Macmillan
Work Out
Series

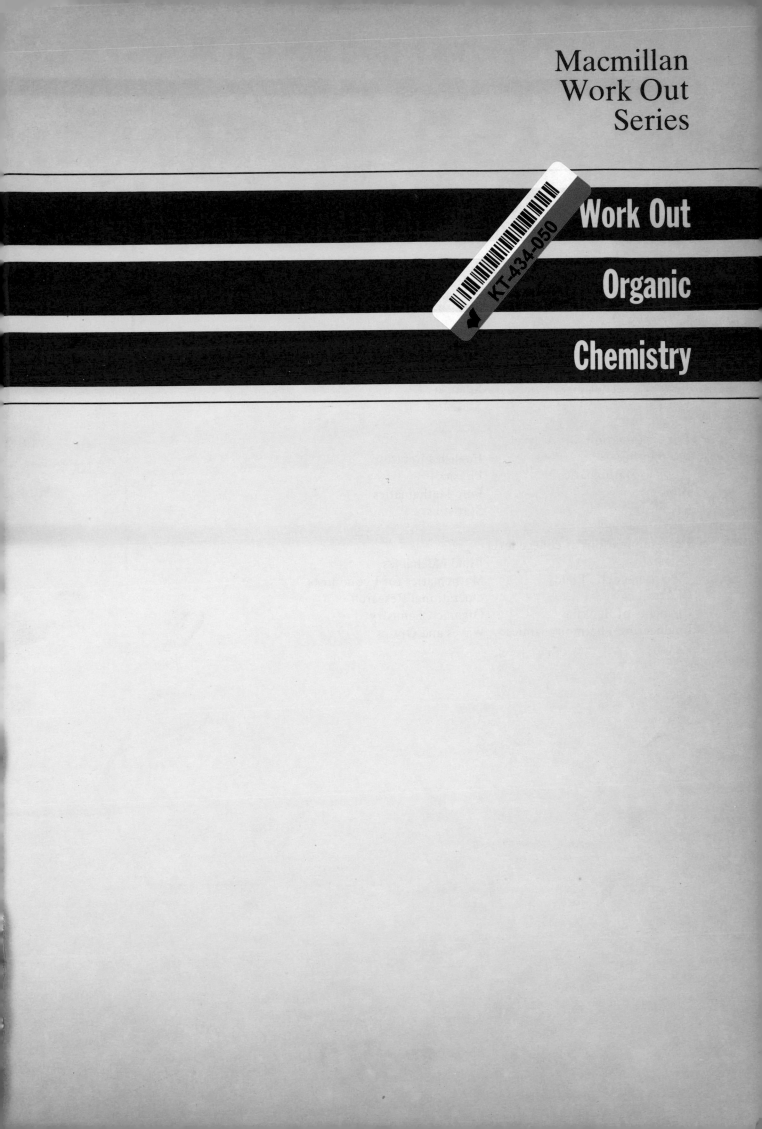

Work Out

Organic

Chemistry

# The titles in this series

MACMILLAN
WORK OUT
SERIES

# Work Out

# Organic

# Chemistry

## C. Went

MACMILLAN
EDUCATION

First published 1988

Published by
MACMILLAN EDUCATION LTD
Houndmills, Basingstoke, Hampshire RG21 2XS
and London
Companies and representatives
throughout the world

Printed in Hong Kong

British Library Cataloguing in Publication Data
Went, Charles
Work out organic chemistry.—
(Macmillan work out series).
1. Chemistry, Organic—Problems,
exercises, etc.
I. Title
547'.0076          QD257
ISBN 0-333-44772-7

# Contents

# Introduction

## How to Use this Book

This book is intended for use both as a 'learning aid' as you complete each topic during your course, and also to assist with your revision and examination preparation at the end of the course. Each chapter has three main sections:

(1) A summary of the major facts and concepts.
(2) Worked examples of representative University and Polytechnic examination questions.
(3) Self-test questions — further examination questions for you to attempt on your own, with comments and outline answers at the back of the book.

For ease of reference, the book is arranged according to the traditional classification of organic compounds, but chapters are also included on organic structure, stereochemistry, mechanisms, and also on 'guided route' and 'short-answer' questions. Chapters are self-contained and may be studied in any order to suit your own programme of study.

The summaries at the beginning of each chapter are intentionally neither greatly detailed, nor completely comprehensive. They are simply reminders of the more important facts or concepts with which you should be familiar. If this is not the case for a particular topic, you should refer back to your lecture notes and/or textbooks before going on to the worked examples. Some space has been left for any additional notes you might wish to include, in order to 'personalise' each chapter for revision purposes.

Bear in mind that there is usually not just one 'correct' answer to an examination question. Most questions could be answered in one of several different ways, involving alternative arrangements of material and the use of different illustrative reactions. Thus the 'worked examples' are not intended to be 'model answers', but are more of a guide to what the examiner is looking for in an answer. They also contain additional comments and discussion of particular points of interest or difficulty which may help to clear up some of the more common misconceptions or misunderstandings.

Depth of treatment of individual topics may vary from one institution to another, so don't worry if a particular 'worked example' contains greater detail than you expected. Use your own notes as a guide — but do just check that any omissions are not the result of a 'missed' lecture, or failure to follow up a suggested reading reference!

Many first year courses now include an introduction to the applications of spectroscopic methods, but the topic is too wide to be adequately treated in one chapter of this book, so with a few simple exceptions, questions involving the use of spectral data have been omitted.

# Nomenclature

Despite the efforts of the ASE and other interested bodies, many institutions still prefer to use traditional names (such as aniline, acetic acid, etc.) for common compounds and reagents. The policy adopted in the worked examples has been to refer to substances by the names given in the question. Where systematic names are used for the first time, the common or traditional name is given in brackets. A reference list of commonly used abbreviations and names with IUPAC equivalents is given in Appendix 1.

# Study Technique

It is a common mistake to attempt to get down virtually every word of a lecture instead of concentrating on what is being said. Try to follow an argument or line of thought and simply jot down key points as later 'reminders', leaving plenty of space for subsequent additions. Above all, avoid the situation where you miss a crucial statement or lose track of an argument because you are copying down relatively unimportant detail which can be obtained from a textbook later on.

As soon as possible after each lecture, read through and consolidate your notes. It really is important to do this while the lecture is still fresh in your memory, i.e. the same evening, or anyway, not later than the next day! Consult relevant sections of your textbook and make additional notes to 'fill in the gaps'. Do not spend time merely rewriting your lecture notes as it is very easy to substitute this activity for the more demanding task of getting to grips with actually understanding and learning the material.

One way of testing your comprehension and memory recall is to work through old examination questions, and this is where the present book should prove useful.

When you complete a particular lecture series, review your notes for the whole topic and prepare a set of summary notes for revision purposes. It may help at this stage to page-number your notes for ready reference should you need to refer back to specific points in more detail when revising.

Incidentally, you will not normally be expected to memorise detailed experimental conditions. Thus, '15% aqueous NaOH at 200°C for 10–12 h' might reasonably be remembered as 'prolonged heating with dilute NaOH at high temperature'.

It is inevitable that from time to time there will be points that you are unsure about. Do not be tempted to shelve any such difficulties for sorting out 'later on', but tackle them as and when they arise.

Try initially to resolve uncertainties for yourself. Reading from an alternative textbook can help: what perhaps seemed obscure in one textbook may appear clearer when differently expressed by another author. Discuss matters that remain unclear with another student. He or she may be similarly puzzled, in which case you should both consult the lecturer or tutor concerned. Lecturers would far rather discuss points of difficulty with you during term-time than encounter them in your examination answers!

For more detailed guidance on 'how to study', you might like to read one or more of the books listed in Appendix 2.

# Preparation for Examinations

It cannot be overstressed that the secret of success in examinations lies in adequate preparation, both long and short term. Most students experience 'examination nerves' to some extent, but outright panic is only likely to result from the dawning realisation too late that not enough work has been done prior to the examination.

Do devote some effort (such as running, walking, swimming, playing tennis) towards maintaining a reasonable level of physical health. There is a definite relationship between physical fitness and mental alertness!

Plan your revision well in advance, and having made a timetable, stick to it. Allocate revision time according to the length and perceived difficulty of each individual topic.

Don't attempt concentrated revision for more than 45–50 min at a time. Have adequate (but not prolonged!) coffee* or tea breaks, and frequently switch topics to sustain interest and stave off boredom. Avoid working so late into the night that you awake feeling tired. A backlog of lost sleep can quickly lead to mental fatigue.

An essential part of your programme must be frequent self-testing. Having revised some material one day, see how much of it you can remember the next. Pinpoint (e.g. by underlining in red) those sections which require further work, and in this way you can avoid spending time reading over material which has already been successfully committed to memory.

Although probably few students ever attempt to learn every single word of wisdom uttered in the lecture room, you should think very carefully about the implications of 'selective revision'. A review of past examination papers is a legitimate way of estimating the frequency with which certain topics occur, but to rely solely on 'question spotting' is likely to be somewhat risky. Many questions do not fall into neat little categories, and it is very frustrating to find a favourite topic combined with a half question on another one which you decided to omit from your revision!

There is another, more fundamental, reason for caution. Much of the treatment in the first year will be laying the foundations for more advanced topics later on. Selective revision of a number of 'banker' questions may, if you are lucky, see you through the examination, but you could then find that an incomplete knowledge of the basic material leads to difficulties in subsequent years of the course.

Do take the trouble to prepare your own individual examination timetable. Carefully check the date, time and location of each examination, and then cross-check with somebody doing the same course. This may sound rather obvious, but if you have a number of examinations to take, it is all too easy to slip up, and every year, sadly, some students do just that!

Most students feel the necessity to look through their revision notes the night before an examination. However, it is unwise either to attempt to learn new material at this stage, or to work too late into the night. If that is what you would normally do, socialise for a while at the end of the evening, but get to bed in good time and don't forget to set the alarm clock!

# The Examination

First of all, do arrive in good time. Check the seating arrangements, especially if other examinations are being held in the same room. It is better to arrive with time to spare, than out of breath, 5 min late, wishing you had caught an earlier bus!

Read through the whole paper, noting any special instructions such as 'start each new answer on a new side of paper', or 'use a separate answer book for section B', etc.

*Preferably decaffeinated!

Decide on the questions you are going to answer, and in which order. A compulsory question or section does not necessarily have to be answered first. There is much to be said for attempting your 'best' or second best question first. Having one good answer down on paper can settle your nerves and put you in a more confident mood for tackling initially less appealing questions later on.

Resist the temptation to spend too much time on your best answer(s). It is very difficult to obtain completely full marks for a question, and you are more likely to pick up extra marks by giving adequate time to your weakest answers than by devoting extra time to your better ones. For the same reason, you should always attempt the full number of questions specified on the paper.

Finally, try to allow sufficient time at the end of the examination to read through all your answers. This may enable you to improve your presentation (spelling, punctuation, headings) or to correct an unintentional error, for example writing electrophilic when you meant nucleophilic.

If you are well prepared and have developed a good examination technique, you will not only give yourself the maximum chance of success, but the examination can become an enjoyable challenge rather than a dreaded barrier. Good luck!

# Acknowledgements

Grateful acknowledgement is made by the author and publishers to the following Institutions who kindly gave permission for the use of questions from their first year examination papers:

University of Bradford
University of Bristol
Brunel, The University of West London
The City University
University of Durham
The University of Dundee
University of Essex
University of Exeter
University of Keele
The University of Leeds
University of Leicester

University of London
Loughborough University of Technology
University College of North Wales
University of Nottingham
The University of Salford
The University of Southampton
The University of Sussex
University College of Swansea
University of Warwick
University of York

Coventry Polytechnic
The Polytechnic, Huddersfield
Kingston Polytechnic
The Polytechnic of North London
Oxford Polytechnic
Plymouth Polytechnic
Portsmouth Polytechnic

Sheffield City Polytechnic
Sunderland Polytechnic
Teesside Polytechnic
Thames Polytechnic
Trent Polytechnic Nottingham
The Polytechnic, Wolverhampton

The author would also like to thank the many members of staff who responded to a request for details of their first year syllabuses and specimen examination papers, and who subsequently provided further helpful information.

Thanks are also due to Dr Paul Lloyd-Williams for his helpful comments on the manuscript. However, the worked examples and answers to self-test questions are entirely due to the author who accepts responsibility for any errors, omissions or misconceptions.

# 1 Structure and Physical Properties: Inductive and Mesomeric Effects

## 1.1 Introduction

The physical and chemical properties of a compound are intimately related to its structure. Electron distribution may differ somewhat from that shown by the classical Lewis formula, and electron shifts are described in terms of two 'effects', the **inductive effect** and the **mesomeric (or resonance) effect**.

## 1.2 The Inductive Effect

This is defined as the *displacement (or unequal sharing) of the electron pair in a sigma bond*. The standard for comparison is the C—H bond which is considered essentially non-polar (i.e. no charge separation).

(1) Electron attracting ligands (atoms or groups) are said to exert a **negative inductive effect**, symbol −I.

(2) Electron releasing ligands are said to exert a **positive inductive effect**, symbol +I.

The direction of displacement is shown by placing an arrowhead midway along the line representing the sigma (electron pair) bond:

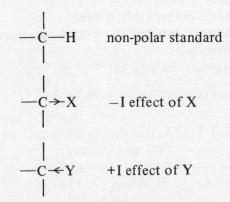

Fractional charges resulting from inductive (or mesomeric) displacements are shown by the symbols δ+ and δ−.

−I ligands include

$$-\overset{\cdot\cdot}{N}H_2,\ -\overset{\cdot\cdot}{N}HR,\ -\overset{\cdot\cdot}{O}H,\ -\overset{\cdot\cdot}{O}R,\ -\overset{\cdot\cdot}{X}:\ \text{(halogen)},\ \overset{\delta+}{-C}\overset{\delta-}{=\overset{\cdot\cdot}{O}},\ \overset{\oplus}{-N}-\overset{\ominus}{\overset{\cdot\cdot}{O}}:,\ -\overset{\oplus}{N}R_3$$

with H below the C, and =Ö below the N.

+I ligands include

$$-\overset{\cdot\cdot}{N}\overset{\ominus}{H},\ -\overset{\cdot\cdot}{\overset{\cdot\cdot}{O}}:^{\ominus},\ -C\overset{\cdot\cdot}{\overset{\cdot\cdot}{O}}\overset{\cdot\cdot}{\overset{\cdot\cdot}{O}}:^{\ominus},\ -R\ \text{(alkyl)}$$

Although sigma electrons are drawn towards the more electronegative of two bonded atoms, *they remain in a bonding position*, i.e. *localised* between the two atoms.

The inductive effect can be transmitted (relayed) along a chain of carbon atoms, but rapidly 'dies out' beyond about the third carbon atom.

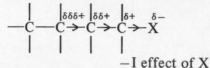

$$-I \text{ effect of } X$$

### (a) Field Effects

Electrostatic (Coulombic) interactions through space or through a solvent are referred to as **field effects**. In practice it may be difficult to distinguish between inductive and field effects.

## 1.3 Polarity and Physical Properties

### (a) Electric Dipole Moments

Defined as the *product of charge and distance of separation*.

**CGS units**
 charge in esu, distance in cm
 Debye unit, D, = $10^{-18}$ esu cm
**SI units**
 charge in Coulomb, distance in m
 usually quoted as $10^{-30}$ C m for convenience
Relationship:
 1 Debye = $3.34 \times 10^{-30}$ C m

Some examples are given in Table 1.1.

**Table 1.1** Electric dipole moments of some methyl compounds expressed in CGS and SI units

| Compound | $CH_3-H$ | $CH_3-NH_2$ | $CH_3-OH$ | $CH_3-Cl$ |
|---|---|---|---|---|
| $\mu$ (D) | 0.0 | 1.32 | 1.69 | 1.82 |
| $\mu$ ($10^{-30}$ Cm) | 0.0 | 4.41 | 5.65 | 6.08 |

**(b) Association**

Two types:

(i) *Dipole–dipole Interaction*

e.g.

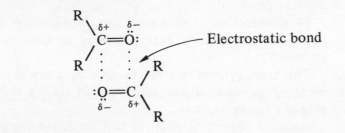

(ii) *Hydrogen Bonding*

1. *Inter*molecular

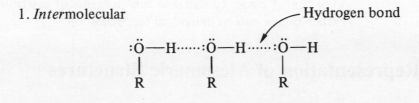

2. *Intra*molecular

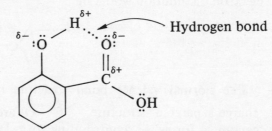

Association influences solubility and volatility, and the ability of solvents to interact with (solvate) molecules and ions. For example, the lower alcohols are water soluble, and alcohols are less volatile than isomeric ethers. *Ortho*-hydroxybenzoic acid (above) is more volatile than the *meta*- or *para*- isomers which are *inter*molecularly hydrogen bonded.

**(c) Acidity of Carboxylic Acids**

Substituents can markedly affect the acidity of carboxylic acids (see Chapter 4).

# 1.4 The Mesomeric (Resonance) Effect

In its simplest form, this relates to the *displacement (unequal sharing) of the pi electron pair in a double bond*. Analogous symbolism is used, −M for electron withdrawal and +M for electron release. The direction of displacement is shown by a 'curved' ('curly') arrow. For example, the carbonyl group in aldehydes and ketones has a polar structure resulting from a combination of −I and −M effects of the oxygen atom:

$$\diagdown C \overset{\curvearrowright}{=\!\!=} \ddot{O}:$$

−M effect

−I effect

The use of a curved arrow in this context is rather different from its use in mechanistic equations.

In mesomerism (resonance), the curved arrow shows the *direction in which electrons have already been displaced* in the 'real' (mesomeric, or resonance) structure.

The same symbol in a mechanistic equation is used to show the *actual movement of an electron pair* which occurs during the course of a reaction (or one step in a reaction).

Another important point is that inductive displacement shifts the sigma electrons towards the oxygen atom, but the electron pair *remains in a bonding position between the two atoms*. Mesomeric displacement of the pi electrons partially shifts them into a *new non-bonding orbital* on the oxygen atom. You may find it easier to visualise this in terms of distribution of *pi electron density* rather than a pair of individual electrons.

## 1.5  Representation of Mesomeric Structures

If the displacement of the pi electron pair went to completion, the resulting electron distribution would be

$$\diagdown \overset{\oplus}{C} — \overset{..\ominus}{\underset{..}{O}}:$$

The normal valence bond structure ('classical' structure) $\diagdown C{=\!\!=}\ddot{O}$ and the 'charge separated' structure, $\diagdown \overset{\oplus}{C} — \overset{\ominus}{O}:$ are described as **limiting forms** (the terms 'canonical forms' or 'contributing forms' are also used) and the actual structure is called a **mesomer** or **resonance hybrid** of the two:

$$\diagdown C{=\!\!=}\underset{..}{\ddot{O}}: \quad \longleftrightarrow \quad \diagdown \overset{\oplus}{C} — \overset{..\ominus}{\underset{..}{O}}:$$

Classical                    Charge separated
limiting form                limiting form

**Resonance symbol**: implies that the actual electron distribution is somewhere between the extremes shown in the two limiting forms

The mesomer (resonance hybrid) can be represented by one composite formula as

$$\diagdown \overset{\delta+}{C}{=\!\!=}\overset{\delta-}{\underset{..}{\ddot{O}}}:$$

or even more simply as

$$\diagdown \overset{\delta+}{C}{=\!\!=}\overset{\delta-}{\underset{..}{O}}:$$

# 1.6 Conjugation

A conjugated system is one of *alternating double and single bonds*, e.g.

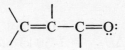

is a conjugated **enone**

Atoms with non-bonded electrons can also form part of a conjugated system. For example, the amino group in an acid amide is said to be conjugated with the carbonyl group

Non-bonded electron
pair equivalent to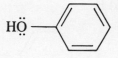
a pi electron pair

and the hydroxyl group in a phenol is conjugated with the aromatic ring — itself a conjugated system par excellence!

HÖ

Relay of the mesomeric effect along a conjugated system, e.g.

composite formula

The shift of electrons from localised bonding positions in classical Lewis formulae is called **delocalisation**, and is characteristic of conjugated structures, which are described as **delocalised systems**.

# 1.7 The Concept of Resonance Energy

*Delocalised systems are at a lower energy* (i.e. more stable) *than they would be if they had the classical Lewis electron distribution.* The difference in stability between the classical and delocalised systems is called **resonance energy** or **delocalisation energy**.

### (a) Resonance Energy of Benzene

This has been determined in two ways.

### (i) *Enthalpy of Combustion*

In the process

$$C_6H_6 \text{ (l)} + 7\tfrac{1}{2}O_2 \longrightarrow 6CO_2 \text{ (g)} + 3H_2O \text{ (l)}$$

benzene evolves 150 kJ mol$^{-1}$ less energy ($\Delta H_{obs} = -3300$ kJ mol$^{-1}$) than calculated ($\Delta H_{calc} = 3450$ kJ mol$^{-1}$) for the Kekulé structure (cyclohexatrienyl structure).

### (ii) *Enthalpy of Hydrogenation*

$$C_6H_6 \text{ (l)} + 3H_2 \text{ (g)} \longrightarrow C_6H_{12} \text{ (l)}$$

For hydrogenation of cyclohexene,

$$\Delta H_{hydrog} = -120 \text{ kJ mol}^{-1}$$

Therefore if benzene had the Kekulé structure the expected value for hydrogenation would be

$$3 \times = 3 \times -120 \text{ kJ mol}^{-1} = -360 \text{ kJ mol}^{-1}$$

but the observed (experimental) value is $-210$ kJ mol$^{-1}$, i.e. $-150$ kJ mol$^{-1}$ less than calculated.

The argument in both cases is that if benzene *evolves* 150 kJ mol$^{-1}$ less energy than calculated for the Kekulé structure, then it must *contain* that much less energy. In other words, benzene is 150 kJ mol$^{-1}$ more stable than cyclohexatriene, and this is a measure of the resonance energy.

### (b) Resonance and Bond Lengths

Delocalisation of the pi electrons in propenal gives a structure (electron distribution) between the two limiting forms

summarised by the composite formula

The C to C and C to O double bonds are said to have *lost some double bond character* and the central C to C position has *gained some double bond character*. This increase or decrease in double bond character is reflected in the observed bond lengths (in pm)*

---

*Some shortening of the C2–C3 bond is to be expected as both the carbon atoms involved are sp$^2$ hybridised, but some part of the 8pm shortening can be attributed to a gain in double bond character.

6

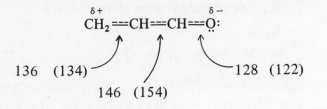

$$136 \quad (134) \qquad\qquad 128 \quad (122)$$
$$146 \quad (154)$$

The figures in brackets refer to 'normal' double or single bond lengths in non-conjugated systems (see Table 1.2).

**Table 1.2** Some reference bond lengths in pm

| | Bond length/pm | |
|---|---|---|
| *Single* | *Double* | *Triple* |
| $-\overset{\mid}{C}-\overset{\mid}{C}-$, 154 | $\overset{}{C}=C$, 134 | $-C{\equiv}C-$, 121 |
| $-\overset{\mid}{C}-\ddot{N}{\big\langle}$, 147 | $C=\ddot{N}-$, 130 | $-C{\equiv}\ddot{N}$, 116 |
| $-\overset{\mid}{C}-\ddot{\underset{\cdot\cdot}{O}}-$, 143 | $C=\ddot{\underset{\cdot\cdot}{O}}$, 122 | |
| $-\overset{\mid}{C}-\ddot{\underset{\cdot\cdot}{C}l}{\colon}$, 176 | | |

## (c) Resonance and Electric Dipole Moments

Delocalisation results in an increase (or decrease if inductive and mesomeric dipoles are in opposition) in electric dipole moment of a molecule when compared with a non-conjugated standard. As an example, values are quoted for butanal and but-2-enal:

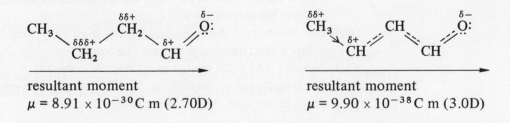

resultant moment
$\mu = 8.91 \times 10^{-30}$ C m (2.70D)

resultant moment
$\mu = 9.90 \times 10^{-38}$ C m (3.0D)

## (d) Resonance and Spectra

In both UV and IR spectra, delocalisation leads to *more intense absorption at longer wavelengths.*

E.g. UV spectrum ($\pi$ to $\pi^*$ transition)

$$CH_3{-}CH_2{-}\overset{\delta+}{CH}{=}\overset{\delta-}{\ddot{O}}\colon \qquad \overset{\delta+}{CH_2}{=}CH{=}CH{=}\overset{\delta-}{\ddot{O}}\colon$$

$\lambda_{max}$ = 188 nm $\qquad\qquad\qquad$ $\lambda_{max}$ = 217 nm, more intense

E.g. IR spectrum ( $\diagdown$ C$=$O$_{str}$)

$$CH_3{-}CH_2{-}CH{=}\ddot{O}\colon \qquad \overset{\delta+}{CH_2}{=}CH{=}CH{=}\overset{\delta-}{\ddot{O}}\colon$$

$\lambda_{max}$ = 1735 cm* $\qquad\qquad\qquad$ $\lambda_{max}$ = 1637 cm*

(*Wavenumbers, reciprocal cm, are a *frequency* unit. Thus a smaller wavenumber means a longer wavelength.)

# 1.8 Some Guidelines on Writing Limiting Forms

1. Begin with the normal Lewis structure, preferably putting in all the non-bonded electron pairs.
2. Derive the other limiting forms in a stepwise fashion, showing electron pair movements with the aid of curved arrows. Remember that atoms stay in the same relative positions.
3. The types of electron movement which occur are

   (a) bonding $\rightarrow$ non-bonding, e.g.

   $$\diagup^{\diagdown}C{=}\ddot{O}\colon \longrightarrow \diagup^{\diagdown}\overset{\oplus}{C}{-}\ddot{\ddot{O}}\colon^{\ominus}$$

   (b) non-bonding $\longrightarrow$ bonding, e.g.

   $$-\overset{|}{\underset{|}{\ddot{N}}}{-}\overset{\oplus}{\underset{|}{C}}{-} \longrightarrow -\overset{\oplus}{\underset{|}{N}}{=}\overset{}{\underset{|}{C}}{-}$$

   (c) bonding $\longrightarrow$ new bonding, e.g.

   $$\diagup^{\diagdown}C{=}C{-}\overset{\oplus}{C}\diagdown_{\diagup} \longrightarrow \diagup^{\diagdown}\overset{\oplus}{C}{-}\underset{|}{C}{=}C\diagdown_{\diagup}$$

4. Note particularly that:
   (i) an atom can be electron deficient (e.g. $C^{\oplus}$ has only 6 valence electrons), but must not be shown with more valence electrons than it can accommodate (usually 8);
   (ii) the number of electron pair bonds may vary, but the total number of valence electrons must remain the same;
   (iii) likewise, the total charge must remain constant;
   (iv) limiting forms are separated by the **resonance symbol**, $\leftrightarrow$; never use the equilibrium sign, $\rightleftharpoons$, for this purpose.

# 1.9 Assessing the Relative Stabilities of Limiting Forms

Factors to consider are:
1. The number of electron pair bonds. (Bonded electrons are at a lower energy level than those in non-bonding positions.)

2. The electronegativity of atoms carrying charges. (Thus $O^\ominus$ is more stable than $N^\ominus$, which is in turn more stable than $C^\ominus$.)
3. Positive carbon, $C^\oplus$, is much less stable than $O^\oplus$ or $N^\oplus$, because $C^\oplus$ has only 6 valence electrons.

# 1.10 Application of Limiting Forms

When using limiting forms to assess the extent of delocalisation, the following generalisations apply.
1. The more stable a particular limiting form, the greater its 'contribution' to the mesomer (i.e. the more closely it resembles the electron distribution in the mesomer).
2. Highly unstable limiting forms may be ignored (make 'little contribution').
3. The larger the number of stable limiting forms that can be drawn, the greater the extent of delocalisation.
4. Delocalisation results in a more stable structure. Hence, the more effective is delocalisation, the larger the resonance energy.

# 1.11 Resonance and Reactivity

Inductive and mesomeric effects determine the polarity of molecules which in turn influences their behaviour in terms of:
1. Physical properties:
   e.g. volatility, solubility, ability to solvate or be solvated by other molecules.
2. Reactivity:
   (a) polarisation, permanent and/or induced, creates nucleophilic and electrophilic sites within a molecule.
   (b) differential stabilisation of transition states, intermediates or products may influence the course and rate of reaction, or the position of an equilibrium. Examples will be found throughout the text, and in particular in the chapters on mechanisms (3), acids and bases (4) and aromatic compounds (10).

This is necessarily a rather brief account. For a more detailed discussion of inductive and mesomeric effects and their significance in organic chemistry, you might like to consult the author's book *Ionic Organic Mechanisms* (Macmillan Education, 1986).

# 1.12 Worked Examples

The following questions illustrate some of the basic principles, but many more examples of the application of inductive and mesomeric effects will be found throughout the text.

**Example 1.1**

1. Show by means of curved arrows the electron shifts necessary to bring about the following transformations:

(a) $CH_3-\overset{..}{\underset{..}{O}}-\overset{\overset{\displaystyle ||}{\underset{\displaystyle :O:}{}}}{C}-CH_3 \longrightarrow CH_3-\overset{\oplus}{\underset{..}{O}}=\overset{\overset{\displaystyle |}{\underset{\displaystyle :\overset{..}{\underset{..}{O}}:\ominus}{}}}{C}-CH_3$

(2 marks)

9

(b) $\ddot{N}H_2-CH=\overset{\oplus}{\ddot{O}}-CH_3 \longrightarrow \overset{\oplus}{N}H_2=CH-\ddot{O}-CH_3$      (*2 marks*)

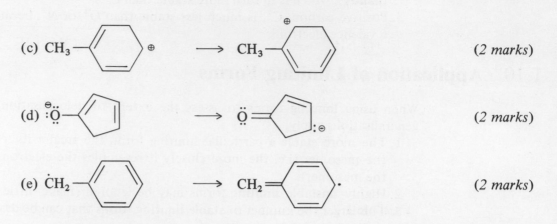

(c) $CH_3-$ ◯ $\oplus \longrightarrow CH_3-$ ◯      (*2 marks*)

(d) $:\overset{\ominus}{\ddot{O}}-$ ◯ $\longrightarrow \ddot{O}=$ ◯ $:\ominus$      (*2 marks*)

(e) $\dot{C}H_2-$ ◯ $\longrightarrow CH_2=$ ◯ $\cdot$      (*2 marks*)

2. For the following species, draw the Lewis structures which would result from the indicated electron movements. In each case show any charges that develop or change places as a consequence of such movements.

(f) $\overset{\frown}{N}H_2-C\overset{\frown}{\equiv}\ddot{N}$      (*2 marks*)

(g) $:\ddot{C}l-CH\overset{\frown}{=}CH_2$      (*2 marks*)

(h) $CH_3-\ddot{O}-\ddot{N}=\ddot{O}:$      (*2 marks*)

(i) $\overset{\ominus}{\ddot{N}}=\overset{\oplus}{N}=\ddot{N}-CH_3$      (*2 marks*)

(j) ◯$-\overset{\oplus}{N}\equiv N$      (*2 marks*)

(Coventry Polytechnic, compiled from
(Mid-Sessional Examinations, 1986/7)

*Solution 1.1*

1. Compare the two structures carefully to ascertain what electron movements have occurred. Make sure that your curved arrows clearly indicate where the electrons originate, and where they are moved to.

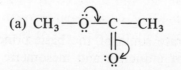 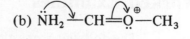

(a) $CH_3-\ddot{O}-C-CH_3$
            $\|$
            $:\ddot{O}$

(b) $\ddot{N}H_2-CH-\ddot{O}-CH_3$

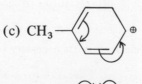

(c) $CH_3-$

(d) $\overset{\ominus}{:}\ddot{O}-$

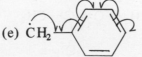

(e) $\dot{C}H_2-$      Note the use of half-headed arrows to indicate one-electron movements.

2. Be particularly careful here to make sure that you allocate any charges correctly.

(f) $\overset{\oplus}{N}H_2\!=\!C\!=\!\overset{\ominus}{\underset{..}{N}}$

(g) $:\overset{..}{\underset{.}{C}l}\!=\!CH\!-\!\overset{\ominus}{\underset{..}{C}H_2}$

(h) $CH_3\!-\!\overset{\oplus}{\underset{..}{O}}\!=\!\overset{}{N}\!-\!\overset{\ominus}{\underset{..}{O}}:$

(i) $\overset{..}{N}\!\equiv\!N\!-\!\overset{\oplus}{\underset{..}{N}}\!-\!CH_3$

(j)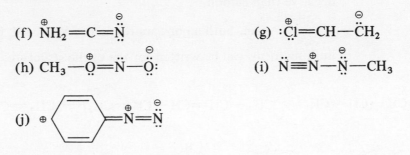

### Example 1.2

State, with reasons, which you would expect to be the more stable anion or cation in each of the following pairs:

(a) $CH_3\!-\!CH_2\!-\!\overset{\oplus}{C}H_2$     or    $CH_2\!=\!CH\!-\!\overset{\oplus}{C}H_2$     *(2 marks)*

(b) $CH_2\!=\!CH\!-\!\overset{\ominus}{\underset{..}{C}}H_2$     or    $\overset{..}{\underset{..}{O}}\!=\!CH\!-\!\overset{\ominus}{\underset{..}{C}}H_2$     *(2 marks)*

(c) $CH_2\!=\!CH\!-\!\overset{\ominus}{\underset{..}{C}}H\!-\!CH\!=\!CH_2$    or    ⬠ $:\ominus$     *(2 marks)*

(d) $CH_3\!-\!\overset{..}{\underset{..}{O}}\!-\!\overset{\oplus}{C}H\!-\!CH_3$    or    $CH_3\!-\!CH_2\!-\!\overset{\oplus}{C}H\!-\!CH_3$   *(2 marks)*

(e) $\overset{\ominus}{\underset{..}{C}}H_2\!-\!CH_3$     or    $\overset{\ominus}{\underset{..}{C}}H_2\!-\!C\!\equiv\!\overset{..}{N}$     *(2 marks)*

(Coventry Polytechnic, 1986)

### Solution 1.2

(a) $CH_2\!=\!CH\!-\!\overset{\oplus}{C}H_2$. This carbocation is resonance stabilised:

$$CH_2\!=\!CH\!-\!\overset{\oplus}{C}H_2 \longleftrightarrow \overset{\oplus}{C}H_2\!-\!CH\!=\!CH_2$$

and the positive charge is therefore shared by *two* carbon atoms. The saturated cation has a primary structure and is relatively unstable. There is some inductive stabilisation by the +I effect of the ethyl group,

$$CH_3CH_2\!\rightarrow\!\overset{\oplus}{C}H_2$$

but the positive charge is essentially localised on one carbon atom.

(b) $\overset{..}{\underset{..}{O}}\!=\!CH\!-\!\overset{\ominus}{C}H_2$. Both anions are resonance stabilised,

$$\overset{\ominus}{\underset{..}{C}}H_2\!-\!CH\!=\!CH_2 \leftrightarrow CH_2\!=\!CH\!-\!\overset{\ominus}{\underset{..}{C}}H_2 \quad \text{and} \quad \overset{\ominus}{\underset{..}{C}}H_2\!-\!CH\!=\!\overset{..}{\underset{..}{O}}: \leftrightarrow CH_2\!=\!CH\!-\!\overset{..}{\underset{..}{O}}:^{\ominus}$$

$$\;\;\;\text{(I)} \qquad\qquad\qquad\qquad \text{(II)} \qquad\qquad\qquad \text{(III)} \qquad\qquad\qquad \text{(IV)}$$

11

but limiting form (IV) is more stable (and therefore makes a larger contribution) as the negative charge is on an oxygen atom (which is more electronegative than carbon).

(c)  Again, both anions are resonance stabilised, but a larger number of limiting forms can be written for the cyclic structure:

$$CH_2\!\!=\!\!CH\!-\!\overset{\ominus}{\ddot{C}H}\!-\!CH\!\!=\!\!CH_2 \leftrightarrow \overset{\ominus}{\ddot{C}H_2}\!-\!CH\!\!=\!\!CH\!-\!CH\!\!=\!\!CH_2 \leftrightarrow CH_2\!\!=\!\!CH\!-\!CH\!\!=\!\!CH\!-\!\overset{\ominus}{\ddot{C}H_2}$$

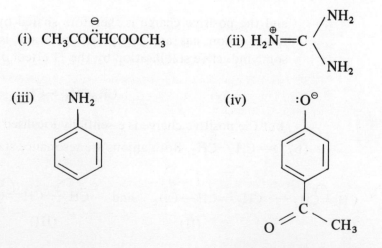

Thus the negative charge is shared equally by the five ring carbon atoms.

(d) $CH_3\!-\!\overset{..}{\ddot{O}}\!-\!\overset{\oplus}{C}H\!-\!CH_3$. This cation is resonance stabilised

$$CH_3\!-\!\overset{..}{\ddot{O}}\!-\!\overset{\oplus}{C}H\!-\!CH_3 \longleftrightarrow CH_3\!-\!\overset{\oplus}{\ddot{O}}\!\!=\!\!CH\!-\!CH_3$$
$$\qquad\qquad\text{(I)} \qquad\qquad\qquad\qquad \text{(II)}$$

and limiting form (II) is particularly stable as all the atoms have complete valence shells. The other cation is inductively stabilised by the two alkyl groups, but most of the positive charge remains localised on one carbon atom.

(e) $\overset{\ominus}{\ddot{C}H_2}\!-\!C\!\!\equiv\!\!\ddot{N}$. Delocalisation again accounts for the difference in stabilities:

$$\overset{\ominus}{\ddot{C}H_2}\!-\!C\!\!\equiv\!\!\ddot{N} \longleftrightarrow CH_2\!\!=\!\!C\!\!=\!\!\overset{\ominus}{\ddot{N}}$$
$$\qquad\text{(I)} \qquad\qquad\qquad\qquad \text{(II)}$$

Limiting form (II) has the negative charge on the more electronegative nitrogen atom.

**Example 1.3**

(a) Draw the principal contributing canonical forms for each of the following species.

(i) $CH_3CO\overset{\ominus}{\ddot{C}H}COOCH_3$

(ii) $H_2\overset{\oplus}{N}\!\!=\!\!C\overset{\displaystyle NH_2}{\underset{\displaystyle NH_2}{}}$

(iii) phenyl–$NH_2$

(iv) 4-acetyl phenolate, $:O^{\ominus}$ ... $O\!\!=\!\!C\!-\!CH_3$

(Sheffield City Polytechnic, 1985)

12

(b) Write the formulae of the other contributors to the resonance hybrids of each of the following:

(i)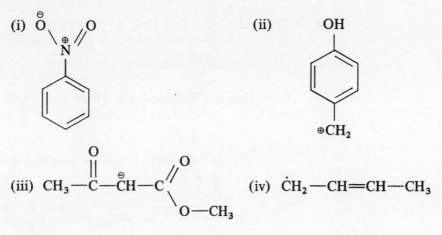

(ii)

(iii) $CH_3-\overset{O}{\overset{\|}{C}}-\overset{\ominus}{CH}-\overset{O}{\overset{\|}{C}}{\underset{O-CH_3}{}}$

(iv) $\dot{C}H_2-CH{=}CH-CH_3$

(University of Leeds, 1983)

(c) Write the formulae of the other contributors to the resonance hybrid of each of the following:

(i) $CH_3-\overset{\ominus}{CH}-\overset{\oplus}{N}{\equiv}N$

(ii)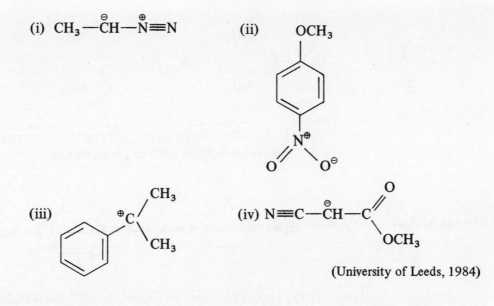

(iii)

(iv) $N{\equiv}C-\overset{\ominus}{CH}-\overset{O}{\overset{\|}{C}}{\underset{OCH_3}{}}$

(University of Leeds, 1984)

*Solution 1.3*

In the experience of the author, many of the errors which occur in drawing limiting forms arise from a misunderstanding (or forgetfulness!) about the numbers of lone pairs of electrons on the atoms involved. Thus, although it is conventional (as in the present examples) to omit such electrons, you are strongly advised to show *all* the non-bonded electrons in the structures that you draw.

(a)

(i)

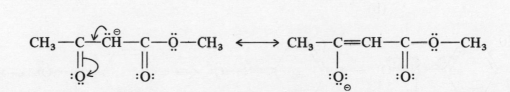

13

Similarly with the other carbonyl group,

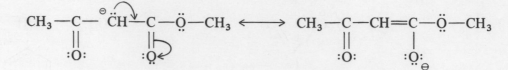

Note that the limiting form which you might draw for resonance in an ester,

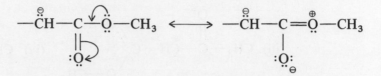

will make very little contribution in this case because it leaves the negative charge localised on the carbon atom.

(ii) This is the conjugate cation of guanidine, a very strong base. The positive charge is shared equally by the three nitrogen atoms:

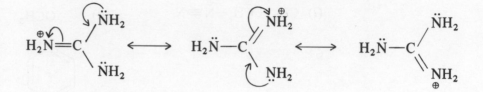

(iii) It is often convenient, although not strictly necessary, to re-draw certain structures such as phenylamine:

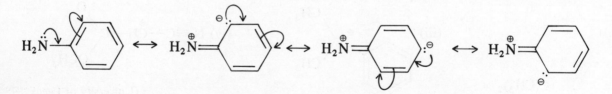

(iv) This applies also to the substituted phenoxide ion:

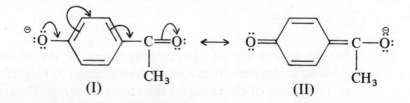

Other limiting forms may be drawn, but (II) is the major additional contributor as it shows the charge on the (electronegative) oxygen atom.

(b)

(i)

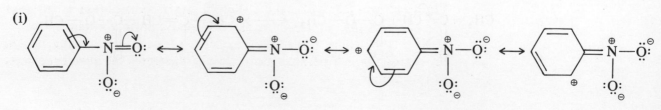

14

(ii)

(I)   (II)

Again, other limiting forms can be drawn, but (II) is more stable than the alternatives because, unlike $C^\oplus$, positively charged oxygen has a complete octet of valence electrons.

(iii) See (a) (i) above.

(iv) This is an example of a resonance stabilised (allylic) free radical. Remember that movement of individual electrons is shown by means of a half-headed arrow.

$$\overset{\cdot}{C}H_2\!-\!CH\!=\!CH\!-\!CH_3 \longleftrightarrow CH_2\!=\!CH\!-\!\overset{\cdot}{C}H\!-\!CH_3$$

(c)

(i)   $CH_3\!-\!\overset{\ominus}{\overset{\cdot\cdot}{C}H}\!-\!\overset{\oplus}{N}\!\equiv\!\overset{\cdot\cdot}{N}: \longleftrightarrow CH_3\!-\!CH\!=\!\overset{\oplus}{N}\!=\!\overset{\cdot\cdot}{\overset{\ominus}{N}}$

(ii)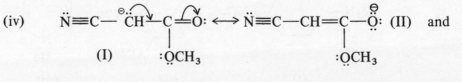

(iii)

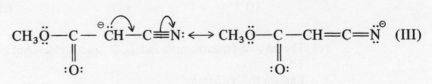

(iv)  $\overset{\cdot\cdot}{N}\!\equiv\!C\!-\!\overset{\ominus\,\cdot\cdot}{C}H\!-\!C\!=\!\overset{\cdot\cdot}{O}: \longleftrightarrow \overset{\cdot\cdot}{N}\!\equiv\!C\!-\!CH\!=\!C\!-\!\overset{\cdot\cdot}{\overset{\ominus}{O}}:$ (II)   and

(I)   $:\overset{\cdot\cdot}{O}CH_3$        $:\overset{\cdot\cdot}{O}CH_3$

$CH_3\overset{\cdot\cdot}{O}\!-\!\overset{\parallel}{C}\!-\!\overset{\ominus\,\cdot\cdot}{C}H\!-\!C\!\equiv\!N: \longleftrightarrow CH_3\overset{\cdot\cdot}{O}\!-\!\overset{\parallel}{C}\!-\!CH\!=\!C\!=\!\overset{\cdot\cdot\ominus}{N}$ (III)
$\quad\;\;:O:$                              $:O:$

Limiting form (II) will make a slightly larger contribution as the negative charge is then on the more electronegative atom, oxygen.

Note that some tutors may prefer a symmetrical drawing for the nitro group which perhaps more closely represents the actual geometry of the group,

15

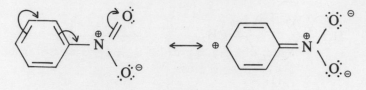

etc.

However, to others this may seem a bit pedantic, and if followed through should also lead to structures such as

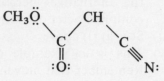

which are clearly more cumbersome to draw.

# 1.13 Self-Test Questions

### Question 1.1

(a) Give the structure that results from the movement of electrons in each of the following two diagrams.

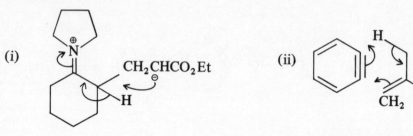

(i)

(ii)

(University of Salford, 1985)

(b) Draw the major resonance contributors to each of the following structures.

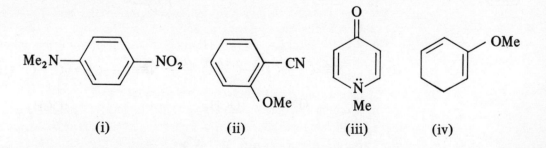

(i)                  (ii)                  (iii)                  (iv)

(c) The $pK_a$ of butanoic acid and 2-, 3-, and 4-chlorobutanoic acids are 4.81, 2.9, 4.1 and 4.5.
Explain this variation.

(d) $EtH_2\overset{\oplus}{N}—\overset{\ominus}{B}Me_3$ is more stable than $H_3\overset{\oplus}{N}—\overset{\ominus}{B}Me_3$, which in turn is more stable than $Et_3\overset{\oplus}{N}—\overset{\ominus}{B}Me_3$.
Comment on these observations.

(e) Describe the main solute–solvent interactions with lithium chloride dissolved in (i) ethanol, and (ii) acetonitrile, MeCN.

(b) to (e) University of Nottingham, 1984)

## Question 1.2

State, with reasons, which is likely to be the major contributor in each of the following pairs of limiting forms.

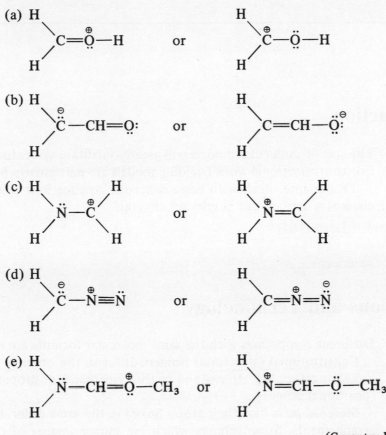

(Coventry Polytechnic, 1986/87)
(Compiled from Progress Tests)

## Question 1.3

Use the concepts of inductive and/or mesomeric (resonance) stabilisation of cations to explain the following observations:

(a) Propene readily reacts with HBr to give 2-bromopropane, but propenoic acid reacts only very slowly to give 3-bromopropanoic acid.  (*6 marks*)

(b) Addition of HI to chloroethene (vinyl chloride, $CH_2{=}CHCl$) is slower than addition to ethene, and gives exclusively 1-chloro-2-iodoethane.  (*7 marks*)

(c) Bromomethoxymethane ($CH_3OCH_2Br$) undergoes unimolecular hydrolysis ($S_N1$) at a rapid rate in alkaline solution. Under similar conditions, bromopropanone tends to form the anion, $CH_3CO\overset{\ominus}{C}HBr$.  (Coventry Polytechnic, 1986)

(Mid-Sessional Examination)

## Question 1.4

Give an account of how inductive and resonance properties of a substituent group R in a benzene nucleus can affect the position of substitution by an electrophilic reagent $E^{\oplus}$.

(City University, 1985)

# 2 Basic Stereochemistry

## 2.1 Introduction

The use of molecular models will greatly facilitate your study of stereochemistry. For conformational work Dreiding models are particularly helpful.

This chapter deals with basic concepts. Applications to other areas of organic chemistry are covered in relevant chapters.

## 2.2 Definitions and Terminology

Different compounds with the same molecular formula are **isomers**.

**Constitutional** (structural) **isomers** differ in the order in which their atoms are linked together, e.g. ethanol and methoxymethane, or propan-1-ol and propan-2-ol (positional isomers).

**Stereoisomers** have their atoms linked in the same order, but in different spatial arrangements. Stereoisomers which are mirror images of one another are called **enantiomers**. Non-enantiomeric stereoisomers are called **diastereoisomers** (or **diastereomers**).

Enantiomers occur only in compounds whose molecules are chiral. A chiral molecule is one that is not superposable (superimposable) on its mirror image. Molecules that are superposable on their mirror images are called **achiral**.

A common, but not the only, cause of chirality is the possession of a **chiral carbon atom**\* (formerly called an asymmetric carbon atom). This is a carbon atom to which are attached four different atoms or groups ('ligands').

Molecules with two or more chiral atoms may or may not be chiral. General criteria are that a molecule will not be chiral if it has (i) a plane of symmetry, (ii) a centre of symmetry, or (iii) an $n$-fold alternating axis of symmetry. For most purposes the possession or non-possession of a plane of symmetry (an imaginary plane that cuts the molecule into two 'mirror image' halves) is sufficient to distinguish between achiral and chiral molecules.

Molecules with no element of symmetry are called **asymmetric**. Some molecules may have an element of symmetry (commonly an alternating axis of symmetry, e.g. certain derivatives of allene) and still be chiral. Such molecules are described as **dissymmetric**.

---

\*The term **stereogenic carbon atom** has been suggested (see Mislow, K., *et al.*, *J. Am. Chem. Soc.*, 1984, **106**, 3319–28) but is not yet in general use.

## 2.3 Conventions for Showing Three-Dimensional Structures on Paper

### (a) Sawhorse Formulae

For example, for ethane

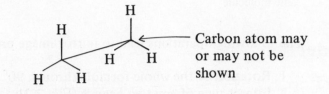

Carbon atom may or may not be shown

### (b) 3-D ('flying wedge') formulae

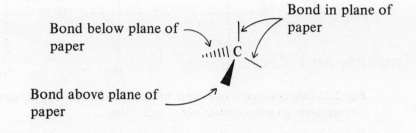

Bond below plane of paper

Bond in plane of paper

Bond above plane of paper

(Bonds to atoms or groups (ligands) below the plane of the paper may also be shown as ·······; however, the use of a wedge, |||ıı. , avoids any confusion with the dashed line convention (-----) for representing bonds being formed or broken in drawings of transition states).

### (c) Fischer Projections

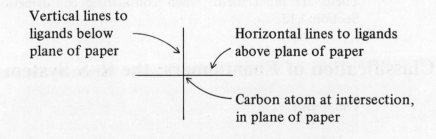

Vertical lines to ligands below plane of paper

Horizontal lines to ligands above plane of paper

Carbon atom at intersection, in plane of paper

### (i) *Manipulation of Fischer Projections*

The following operations result in an alternative projection of the *same* molecule:
1. Rotation of the whole molecule through 180°.
2. Rotation of any three ligands by one or two places clockwise or anti-clockwise, whilst the fourth ligand remains in the same position (Fig. 2.1).

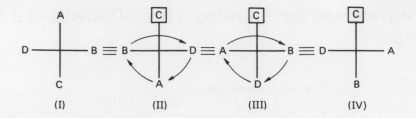

**Fig. 2.1** Manipulation of Fischer projection formulae. I–IV are different projections of the same molecule

The following operations lead to mirror image projections;

1. Rotation of the whole formula through 90°;
2. Interchange of any two ligands (Fig. 2.2).

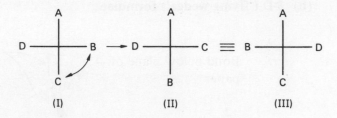

**Fig. 2.2** Interconversion of Fischer projection formulae. Projections I and III are mirror images and represent an enantiomeric pair of molecules

When comparing two Fischer projection formulae to assess whether they represent two views of the same molecule or an enantiomeric pair of molecules, carry out the permitted manipulations on one projection until any two ligands occupy similar positions. Then if the other two ligands also coincide, the two formulae represent the same molecule. If not, the two projections represent enantiomeric isomers.

### (d) Newman Projections

These are most useful when considering conformations of a molecule – see Section 2.15.

## 2.4 Classification of Enantiomers: the R–S System

The steps are:

1. Assign an order of priority to each ligand on the chiral carbon atom by the **Sequence Rules**.
2. Orient the drawing or projection of the molecule so that the ligand of lowest priority faces away from the viewer. When using Fischer projections this means manipulating the formula so that the ligand of lowest priority is at the top or bottom of the projection.
3. Observe the order of decreasing priority of the other three ligands: if this is clockwise, the configuration is R (rectus); if it is anticlockwise the configuration is S (sinister) (see Fig. 2.3).

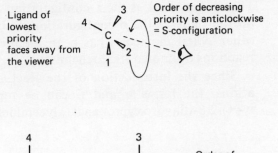

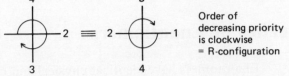

**Fig. 2.3** Assignment of configuration by the R–S system

## 2.5 The Sequence Rules (Cahn–Ingold–Prelog Rules)

1. Atoms directly attached to the chiral carbon atom are arranged in order of *decreasing atomic number* (lower atomic number, lower priority).
2. Isotopes when present are arranged in decreasing *mass number*.
3. For ligands attached by the same kind of atom (e.g. methyl and ethyl groups are both attached by a carbon atom), the next atoms are considered and the process continued until a decision can be made. Thus ethyl, with C–C has priority over methyl with C–H.
4. Doubly or triply bonded atoms are treated as though both atoms were duplicated or triplicated. For example

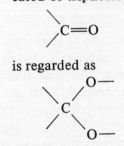

## 2.6 Absolute Configuration

In 1951 dextrorotatory (+)-2,3-dihydroxybutanedioic acid ((+)-tartaric acid) was shown by X-ray diffraction to have an absolute (i.e. actual) configuration with the 3-hydroxyl group on the left as seen in the Fischer projection:

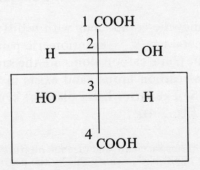

This is described as the L-**configuration** and the mirror image (with the OH group on the right) is the D-**configuration**. Subsequently, the absolute configuration of other chiral molecules has been established by chemical interconversions using reactions of known stereochemistry.

Since the introduction of the Rectus–Sinister system for describing configurations, the terms D and L can be replaced appropriately by R and S. Thus, D-(+)-2,3-dihydroxypropanal (glyceraldehyde) has the R-configuration.

## 2.7 Properties of Enantiomers

Enantiomers have identical physical and chemical properties except for:
1. Their ability to rotate the plane of polarisation of polarised light clockwise ((+), dextrorotatory) or anticlockwise ((−), laevorotatory). (Referred to as '**optical activity**'.)
2. Their behaviour towards other chiral substances, e.g. enantiomers vary in their solubility in a chiral solvent, and react at different rates with chiral reagents.

## 2.8 Separation of Enantiomers

An optically inactive assembly of equal numbers of a pair of enantiomeric molecules is called a **racemic modification**.* The formation of such a racemate from an optically pure enantiomer is termed **racemisation**.

The separation of one or both pure enantiomers from a racemic modification is called **resolution**.

Preferential decomposition of one enantiomer from a racemic modification by a specific enzyme has obvious disadvantages (e.g. at least half the material is lost) and is of limited applicability. Chromatographic resolution using a chiral stationary phase has had some success but the method of general applicability is to react the R–S racemic modification with a chiral reagent X to give a mixture of diastereoisomers RX and SX.† Unlike enantiomers, these are different compounds with different physical and chemical properties and can therefore be separated by conventional means such as fractional crystallisation or column chromatography. The individual enantiomers are then recovered from the separated diastereoisomers. For example, a racemic acid may be resolved by salt formation with a (+) or (−) base, or by esterification with a (+) or (−) chiral alcohol (see Worked Example 2.2).

## 2.9 Molecules with Two or More Chiral Carbon Atoms

In alicyclic compounds with $n$ different chiral carbon atoms, there are $2^n$ stereoisomers, i.e. $2^{n-1}$ enantiomeric pairs. There are fewer isomers when two or more of the chiral carbon atoms are the same. Tartaric acid, for example, has two similar chiral carbon atoms, and exists in three stereoisomeric forms, 2R3R, 2S3S, and 2R3S. The latter has a plane of symmetry and is an optically inactive meso form (see Fig. 2.4).

*The terms racemate and racemic mixture are also commonly used.
†You might like to verify this by the use of models.

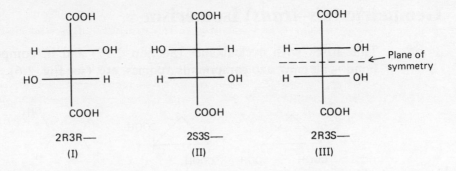

**Fig. 2.4** The three stereoisomers of tartaric acid. Note that (III) (*meso* tartaric acid) is a diastereoisomer of (I) and (II)

## 2.10 Chirality in Compounds with no Chiral Carbon Atoms

Cyclic and bicyclic compounds, biphenyls, allenes, spiranes and helical compounds are all examples of substances that can exist in enantiomeric forms although no chiral carbon atoms are present (see Fig. 2.5).

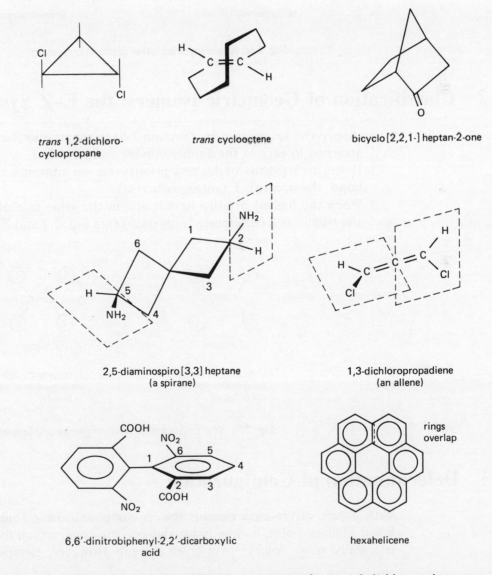

**Fig. 2.5** Some chiral molecules. With the exception of *trans* 1,2-dichlorocyclopropane, none of the molecules has a chiral carbon atom

## 2.11 Geometric (*cis–trans*) Isomerism

This arises with cycloalkanes (Section 2.17) and in compounds with a double bond — alkenes, azo compounds, oximes, etc. (see Fig. 2.6).

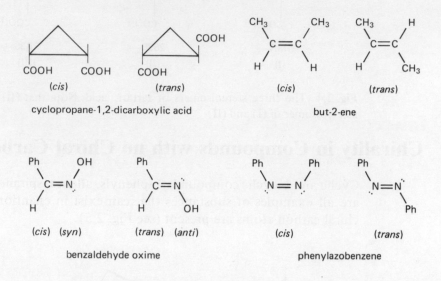

cyclopropane-1,2-dicarboxylic acid

but-2-ene

benzaldehyde oxime

phenylazobenzene

**Fig. 2.6**  Some representative geometric isomers

## 2.12 Classification of Geometric Isomers: the E–Z System

1. Apply the sequence rules (Section 2.5) to assign priorities to the two ligands attached to each of the doubly bonded carbon atoms.
2. If the two ligands of highest priority are on opposite sides of the double bond, the isomer is **E** (**entgegen**, across).
3. When the highest priority ligands are on the same side of the double bond, the isomer is **Z** (**zusammen**\*, together) (see Fig. 2.7 and Example 2.3).

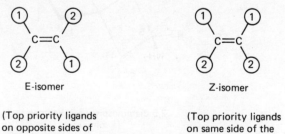

E-isomer

(Top priority ligands
on opposite sides of
the pi bond)

Z-isomer

(Top priority ligands
on same side of the
pi bond)

**Fig. 2.7**  The E–Z classification of geometric isomers.

## 2.13 Determination of Configuration

With a pair of *cis–trans* isomers the *cis* compound often (but not always) has a lower melting point, boiling point and specific gravity than the *trans* isomer, and is usually more soluble in a given solvent. However, electric dipole moments,

---

\*A useful nmemonic here is that zusammen contains the word 'same'.

when applicable, and various spectroscopic methods are usually more reliable. The coupling constant ($J$) for alkenic protons in $^1$H NMR spectroscopy is higher for E compounds (12–18 Hz) than for isomeric Z compounds (7–11 Hz).

For compounds of the form RCH=CHR′, the C—H out of plane bending (IR spectrum) occurs at 675–730 cm$^{-1}$ for the Z-isomer, and at 965–975 cm$^{-1}$ for the E-isomer.

Further evidence may be obtained from chemical reactivity. The two isomers may react at different rates because of the different spatial arrangement of ligands, and an E and Z alkene for example may give rise to products with different stereochemistry.

## 2.14 Interconversion of Isomers

Interconversion requires 'breaking' the pi bond to allow rotation about the sigma bond. With alkenes, this may be achieved by temporary formation of a radical or a carbocation (see Fig. 2.8).

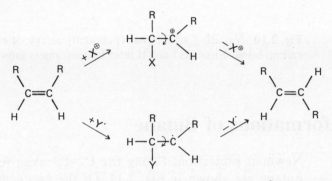

**Fig. 2.8** Interconversion of *cis-trans* isomers by formation of an intermediate radical or carbocation

## 2.15 Conformation

The **conformations** of a molecule are the different spatial arrangements of the atoms which result from rotations of groups about a single (sigma) bond. A study of the energy changes which accompany these rotations is called **conformational analysis**.

### (a) Newman Projections

A convenient representation of a molecule viewed along one of its carbon–carbon axes is the Newman projection. Bonds on the front carbon atom are shown as

and those on the rear carbon atom as . Newman projections of

the two extreme conformations of ethane are shown as a reminder in Fig. 2.9.

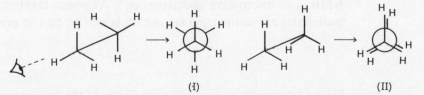

**Fig. 2.9** Newman projections of two conformations of ethane, **fully eclipsed (II)** and **staggered** or **anti (I)**

25

In the **staggered** (or **anti**) conformation (I), the H atoms are as far apart as possible, and electrostatic repulsion between the C–H bonding electron pairs is at a minimum. In the least stable **eclipsed** conformation (II), electron pair repulsion is maximised. The changes in internal (potential) energy which accompany rotation about the carbon–carbon bond in ethane are shown diagrammatically in Fig. 2.10.

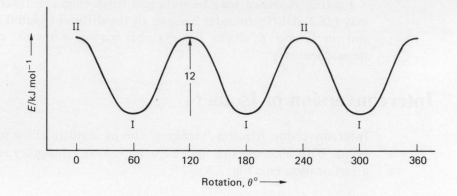

**Fig. 2.10** Variation of internal (potential) energy of ethane with rotation about the carbon-carbon bond. Numbers I and II refer to conformers shown in Fig. 2.9

## 2.16  Conformations of Butane

Newman projections (along the C2–C3 axis) for the principal conformations of butane are shown in Fig. 2.11. Of the two eclipsed conformations, (I) (referred to as **synperiplanar**)* and (III) (**anticlinal**), (I) is the least stable because of additional steric strain due to the close proximity of the two bulky methyl groups.

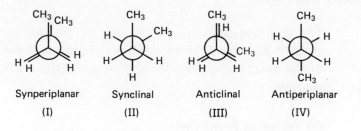

**Fig. 2.11**  Principal conformations of butane

Similarly, the staggered conformer (IV) (**antiperiplanar**) is more stable than (II) (**synclinal**) where some steric strain is also present. Energy differences for the various conformations are shown in Fig. 2.12.

The proportions of the different conformations present and energy differences between them may be determined by various spectroscopic methods (Raman, IR, NMR and microwave spectroscopy). At room temperature about 70% of butane molecules are antiperiplanar, which is said to be the **preferred conformation**.

*These terms refer to the relative positions of the two methyl groups. syn = together, anti = opposite, periplanar = about planar, and clinal = inclined.

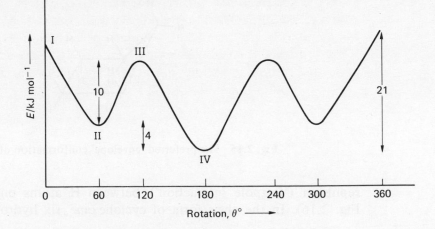

**Fig. 2.12** Variation in internal (potential) energy of butane with rotation about the C2–C3 carbon–carbon bond. Numbers I–IV relate to conformers shown in Fig. 2.11

## 2.17 Conformations of Cycloalkanes

Cyclopropane is relatively unstable because it has both ring strain from a shortening of the C–C–C bond angle (109.5° down to 60°), and torsional (Pitzer) strain due to eclipsing of the hydrogen atoms (see Fig. 2.13).

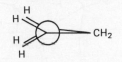

**Fig. 2.13** A Newman projection of cyclopropane

The folded ('butterfly') shape of **cyclobutane** (Fig. 2.14) is a compromise in which a slight increase in ring strain (88° rather than 90°) allows relief from torsional strain by non-eclipsing of the hydrogen atoms.

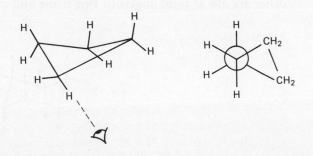

**Fig. 2.14** The preferred 'butterfly' conformation of cyclobutane

**Cyclopentane** is a flexible molecule with rapidly interchanging conformations. The most stable ('preferred conformation') of these is the 'envelope' in which torsional strain is minimised by having one carbon atom above (or below) the plane of the other four (see Fig. 2.15).

Two important conformations of **cyclohexane** are the semi-rigid chair form which is free of both ring strain and torsional strain, and the flexible boat form, which, although free of ring strain, has both torsional strain and van der Vaals

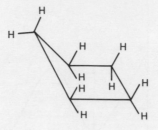

Fig. 2.15   The preferred 'envelope' conformation of cyclopentane

repulsion ('flagpole interaction') between H atoms on carbons 1 and 4 (see Fig. 2.16). In the chair form of cyclohexane, six hydrogen atoms, one on each

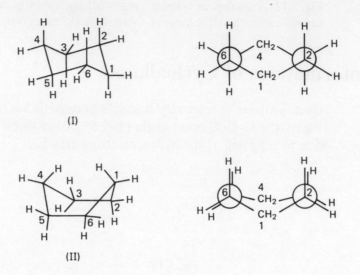

Fig. 2.16   Chair (I) and boat (II) conformations of cyclohexane

carbon atom, lie in the general plane of the ring and are called **equatorial** (e). The other six are at right angles to this plane and are called **axial** (a) (see Fig. 2.17).

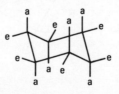

Fig. 2.17   Axial (a) and equatorial (e) positions in the chair conformation of cyclohexane

At normal temperatures the cyclohexane ring rapidly flips back and forth between two equivalent chair conformations and in so doing all of the bonds that were axial become equatorial and vice versa. When a substituent such as a methyl group is present (Carbon 1) van der Vaals repulsion between it and axial H atoms on carbons 3 and 5 ('steric crowding' or 'non-bonded interaction') is minimised if the substituent is in an equatorial position (see Fig. 2.18). Conformations of di-substituted cyclohexanes are summarised in Table 2.1.

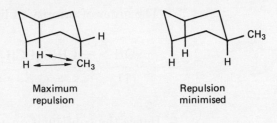

Maximum repulsion    Repulsion minimised

**Fig. 2.18** 1,3-diaxial interaction in methylcyclohexane (only relevant H atoms are shown)

**Table 2.1** Equatorial (e) and axial (a) position of substituents in chair conformations of *cis* and *trans* disubstituted cyclohexanes

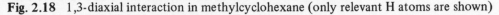

| Isomer | Cis or trans | Position of substituents |
|--------|--------------|--------------------------|
| 1,2 | *cis* | ea and ee |
| | *trans* | ee and aa |
| 1,3 | *cis* | ee and aa |
| | *trans* | ea |
| 1,4 | *cis* | ea |
| | *trans* | ee and aa |

## 2.18  Worked Examples

**Example 2.1**

Write down Fischer–Rosanov* structures for the following compounds. State whether each is a *meso*-form, or an enantiomer. Where the compound is an enantiomer, show the enantiomeric form. Will any of these have a diastereoisomer? If so, show the structure.

(a) (2S,3R)-butan-2,3-diol
(b) (S)-2-amino-3-thiolpropionic acid
(c) (2R,3S)-2,3,4-trihydroxybutanal
(d) (2S,3R)-2-chloro-3-fluorobutanoic acid.    (University of Bradford, 1982)

*Solution 2.1*

Begin this question by drawing Fischer projection formulae for each compound and assigning an order of priority to the ligands on each of the chiral carbon atoms present. Classify each chiral centre as R- or S- and then modify the formulae as necessary to conform with the stipulated configurations.

(a)

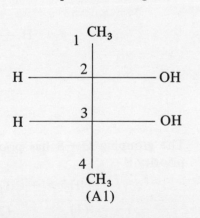

(A1)

\* Usually referred to simply as Fischer structures or projections.

The order of priority of ligands on carbon 2 is

$$-OH > -CH(OH)CH_3 > -CH_3 > -H$$
$$\quad\text{(1)} \qquad\qquad \text{(2)} \qquad\qquad \text{(3)} \qquad \text{(4)}$$

Hence the configuration about carbon 2 is

The descending order of priority is anticlockwise, i.e. Sinister
Since the arrangement of ligands about carbon 3 is the mirror image of that about carbon 2, the configuration at carbon 3 must be R. (If this idea is new to you, verify by the above procedure.)

The configuration 2S, 3R is that stipulated by the question, and you can see that with a plane of symmetry between carbons 2 and 3, the Fischer projection (A1) represents an optically inactive *meso* form. Two enantiomers will exist with Fischer projection formulae

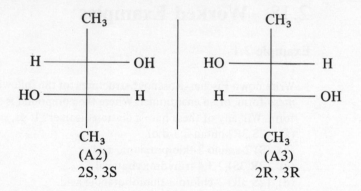

Structures (A2) and (A3) are stereoisomers of (A1) but do not have a mirror image relationship with (A1), i.e. they are diastereoisomers of (A1).

(b)

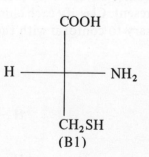

The grouping C—S has priority over C—O, so the order of decreasing priority is

$$-NH_2 > -CH_2SH > -COOH > -H$$
$$\quad\text{(1)} \qquad\qquad \text{(2)} \qquad\qquad \text{(3)} \qquad \text{(4)}$$

Hence

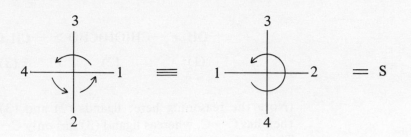

Again, projection (B1) has the required configuration. There will be a mirror image enantiomer (B2),

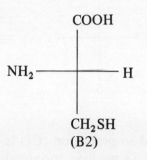

(B2)

with the 2R configuration, but there are no *meso* or other diastereoisomeric forms.

(c)

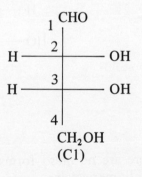

(C1)

For ligands on carbon 2, the order of priority is

$$-OH > -CHO > -CH(OH)CH_2OH > -H$$
$$\quad(1)\qquad\ (2)\qquad\qquad(3)\qquad\qquad(4)$$

Hence the configuration is

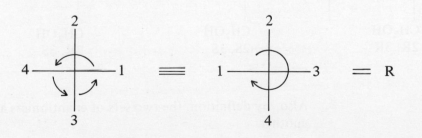

For ligands on carbon 3, the order of priority is

$$\text{—OH} > \text{—CH(OH)CHO} > \text{—CH}_2\text{OH} > \text{—H}$$
$$(1) \qquad\qquad (2) \qquad\qquad\quad (3) \qquad\quad (4)$$

(Note the reasoning here: ligands (2) and (3) have C—O, but ligand (2) then has C—C, whereas ligand (3) has only C—H.)
The configuration about carbon 3 is

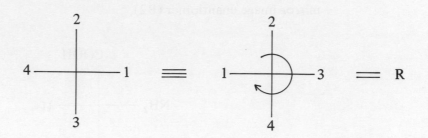

and so projection (C1) is 2R,3R. To produce the required projection with 2R,3S we have to interchange the position of any two ligands on carbon 3, e.g.

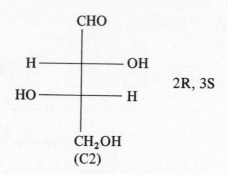

There are no *meso* forms, but (C1) and (C2) are each one isomer of an enantiomeric pair:

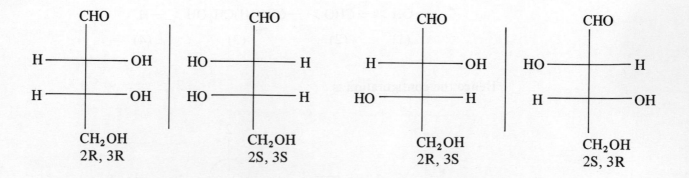

Also, by definition, the two sets of enantiomers are diastereoisomers of one another.

(d)

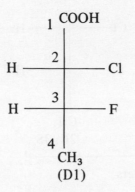

1 COOH
2
H ——|—— Cl
3
H ——|—— F
4
CH₃
(D1)

For carbon 2,

—Cl > —CHFCH₃ > —COOH > —H
(1)        (2)           (3)       (4)

(Again note that C—F has priority over C—O)

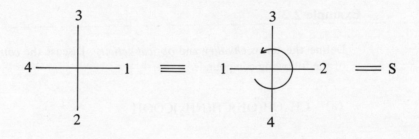

And for carbon 3

—F > —CHClCOOH > —CH₃ > —H
(1)         (2)              (3)      (4)

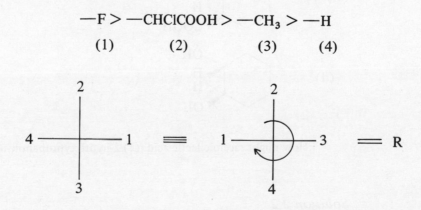

Projection (D1) is 2S,3R, which is the required configuration. As in the previous example, there is no *meso* form, but there are two enantiomeric pairs which are diastereoisomeric:

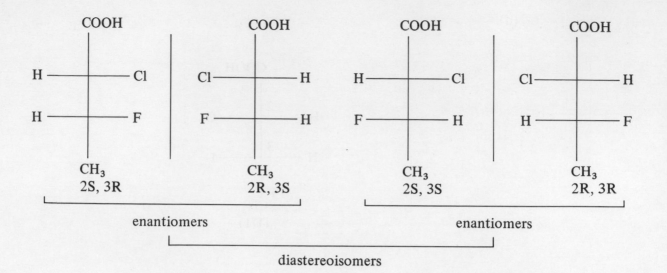

enantiomers                  enantiomers

diastereoisomers

Even if this type of question does not specifically ask you to show your reasoning, always indicate (at least in outline) how you arrive at the various configurations. Should a small slip — for example in transferring priorities to the Fischer projections — lead to the wrong conclusion, you will probably still be awarded a proportion of the marks for showing that you understand the principles involved.

## Example 2.2

Define the terms *chirality* and *optical activity*. Discuss the *configurational stereochemistry* of the following compounds.

(a)   $CH_3CH(OH)CH(NH_2)COOH$

(b)   $CH_3CH(OH)CH(OH)CH_3$

(c)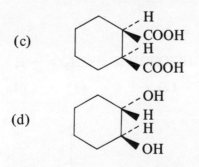

(d)

How might racemic lactic acid ((±)-2-hydroxypropanoic acid) be resolved?

(University of Exeter, 1985)

*Solution 2.2*

Don't be caught out not knowing named concepts and terms such as chirality and optical activity!

(a) This compound has two different chiral centres and will therefore have $2^2 = 4(= 2$ enantiomeric pairs of) stereoisomers:

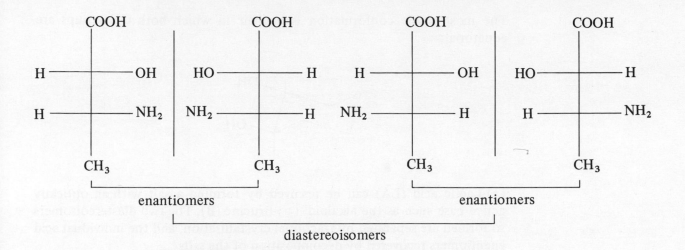

enantiomers            enantiomers

diastereoisomers

The two pairs are diastereoisomers of one another. (Explain any terms that you use, such as enantiomers and diastereoisomers.)

(b) See Example 2.3.

(c) *Cis* cyclohexane-1,2-dicarboxylic acid (c) has a plane of symmetry (show this) and is therefore an optically inactive *meso* form. Although the question does not specifically require this, you could indicate that the most stable conformation for this compound is a *chair* form in which one carboxyl group is axial and the other is equatorial:

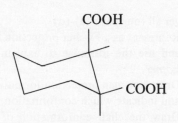

(d) Compound (d) is the *trans* 1,2-diol. This has no plane of symmetry and therefore has two enantiomeric forms:

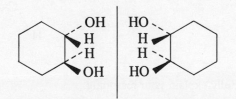

If you are not used to drawing 'flying wedge' formulae, you could show the two isomers in simplified form:

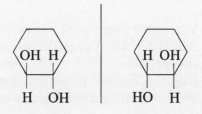

35

The most stable conformation is a chair in which both OH groups are equatorial:

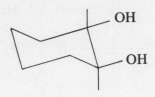

(±)-Lactic acid (LA) can be resolved by forming a salt with an optically active base such as the alkaloid (−)-Brucine (B). The two diastereoisomers so formed are separated by fractional crystallisation, and the individual acid enantiomers recovered by decomposition of the salts:

$$2 (\pm)\text{-LA} + 2(-)\text{-B} \longrightarrow \begin{cases} (+)\text{-LA.} (-)\text{-B} \\ (-)\text{-LA.} (-)\text{-B} \end{cases}$$

$$(+)\text{-LA.} (-)\text{-B} \longrightarrow (+)\text{-LA}$$
$$(-)\text{-LA.} (-)\text{-B} \longrightarrow (-)\text{-LA}$$

**Example 2.3**

Answer all four parts (a)–(d).
(a) Represent as a Fischer projection formula each stereoisomer of 2,3,4-trihydroxybutanal, and use the formulae to explain the meaning of the terms *enantiomer* and *diastereoisomer*. (*9 marks*)
(b) Draw Newman projection formula for the two energy-minimum conformations of butane and indicate which conformation is thermodynamically more stable. (*5 marks*)
(c) Draw the chair conformation of cyclohexane and clearly indicate the axial and equatorial bonds. (*3 marks*)
(d) Designate as E- or Z- the configuration represented by the following structure:

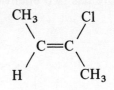

Fully explain your reasoning. (*3 marks*)
(Sheffield Polytechnic, 1985)

*Solution 2.3*

(a) You could begin your answer with a statement such as 'From the formula $HOCH_2CH(OH)CH(OH)CHO$, carbon atoms 2 and 3 are seen to be chiral, and so there are $2^2 = 4 = 2$ enantiomeric pairs of isomers'. Then draw the required Fischer projection formulae, adopting the convention of putting the principal functional group at the top:

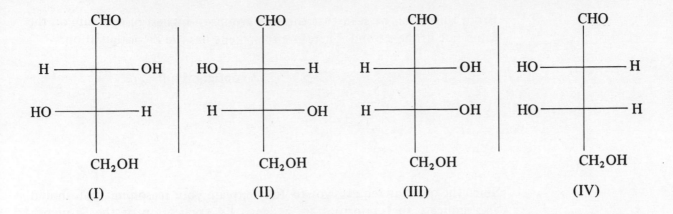

(I)                  (II)                 (III)                 (IV)

I and II, and III and IV are non-superposable mirror images of one another and therefore represent two pairs of enantiomers. I and II are each stereo-isomers of III and IV, but clearly they do not have a mirror image relationship. They are therefore diastereoisomers. You could summarise the relationships thus:

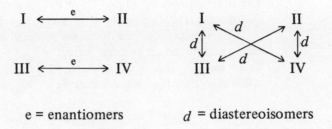

e = enantiomers          $d$ = diastereoisomers

(b) The two required projections are along the C2–C3 axis. Energy-minimum conformations are staggered:

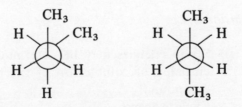

Of the two, the *anti* conformer (where the bulky $CH_3$ groups are as far apart as possible) will be the more stable.

(c) Practice drawing formulae such as this. You will need this ability later on in other parts of your course.

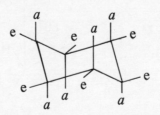

(d) Begin by assigning priorities to the two groups on each of the doubly bonded carbon atoms:

$$-CH_3 > -H \quad \text{and} \quad -Cl > -CH_3$$

37

From this it can be seen that the two groups of highest priority are on the same side of the $\pi$-bond. Therefore the alkene has the Z-configuration:

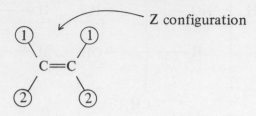

Since the question requires you to 'fully explain your reasoning', you should also indicate that priorities are assigned by application of the Sequence Rules under which orders of priority are based on decreasing atomic number (Section 3.5).

### Example 2.4

(i) The strain energies (in $kJ\ mol^{-1}$) of the simplest cycloalkanes are: cyclopropane, 115; cyclobutane, 110; cyclopentane, 27; and cyclohexane, 0.

Draw the preferred conformation for each of these compounds and comment on each structure and on its strain energy. (*12 marks*)

(ii) There are two isomers of 1,3-di-*tert*-butylcyclohexane and one is $25\ kJ\ mol^{-1}$ more stable than the other. Make a drawing of the conformation of the more stable isomer. (*4 marks*)

(iii) It has been suggested that the energy difference between the isomers in (ii) is a reasonable estimate of the energy difference between chair and twist cyclohexane.

Discuss this argument and make a drawing of a possible structure of the less stable isomer. (*8 marks*)

(University of Bristol, 1985)

*Solution 2.4*

(i) Strain energies may be determined, for example, by comparison of the enthalpy of combustion of the cycloalkane with that of the open chain compound.
**Cyclopropane**

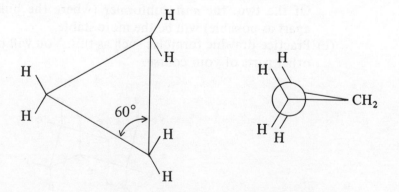

The large value for the strain energy results from a combination of *ring* (or *angle*) strain (C—C—C bond angle reduced from $109.5°$ to $60°$), and *torsional* strain due to eclipsing of hydrogen atoms on adjacent carbon atoms. Poor overlap of the $sp^3$ hybrid atomic orbitals weakens the carbon–

carbon bonds and cyclopropane readily undergoes ring-opening addition reactions, behaving more like an alkene than an alkane.

**Cyclobutane**

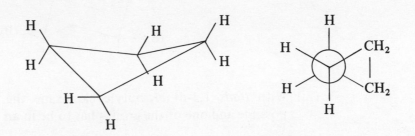

Cyclobutane adopts a slightly folded ('**butterfly**') shape in which a small increase in angle strain (C—C—C bond angle 88°) is more than outweighed by a consequent decrease in torsional strain. Hydrogen atoms on adjacent carbon atoms are not fully eclipsed as they would be if cyclobutane retained a planar structure with a C—C—C bond angle of 90°.

**Cyclopentane**

With a bond angle of 108°, a planar pentagon structure would be virtually free of angle strain, but all the adjacent hydrogen atoms would be eclipsed. The preferred '**envelope**' conformation is one in which torsional strain is minimised by having one carbon atom above (or below) the plane of the other four:

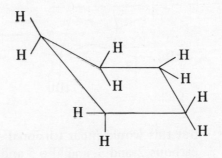

The result is a very flexible, almost strain-free structure, with relatively low strain energy.

**Cyclohexane**

The preferred conformation for this compound is the strain-free, semi-rigid chair form, which has 'normal' C—C—C bond angles and no eclipsing between adjacent hydrogen atoms:

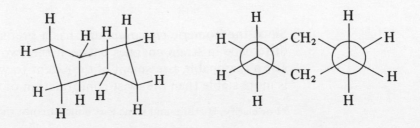

(ii) The *cis* isomer has a preferred conformation in which the two bulky *tert*-butyl groups occupy sterically less crowded equatorial positions:

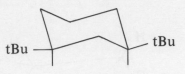

(iii) With *trans* 1,3-di-*tert*-butylcyclohexane the above conformation is not possible and one of the groups has to be in an axial position:

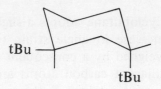

Non-bonded interaction between the bulky axial *tert*-butyl group and diaxial hydrogen atoms on carbons 3 and 5 could be removed by switching to the boat conformation*,

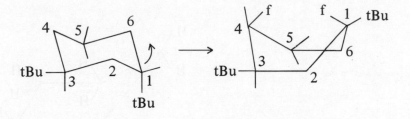

but this would incur torsional strain from eclipsing of hydrogen atoms on carbons 2 and 3, and also 5 and 6. Additional strain would result from the close proximity of the flagpole hydrogens (f). Strain energy is therefore minimised by the adoption of a *skew-boat* or *twist* conformation in which the flagpole hydrogen atoms move further apart, C2–C3 and C5–C6 hydrogens are only partially eclipsed, and the two *tert*-butyl groups are essentially equatorial:

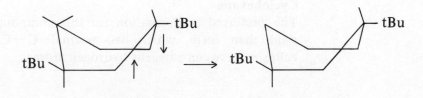

Since the isomeric *cis* compound has a preferred chair conformation, the difference in strain energies between the two isomers (25 kJ mol$^{-1}$) does give a reasonable assessment of the extent to which the chair conformation is more stable than the twist conformation of cyclohexane itself.

*For clarity, H-atoms and some C-H bonds are not shown.

# 2.19  Self-Test Questions

### Question 2.1

Explain concisely the meaning of the terms configuration and conformation.    (*4 marks*)
Indicate, with brief reasoning, the number and type of stereoisomers of the following:
(a)  2,3-dibromobutane;
(b)  2,3-dichloropentane;
(c)  4-bromo-2-hexene;
(d)  2,4-heptadiene.                                                            (*10 marks*)
Giving your reasoning, decide whether the optical isomer (I) whose Fischer projection
formula is shown below has the R or S specification according to the Cahn–Ingold–Prelog
convention.

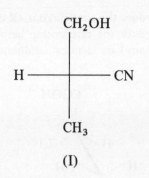

(I)

(*6 marks*)
(Kingston Polytechnic, 1985)


### Question 2.2

Assign R and S designations to each of the following compounds, clearly indicating the
order of priority of groups and how you arrive at your conclusion.

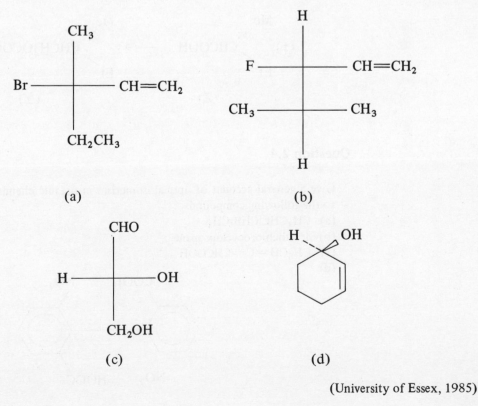

(a)                              (b)

(c)                              (d)

(University of Essex, 1985)

41

## Question 2.3

(a) Express the following molecule (T) as a Newman projection:

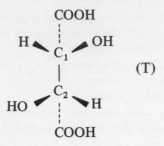

(T)

(b) Deduce the configuration (R or S) of the two asymmetric centres C-1 and C-2.

(c) Deduce the relationship between the following molecules (U, V, W and X), i.e. which (if any) are identical, or diastereoisomers, or enantiomers.

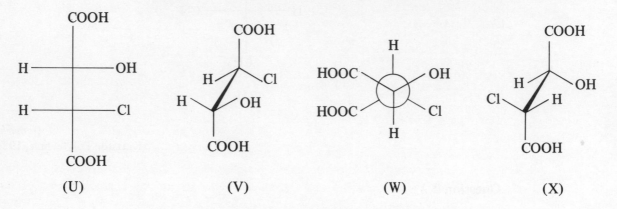

(U)          (V)          (W)          (X)

(d) How would you prepare one of the optically active isomers of the ester, Y, from the racemic acid Z?

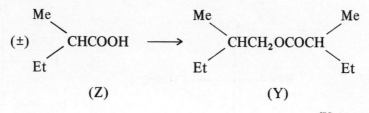

(Z)                    (Y)

(University of Leeds, 1984)

## Question 2.4

Give a general account of optical isomerism in organic chemistry, with particular reference to the following compounds:

(a) $CH_3CHClCHBrCH_3$

(b) 1,2-dichlorocyclopropane

(c) $CH_3CH{=}C{=}CHCOOH$

(d)

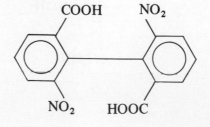

(University of London, 1984)

# 3 Reaction Mechanisms: Basic Principles

## 3.1 Terminology

All organic reactions involve breaking and re-formation of covalent bonds. Bond breaking may be

**homolytic**   $A - B \rightarrow A\cdot + B\cdot$   (*free radical* reactions), or
**heterolytic**   $A - B \rightarrow A^{\oplus} + B^{\ominus}$   (*ionic* reactions).

Reagents are classified as:

(1) **Free radicals**: high energy species with an unpaired electron, e.g. $:\overset{\cdot}{\underset{\cdot\cdot}{Cl}}:$ (abbreviated to $Cl\cdot$) and $\overset{\cdot}{C}H_3$.

(2) **Nucleophiles** (Lewis bases): species with an unshared electron pair which can be donated to form a sigma bond with an electrophilic species.

(3) **Electrophiles** (Lewis acids): species which react by receiving an electron pair from a nucleophile.

Examples of common nucleophiles and electrophiles are given in Table 3.1.

A **reaction mechanism** is a description at the molecular level of the electron shifts necessary to bring about changes in structure during the course of a reaction or one step in a reaction. These changes are summarised in the form of a **mechanistic equation**.

In the reaction

$$\overset{\ominus}{H\overset{\cdot\cdot}{O}}: \quad \overset{\beta CH_3}{\underset{\alpha CH_2}{|}}\text{---}\overset{\cdot\cdot}{\underset{\cdot\cdot}{I}}: \quad \longrightarrow \quad H\overset{\cdot\cdot}{O}\text{---}CH_2 \; \overset{CH_3}{\underset{|}{}} + \; :\overset{\cdot\cdot}{\underset{\cdot\cdot}{I}}:^{\ominus}$$

the hydroxide ion is the **attacking reagent**, and the molecule attacked, iodoethane, is the **substrate**. Within the substrate molecule, the $\alpha$-carbon atom is the **electrophilic centre** and the iodine atom is the (good) **leaving group**. Ethanol is the organic **product** and the iodide ion is the **displaced group**. Similar terminology is used in other reactions.

## 3.2 Some Guidelines for Writing Ionic Mechanistic Equations

### (a) The Correct Use of the Curved Arrow

Movement of an electron pair is shown by means of a **curved (curly) arrow**. The *tail* of the arrow should clearly show the *origin* of the electron pair, and the *arrowhead* should clearly show the *destination* of the electron pair,

**Table 3.1** Some common nucleophiles and electrophiles

| Nucleophiles | Electrophiles |
| --- | --- |
| These may be anions or neutral molecules having an atom with an unshared electron pair, or molecules with one or more multiple bonds | These may be cations or polar molecules with an electron deficient centre, or polarisable molecules which can develop such a centre |

<table>
<tr><td>$H\ddot{O}\colon^{\ominus}$, $R\ddot{O}\colon^{\ominus}$<br>(hydroxide and alkoxide ions)</td><td>$H^{\oplus}$, $R^{\oplus}$<br>(protons, carbocations)</td></tr>
<tr><td>$H_2\ddot{O}$, $R\ddot{O}H$<br>(water, alcohols)</td><td>$\overset{\oplus}{:\ddot{C}l}\colon$, $\overset{\oplus}{:\ddot{B}r}\colon$, $\overset{\oplus}{:\ddot{I}}\colon$<br>(halogen cations)</td></tr>
<tr><td>$\ddot{N}H_2^{\ominus}$, $R\ddot{N}H^{\ominus}$<br>(amide ions)</td><td>$\overset{\delta+}{:\ddot{C}l}{\rightarrow}\overset{\delta-}{\ddot{O}H}$<br>(chloric (1) acid)</td></tr>
<tr><td>$\ddot{N}H_3$, $R\ddot{N}H_2$<br>(ammonia, amines)</td><td>$\overset{\delta+}{:\ddot{B}r}{\rightarrow}\overset{\delta-}{\ddot{B}r}\colon$<br>(polarised bromine molecule)</td></tr>
<tr><td>$:\ddot{C}l\colon^{\ominus}$, $:\ddot{B}r\colon^{\ominus}$, $:\ddot{I}\colon^{\ominus}$<br>(halide ions)</td><td>$ArN_2^{\oplus}$<br>(diazonium cation)</td></tr>
<tr><td>$N\ddot{C}^{\ominus}$, $\ddot{R}^{\ominus}$<br>(cyanide ion, carbanions)</td><td>$NO_2^{\oplus}$<br>(nitryl cation)</td></tr>
<tr><td>alkenes, alkynes, benzene</td><td>$NO^{\oplus}$<br>(nitrosyl cation)</td></tr>
</table>

e.g. for the substitution reaction $R\!-\!I + \overset{\ominus}{OH} \longrightarrow R\!-\!OH + I^{\ominus}$,

Tail close to sigma bonded electron pair

Arrowhead shows that the destination is a non-bonding position on the iodine atom

Tail close to a non-bonded electron pair

Arrowhead shows electron pair is displaced to a bonding position with the α-carbon atom

Remember that curved arrows should *not* be used to show changes in charge distribution which may result from electron movements. Thus, is incorrect, sloppy, and could lead to confusion when writing more complicated mechanistic equations.

### (b) Non-bonded Electrons

Although it takes a little extra time, it is well worth drawing structures in which *all the non-bonded electrons are shown*. This is a great aid in keeping track of electron movements and helps to avoid the common error of showing atoms with incorrect numbers of valence electrons.

### (c) Designation of Charges

To avoid confusion with plus and minus signs, always show charges as ⊕ or ⊖, and omit the plus sign between reagents:

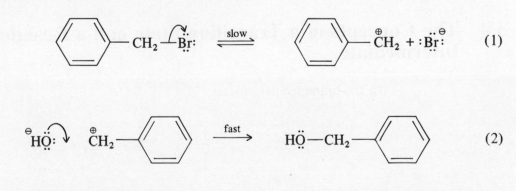

Negative charge close to oxygen, but clear of arrow tail

No + sign between reagent and substrate

### (d) Direction of Electron Movements

Try to arrange equations so that electrons (and hence curved arrows) always *flow from left to right*. This may sometimes involve reversing the 'normal' way of drawing a structure. For example, alkaline hydrolysis of bromoethylbenzene (benzyl bromide) occurs in two stages:

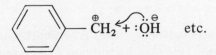

Equation (2) is obviously neater, clearer, and less likely to lead to any misunderstanding than alternative formulations such as

## 3.3 Guidelines for Writing Free Radical Mechanistic Equations

### (a) Curved Arrows

Movement of single electrons is shown by means of *half-headed* curved arrows, ⌒ .

### (b) Non-bonded Electrons

Greater clarity results from omission of non-bonded electrons,

$$\text{e.g. } Cl\!-\!Cl \longrightarrow 2\ \overset{\cdot}{Cl}$$

but remember that the chlorine atom has 7 valence electrons.

### (c) Direction of Electron Movements

Although single electron movements occur in both directions, there is some advantage in writing equations with the majority of half-headed arrows going from left to right. Thus

$$\overset{\cdot}{Cl}\ \ H\!-\!CH_3 \longrightarrow Cl\!-\!H + \overset{\cdot}{C}H_3$$

is more consistent with the convention for writing ionic equations than is the alternative

$$\overset{\cdot}{C}H_3\!-\!H\ \ \overset{\cdot}{Cl} \longrightarrow \overset{\cdot}{C}H_3 + H\!-\!Cl$$

Mechanisms of free radical reactions are discussed in Chapter 12.

## 3.4 The Concepts of a Transition State and a Reaction Intermediate

In the unimolecular process

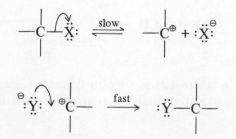

the carbocation, $-\overset{|}{\underset{|}{C}}{}^{\oplus}$, is a short-lived species described as a reaction **intermediate**. This may be thought of as a temporary product of the first step.

### (a) Structure of Carbon Intermediates

(1) **Carbocations**, $R^{\oplus}$, are planar, e.g.

vacant 2p orbital

(2) **Carbanions**, $\overset{..}{R}{}^{\ominus}$, are tetrahedral, e.g.

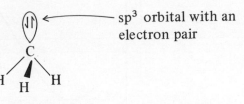

sp³ orbital with an electron pair

(3) **Free radicals**, $R^{\cdot}$, are usually planar, e.g.

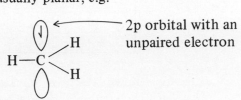

2p orbital with an unpaired electron

## 3.5 Stability of Carbon Intermediates

For carbocations and free radicals the order of increasing stability is

$1° < 2° < 3°$ (inductively stabilised) $<$ allyl, benzyl (resonance stabilised)

Carbanions show less regular variation, but

alkyl $<$ allyl, benzyl

In order to form the carbocation intermediate, the molecule has to momentarily pass through a high energy condition where the R to X bond is stretched, but not yet fully broken. This is the **transition state**, shown as

The broken line represents the bond breaking and the $\delta-$ sign shows that electron density has been displaced on to the departing group X.

In an $S_N2$ reaction

$$:\overset{\ominus}{\underset{..}{Y}}: \quad \overset{|}{\underset{|}{C}}-\overset{..}{\underset{..}{X}}: \longrightarrow :\overset{..}{\underset{..}{Y}}-\overset{/}{\underset{\backslash}{C}} + :\overset{..}{\underset{..}{X}}:^{\ominus}$$

there is no reaction intermediate, but the reactants have to pass through a bimolecular transition state* shown as

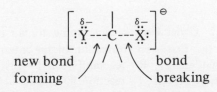

new bond forming          bond breaking

*'Bimolecular' here referring to the number of species involved.

47

## (a) Energy Profile Diagrams

These are simply a qualitative graphical representation of the energy changes involved in going from reactants to products. Most common reactions can be described by one or other of two basic profiles (Figs 3.1 and 3.2).

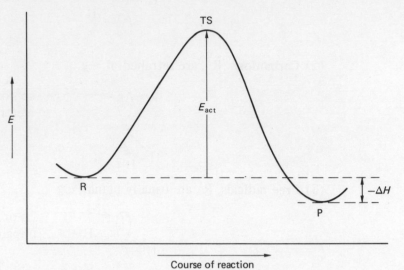

**Fig. 3.1** Generalised energy profile for an exothermic bimolecular reaction. $E$ = potential energy; R = reactant(s); TS = transition state; $E_{act}$ = activation energy; P = product(s); $\Delta H$ = enthalpy change

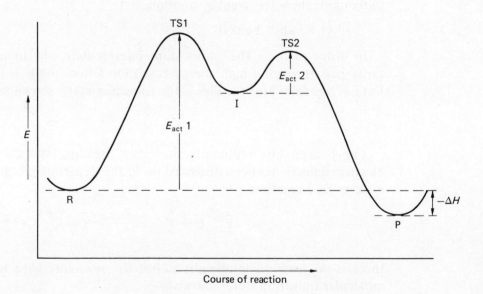

**Fig. 3.2** Generalised energy profile for an exothermic unimolecular process. Since $E_{act}1 > E_{act}2$, the first step (formation of the intermediate) is the slow rate determining step. $E$ = potential energy; R = reactant(s); TS1 = first transition state; $E_{act}1$ = activation energy for formation of TS1; I = reaction intermediate; TS2 = second transition state; $E_{act}2$ = activation energy for formation of TS2; P = product(s); $\Delta H$ = enthalpy change

## 3.6 Molecularity and Kinetic Order of Reaction

**Molecularity** of reaction is a **theoretical** concept, defined as the *number of species involved in the transition state of the rate determining step*.*

The **kinetic order** of a reaction is an *experimentally* determined quantity relating to the *number of species on whose concentration the rate of reaction depends*.

## 3.7 Role of the Solvent

The influence of a **solvent** on organic reactions can be both complex and subtle. The ability to dissolve the reactants and bring them into intimate contact with one another is but one aspect of solvent behaviour.

Solvation involves *electrostatic interaction* between solvent molecules and the solvated species which increases in the order

$$\text{weakly polar substrate} < \text{more polar transition state} < \text{ions with unit charges}$$

Thus, *differential* solvation of a transition state can lower the activation energy for its formation and so increase the rate of reaction.

Conversely, more effective solvation of an ionic reagent relative to a less polar transition state may decrease the rate of a bimolecular reaction.

The effectiveness of solvation may also depend to some extent on the *need* for such stabilisation. The more an ion or transition state is stabilised by, say, inductive or resonance dispersal of charge, the smaller the contribution made by the solvent.

Solvation can also affect the nucleophilic strength of anions. The small fluoride ion is readily solvated by water molecules and is only weakly nucleophilic in aqueous solution. Changing to an aprotic solvent such as dimethylsulphoxide (which is relatively ineffective in solvating anions) greatly increases its nucleophilicity.

## 3.8 The Use of Partial Structures

The shorthand device of *partial structures* is particularly useful when making summary notes. For example, many different bimolecular displacement reactions conform to the general mechanistic equation:

Even a specific type of reaction such as hydrolysis of an iodoalkane can still be generalised for a variety of iodoalkane substrates by putting $\ddot{Y}^\ominus = H\ddot{O}{:}^\ominus$ and $-X = -\ddot{I}{:}$, etc. However, take care to see that the normal valencies of atoms are adhered to, and avoid abbreviations such as

$$^\ominus\ddot{Y} \quad C—X \quad !!$$

*The slowest step in a multistep reaction (Fig. 3.2).

## 3.9 Mechanistic Classification of Ionic Reactions

In the substitution reaction

$$HO^{-} \quad CH_3 {-} I \longrightarrow HO{-}CH_3 + I^{\ominus}$$

the attacking reagent is a nucleophile, and so the reaction is classified as **aliphatic nucleophilic substitution**.

Similarly, in the addition reaction

$$CH_2{=}CH_2 + HBr \longrightarrow CH_3{-}CH_2{-}Br$$

ethene is the substrate and HBr the attacking reagent. The first step is protonation of the alkene

$$CH_2{=}CH_2 \quad H{-}Br \longrightarrow \overset{\oplus}{C}H_2{-}CH_3 + {:}Br^{\ominus}$$

$H^{\oplus}$ is an electrophile, and so the overall reaction is described as **electrophilic addition** to the alkene.

However, like the substitution reaction above, protonation of the alkene can also be described as a nucleophile–electrophile reaction:

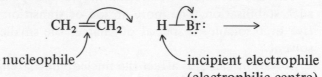

Likewise, the second step is

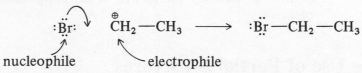

In fact, however they may be *classified*, all simple ionic processes are *mechanistically* similar in that they can be broken down into one or more nucleophile-electrophile interactions. Some examples are given in the following section, and such rationalisation should help you in both understanding and memorising the many ionic reactions that you study.

## 3.10 Summary of Some Common Ionic Processes

### (a) Nucleophilic Substitution

$S_N 1$

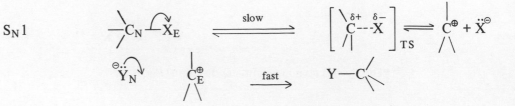

$S_N2$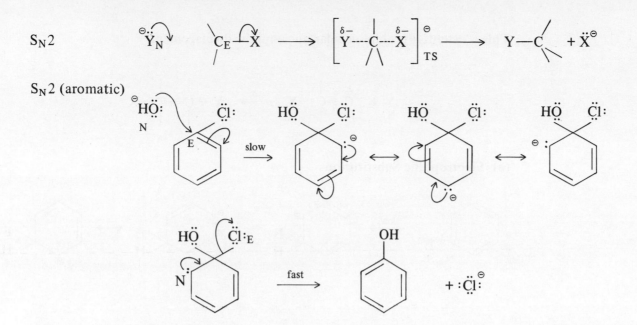

$S_N2$ (aromatic)

N.B. Activating groups (such as *o*- and *p*-$NO_2$) have been omitted for clarity.

### (b) Elimination

E1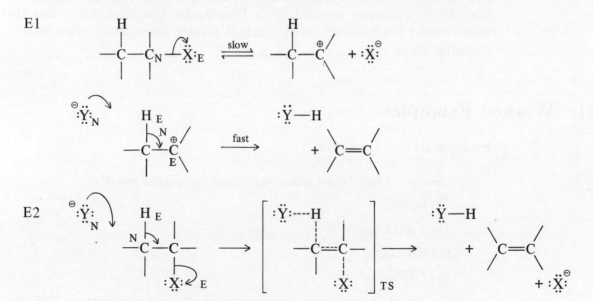

E2

### (c) Nucleophilic Addition

Addition may be completed by protonation of the oxygen atom and/or followed by elimination. In acid catalysed reactions protonation may be the first step.

### (d) Electrophilic Addition

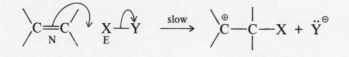

This is essentially a bimolecular nucleophilic displacement

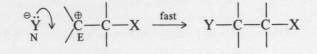

### (e) Electrophilic Substitution

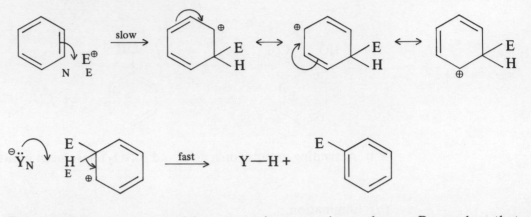

For simplicity, most transition states have not been shown. Remember that bimolecular reactions proceed via a bimolecular transition state, and that in unimolecular processes the intermediate is formed from, and further reacts via, a transition state.

## 3.11 Worked Examples

### Example 3.1

(a) Show how THREE of the following nucleophiles could be generated:

    (i) $HC\equiv\ddot{C}^{\ominus}$

    (ii) $\overset{\ominus}{C}H(CO_2Et)_2$

    (iii) $\overset{\ominus}{C}H_2CO_2Me$

    (iv) $CH_3\ddot{\overset{\ominus}{O}}:$

(b) Use curved arrows to show the expected electron pair movements in THREE of the following systems, and hence deduce the reaction product in each case.

    (i) $CH_3CO_2CH_3 \xrightarrow{\overset{\ominus}{O}H/H_2O}$

    (ii) $PhCO_2CH_3 \xrightarrow[\text{2. } H_2O]{\text{1. LiAlH}_4}$

    (iii) $PhCOCl \xrightarrow{NH_3}$

    (iv) $CH_2\overset{\displaystyle O}{-\!\!\!\diagup\!\!\!\diagdown\!\!\!-}CH_2 \xrightarrow{H^{\oplus}/H_2O}$

(Sheffield City Polytechnic, 1984)

*Solution 3.1*

(a) You can assume here, and in similar questions, that the examiner is looking for knowledge of general conditions rather than exact statements about concentrations or temperatures, etc.

(i) Alkynes are only very weakly acidic, and removal of alkynic hydrogen requires the use of a very strong base. Methods available include:

1. Reaction with sodamide in liquid ammonia:

$$\overset{\ominus}{\ddot{N}}H_2 \quad H-C\equiv CH \longrightarrow \ddot{N}H_2-H + \overset{\ominus}{\ddot{C}}\equiv CH$$

2. Reaction with a metal alkyl such as butyl lithium in ethereal solution:

$$(Li^+)\, CH_3(CH_2)_2\overset{\ominus}{\ddot{C}}H_2 \quad H-C\equiv CH \longrightarrow CH_3(CH_2)_2CH_3 + Li^{\oplus}\,\overset{\ominus}{\ddot{C}}\equiv CH$$

3. Reaction with an alkyl Grignard reagent in ether

$$(MgX^{\oplus})\,\overset{\ominus}{\ddot{R}} \quad H-C\equiv CH \longrightarrow RH + MgX^{\oplus}\,\overset{\ominus}{\ddot{C}}\equiv CH$$

Only one method need be quoted, but illustrate the reaction involved with an outline mechanistic equation as shown. In practice, of course, the method selected will depend on the use to which the nucleophile is to be put.

(ii) The anion of 'diethyl malonate' (diethylpropandioate) is best prepared by treating the ester with sodium ethoxide in ethanol:

$$(Na^{\oplus})\, Et\overset{\ominus}{\ddot{O}}\colon \quad H-\overset{..}{C}H(CO_2Et)_2 \longrightarrow Et\ddot{O}H + Na^{\oplus}\,\overset{\ominus}{\ddot{C}}H(CO_2Et)_2$$

(iii) Methyl hydrogen in the ethanoate ester is a little less acidic than that in diethyl malonate above, but can be displaced in the same way by the powerfully basic ethoxide ion:

$$(Na^{\oplus})\, Et\overset{\ominus}{\ddot{O}}\colon \quad H-CH_2CO_2Me \longrightarrow Et\ddot{O}H + Na^{\oplus}\,\overset{\ominus}{\ddot{C}}H_2CO_2Me$$

(iv) Alkoxide ion is prepared by treating the dry alcohol with sodium metal:

$$2\,Na^{\cdot} + 2\,H-\overset{..}{\ddot{O}}CH_3 \longrightarrow H_2 + 2\,Na^{\oplus}\colon\overset{\ominus}{\ddot{O}}-CH_3$$

(b) (i) This first reaction is base catalysed hydrolysis of an ester. Evidence favours a two-step mechanism with an anionic intermediate, rather than direct displacement of the ethanoate anion:

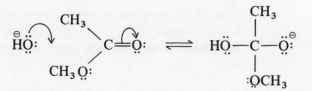

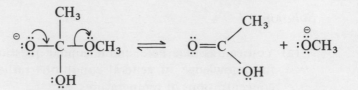

$$CH_3\overset{\ominus}{O} + H_2O \longrightarrow CH_3OH + \overset{\ominus}{O}H \quad \text{(catalyst regenerated)}$$

And in alkaline solution,

$$CH_3COOH + NaOH \longrightarrow CH_3CO\overset{\ominus}{O}\overset{\oplus}{Na} + H_2O$$

These last two reactions effectively prevent the reverse steps and base catalysed hydrolysis goes to completion.

(ii) Reduction of an ester with lithium tetrahydridoaluminate (lithium aluminium hydride) occurs in two stages. If you first explain that donation of hydride ion to the carbon atom of the carbonyl group is accompanied by complexing of the residual $AlH_3$ with the carbonyl oxygen atom,

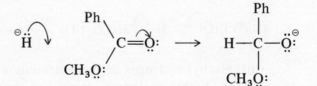

and that in subsequent workup the oxygen atom is protonated by a water molecule to give an alcoholic OH group, you can then simplify your description of the mechanism for reduction of the ester as follows:

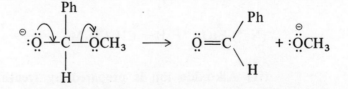

Further reaction reduces the aldehyde in a like manner:

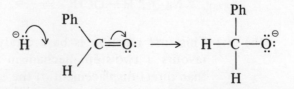

And in the aqueous workup, the complexed anions are protonated by water molecules to give the final reduction products, $CH_3OH$ and $PhCH_2OH$ (together with LiOH and $Al(OH)_3$).

(iii) As with hydrolysis of an ester, discussed above, replacement of halogen in an acid chloride by ammonia is believed to occur in two distinct steps with the involvement of an anionic intermediate:

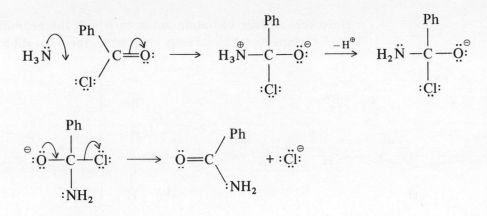

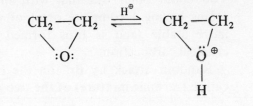

Loss of the proton may occur before, during, or after loss of the chloride ion.

(iv)

$$CH_2—CH_2 \; \underset{}{\overset{H^{\oplus}}{\rightleftharpoons}} \; CH_2—CH_2$$

Protonation weakens the O—C bond and increases the positive charge on the α-carbon atom.

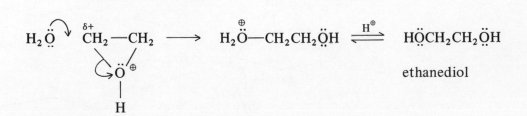

ethanediol

## Example 3.2

1. Describe an experiment which may be taken as evidence for the formation of a cyclic bromonium ion during the ionic addition of bromine to an alkene and explain the significance of the results.
2. In what way do the mechanisms for the addition of hydrogen bromide and bromine to an alkene differ? Explain how this affects the addition of these two reagents to 1,2-dimethylcyclopentene. (Wolverhampton Polytechnic, 1984)

*Solution 3.2*

1. The idea of a bromonium ion was first proposed as an explanation for the exclusive formation of the *trans* dibromide in the ionic addition of bromine to a cyclic alkene such as cyclohexene. If the reaction involved the 'normal' planar carbocation intermediate, then uptake of $Br^{\ominus}$ would be expected to give a mixture with at least *some cis*-isomer present, even if steric hindrance by the large bromine atom favoured formation of the *trans*-isomer:

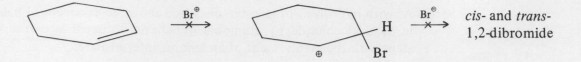

cis- and trans-1,2-dibromide

However, a cyclic bromonium ion in which the bromine atom is bonded to *both* carbon atoms could only give rise to the *trans* dibromide:

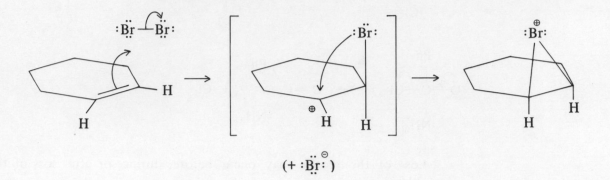

$(+ :\ddot{\underset{..}{Br}}:^{\ominus})$

You could point out that with an extra electron pair bond and with all atoms having a complete octet of valence electrons, the bromonium ion is more stable (and is thus formed more readily) than the corresponding classical carbocation.

Random attack by $Br^{\ominus}$ on the bromonium ion gives a racemic modification (racemic mixture) of the two optical isomers:

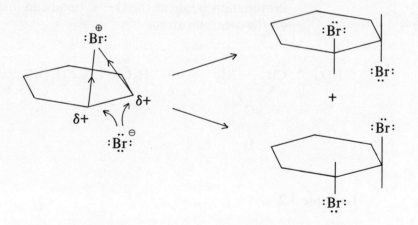

2. Addition of hydrogen bromide occurs via a normal planar carbocation intermediate, giving rise to a mixture of four isomers — two enantiomeric pairs which are diastereoisomers of one another:

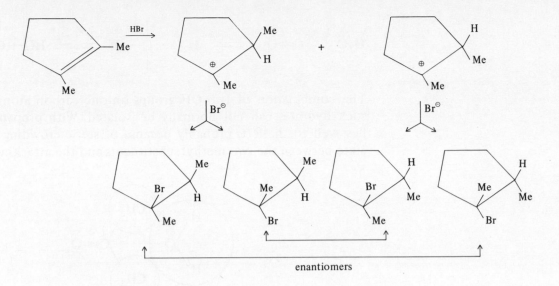

Bromination on the other hand gives an enantiomeric mixture of the two *trans* 1,2-dibromo-1,2-dimethylcyclopentanes:

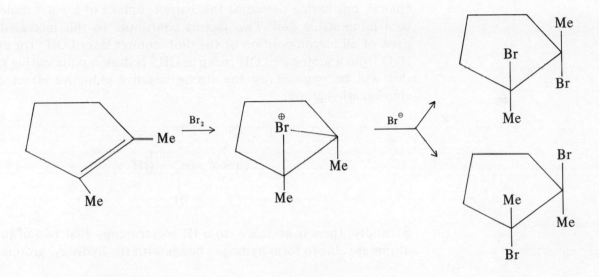

## Example 3.3

By considering the mechanisms of the reactions concerned, explain the following observations:

(a) Chloral (trichloroethanal) most commonly exists as a crystalline hydrate, while acetone forms very little hydrate even in aqueous solution.      (*5 marks*)

(b) One mole of the aldehyde, RCHO, reacts with LiAlH$_4$ (0.25 mole) to give a complex which is hydrolysed to the alcohol, RCH$_2$OH, while one mole of the ester, RCOOR, requires 0.5 mole of LiAlH$_4$ to give, after work-up, the same alcohol.      (*3 marks*)

(c) The base catalysed addition of HCN to methyl ethyl ketone (butan-2-one) gives the cyanohydrin while similar reaction with methyl vinyl ketone (but-3-en-2-one) gives 4-oxopentanenitrile (CH$_3$.CO.CH$_2$.CH$_2$CN).      (*4 marks*)

(d) Acetone is brominated in acid solution to form a monobromo product while in alkaline solution the product is bromoform (tribromomethane).      (*8 marks*)

(The City University, 1984)

*Solution 3.3*

(a) There is spectroscopic evidence (e.g. carbonyl absorption in the UV) that carbonyl compounds react reversibly with water in aqueous solution

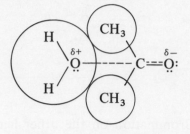

The combination of two OH groups on one carbon atom is unstable, and such hydrates can not normally be isolated. With propanone, equilibrium lies well to the left, probably because of steric crowding in the transition state between the two methyl substituents and the attacking water molecule

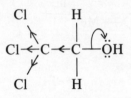

There may well be a similar effect from the large trichloromethyl group in chloral, but having overcome this barrier, uptake of a water molecule leads to a more stable diol. Two factors contribute to this increased stability. First of all, decomposition of the diol requires loss of $OH^{\ominus}$ (or more likely $H_2O$ from a protonated OH group as $OH^{\ominus}$ is itself a poor leaving group) and this will be opposed by the strong negative inductive effect of the trichloromethyl group:

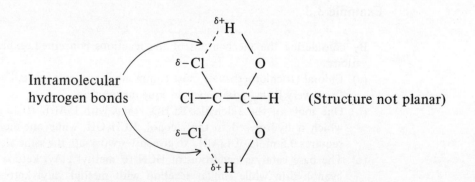

Secondly, there is evidence from IR spectroscopy that two of the chlorine atoms are able to form hydrogen bonds with the hydroxyl groups:

Intramolecular hydrogen bonds

(Structure not planar)

Such bonding lowers the internal energy of the molecule, and so further increases the activation energy for the reverse reaction.

(b) In the reduction of carbonyl compounds, the reagent lithium aluminium hydride (lithium tetrahydridoaluminate (III)) functions by donating hydride ion ($\overset{..}{H}{}^{\ominus}$) to the carbonyl carbon atom:

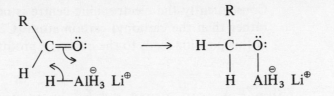

Repetition with further molecules of the aldehyde leads to

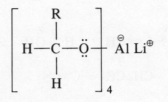

which on hydrolysis with water gives four molecules of the alcohol $RCH_2OH$. Thus reduction of one mole of RCHO requires the use of 0.25 mole of $LiAlH_4$.

Similar reaction with the ester RCOOR gives initially an aldehyde, RCHO, together with the alcohol, ROH.

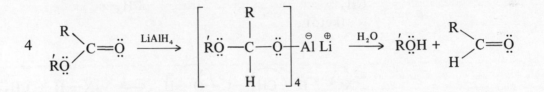

The aldehyde then further reacts as above, so that total reduction of one mole of ester requires 0.25 + 0.25 = 0.5 mole of $LiAlH_4$.

(c) Hydrogen cyanide is only weakly acidic, and the base catalyst increases the concentration of cyanide ion

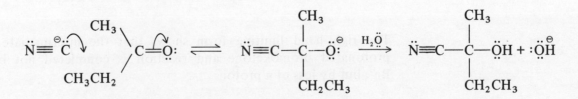

Attack by the nucleophilic cyanide ion on the saturated ketone leads to the cyanohydrin:

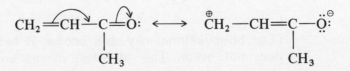

But-3-en-2-one has a mesomeric structure in which the carbonyl group is conjugated with the alkene double bond:

Consequently the electrophilic centre is now the alkene carbon atom (C4) rather than the carbonyl carbon atom (C2), and attack by the cyanide ion at this position leads to the observed product:

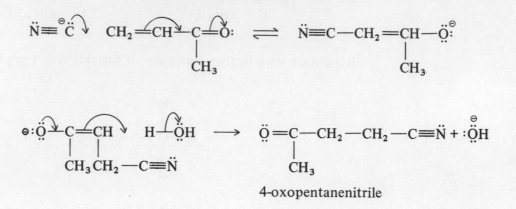

4-oxopentanenitrile

(d) In acid solution, it is the **enol tautomer** of acetone which is brominated:

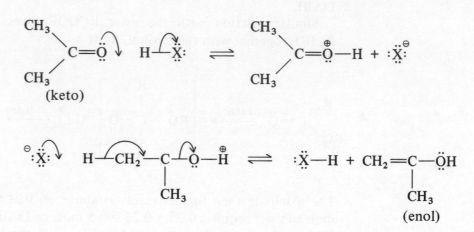

(keto)

(enol)

Bromine then adds in the usual way to the alkene double bond. Uptake of $Br^{\oplus}$ occurs so as to give the more stable of the two possible carbocation intermediates, which is resonance stabilised:

$$HO-C=CH_2 \quad :Br-Br: \quad \rightleftharpoons \quad \left[ HO-C-CH_2-Br: \leftrightarrow HO=C-CH_2-Br: \right] + :Br:^{\ominus}$$

The right-hand limiting form shows that the intermediate is in fact a protonated bromoketone and reaction is completed not by uptake of $Br^{\ominus}$, but by loss of a proton:

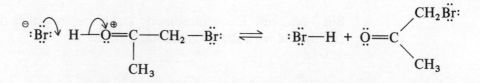

(The bromoketone may well enolise as before, but further bromination does not occur. The −I effect of the bromine substituent lowers the reactivity of the alkene double bond and destabilises the cationic intermediate.)

In alkaline solution the presence of a base results in the formation of a resonance stabilised carbanion:

$$\overset{\ominus}{H\ddot{O}:} \quad H-CH_2-C\overset{..}{=}\ddot{O} \rightleftharpoons H\ddot{O}-H + \left[ \overset{\ominus}{\ddot{C}H_2}-C\overset{CH_3}{=}\ddot{O}: \leftrightarrow CH_2=C\overset{..}{-}\ddot{O}:^{\ominus} \right]$$

$$\underset{CH_3}{\qquad} \qquad \underset{CH_3}{\qquad} \quad \underset{CH_3}{\qquad}$$

which then displaces $Br^{\ominus}$ from a bromine molecule

$$\ddot{O}=C\overset{CH_3}{\underset{|}{-}}\overset{\ominus}{\ddot{C}H_2} \quad :\ddot{B}r-\ddot{B}r: \longrightarrow \ddot{O}=C\overset{CH_3}{\underset{|}{-}}CH_2-\ddot{B}r: + :\ddot{B}r:^{\ominus}$$

(This step can be shown as a reaction of the enolate ion:

$$\overset{\ominus}{:\ddot{O}}-C\overset{CH_3}{=}CH_2 \quad :\ddot{B}r-\ddot{B}r: \longrightarrow \ddot{O}=C\overset{CH_3}{\underset{|}{-}}CH_2-\ddot{B}r: + :\ddot{B}r:^{\ominus})$$

The two remaining hydrogen atoms are made even more acidic by the $-I$ effect of the bromine atom and are therefore rapidly replaced to give the bromoketone $CH_3COCBr_3$.

Reaction is completed by a two-step bimolecular replacement of $\overset{\ominus}{C}Br_3$ by $HO^{\ominus}$:

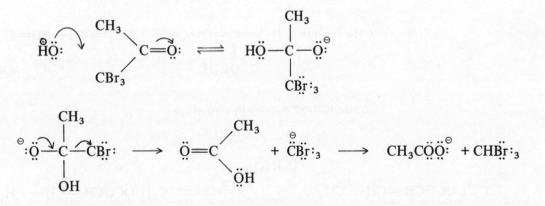

Such displacement is possible because inductive stabilisation of the carbanion by the three electron attracting bromine atoms makes $CBr_3^{\ominus}$ a reasonably good leaving group.

### Example 3.4

How could you account for THREE of the following observations? *(3 × 7; max 20)*

(a) 
$$\overset{CO_2C_2H_5}{\underset{CO_2C_2H_5}{|}} + CH_3CH=CHCO_2C_2H_5 \xrightarrow{C_2H_5O^{\ominus}} C_2H_5O_2CCOCH_2CH=CHCO_2C_2H_5 + C_2H_5OH$$

(b) When 2,2-dimethylpropan-1-ol ($(CH_3)_3CCH_2OH$) is heated with acid, it gives two different alkenes, A and B, each of formula $C_5H_{10}$. A comprises 85% of the mixture, and B 15%.

(c) 2-methylpropene ($(CH_3)_2C=CH_2$) reacts with ethylene (ethene) and hydrogen chloride, under polar conditions, to yield 1-chloro-3,3-dimethylbutane.

(d) Whilst most organic cyanides do not react with C—C cleavage, compound II readily ionises in a polar solvent, and material recovered from this solvent contains compound II and compound III.

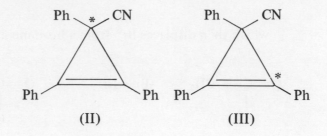

(II)  (III)

(* represents an isotope of carbon)    (University of Sussex, 1984)

*Solution 3.4*

(a) This is a condensation reaction in which a carbanion formed from the unsaturated ester displaces ethoxide ion in the saturated ester. The methyl hydrogen atoms in the unsaturated ester are weakly acidic because their removal (by strongly basic ethoxide ion) gives rise to a resonance stabilised anion:

$$C_2H_5\overset{\ominus}{\ddot{O}}: \quad H\!-\!CH_2CH=CHC\ddot{O}_2C_2H_5 \; \rightleftharpoons \; C_2H_5\ddot{O}H + \overset{\ominus}{C}H_2CH=CHC\ddot{O}_2C_2H_5$$

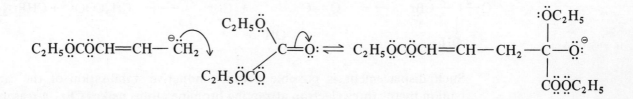

Condensation occurs in two steps:

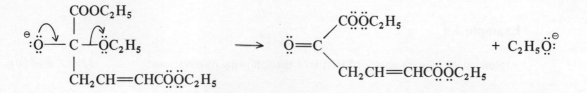

(b) Protonation of the alcohol converts the poor leaving OH group into a good leaving water molecule. Ionisation is facilitated by migration of a methyl group (neighbouring group participation) which converts what would have been an unstable 1° carbocation into a much more stable 3° carbocation:

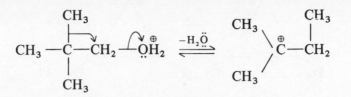

There are two dfferent β-carbon atoms in the cation from which loss of H$^\oplus$ can occur, thus giving rise to two different alkenes:

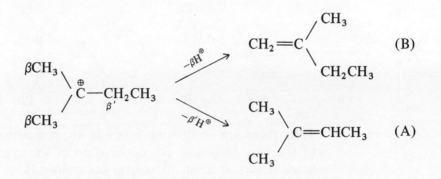

The more stable (more alkylated) alkene, 2-methylbut-2-ene (alkene **A**) is the major product (Saytzeff Rule).

(c) Of the two alkenes, 2-methylpropene is the more readily protonated by hydrogen chloride because it gives rise to a more stable 3° carbocation

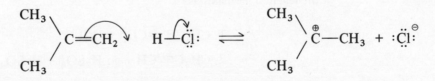

Reaction continues by addition of this carbocation to a molecule of ethene,

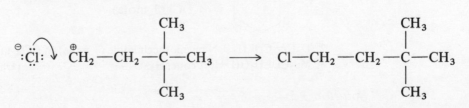

and is completed by uptake of $:\ddot{\text{C}}\text{l}:^\ominus$

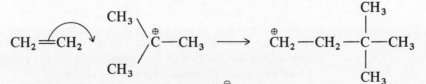

1-chloro-3,3-dimethylbutane

63

(d) Carbon–carbon cleavage can occur in this example because loss of $CN^{\ominus}$ gives a resonance stabilised cation:

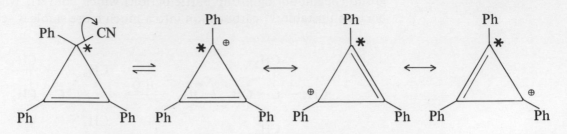

From each of the three limiting forms shown, further limiting forms can be drawn in which the +ve charge appears on ring carbon atoms of the phenyl substituent:

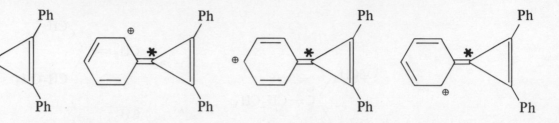

Thus there are 16 limiting forms in all, indicating extensive delocalisation of the positive charge. Recombination of the cyanide ion can occur at the original carbon atom to reform compound II, or at one of the other two ring carbon atoms, either of which will give rise to compound III.

## Example 3.5

(a) Predict the products of the following reactions, explaining your reasoning and showing the reaction mechanisms.

1. $CF_3CH{=}CH_2 + HCl$

2. $CH_3C{\equiv}CH + aq.H_2SO_4 + HgSO_4$

3. 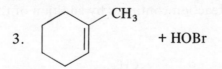 + HOBr

(b) Each of the reactions below gives two products (not necessarily in equal yield) that differ only in the location of deuterium. What are the products? What does their formation tell you about the mechanisms of the reactions?

1. 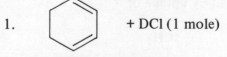 + DCl (1 mole)

2. $D_2C{=}CHCH_3 + $ N-bromosuccinimide (1 mole) in $CCl_4$ in sunlight

3. $Me_2CDCH_2OH + $ hot conc. $H_2SO_4$  (University of Leicester, 1985)

*Solution 3.5*

There are six mechanistic questions here, so clearly you will not be expected to answer each one in great detail.

(a) 1. Ionic addition gives 1,1,1-trifluoro-3-chloropropane. Protonation of the alkene occurs preferentially at C2, rather than C1, since the 1° carbocation in this instance is more stable than the alternative 2° carbocation. This is because protonation of C2 places the positive charge on C1 which is further removed from the positive carbon atom of the trifluoromethyl group:

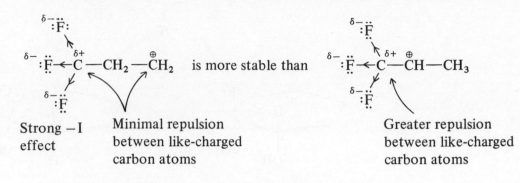

Strong −I effect | Minimal repulsion between like-charged carbon atoms | Greater repulsion between like-charged carbon atoms

Thus the mechanism can be formulated as follows:

$$CH_2\!=\!CHC\ddot{F}_3 \quad H\!-\!\ddot{C}l: \overset{slow}{\rightleftharpoons} \overset{\oplus}{C}H_2\!-\!CH_2C\ddot{F}_3$$

$$^{\ominus}:\ddot{C}l: \quad \overset{\oplus}{C}H_2\!-\!CH_2C\ddot{F}_3 \xrightarrow{fast} :\ddot{C}l\!-\!CH_2\!-\!CH_2C\ddot{F}_3$$

2. Acid catalysed hydration of propyne gives propanone. Reaction is slow, but is further catalysed by $Hg^{2\oplus}$ ions (the role of which is not fully understood). Protonation occurs at C1 to give the more stable of the two possible carbocation intermediates:

$$(H_2O + H_2SO_4 \rightleftharpoons H_3\overset{\oplus}{O} + HSO_4^{\ominus})$$

$$CH_3\!-\!C\!\equiv\!CH \quad H\!-\!\overset{\oplus}{\ddot{O}}H_2 \rightleftharpoons CH_3\!-\!\overset{\oplus}{C}\!=\!CH_2 + \ddot{O}H_2$$

$$CH_3\!-\!\overset{\oplus}{C}\!=\!CH_2 \longrightarrow CH_3\!-\!C\!=\!CH_2 \xrightarrow{-H^{\oplus}} CH_3\!-\!C\!=\!CH_2 \rightleftharpoons CH_3\!-\!C\!-\!CH_3$$
$$\quad\ H_2\ddot{O} \qquad\qquad\qquad \overset{\oplus}{O}H_2 \qquad\qquad\qquad :\ddot{O}H \qquad\qquad\qquad \|$$
$$\qquad\qquad\qquad\qquad\qquad\qquad\qquad\qquad\qquad (enol) \qquad\qquad\qquad :O \ (keto)$$

The first formed enol tautomer rapidly isomerises to the more stable keto form.

3. Bromic(I) acid (HOBr) will add $Br^{\oplus}$ to the alkene to give the intermediate bromonium ion:

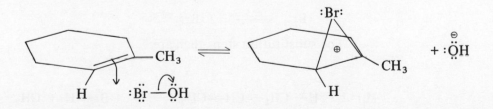

65

Attack by OH$^{\ominus}$ (or a water molecule) gives rise to the two products as shown:

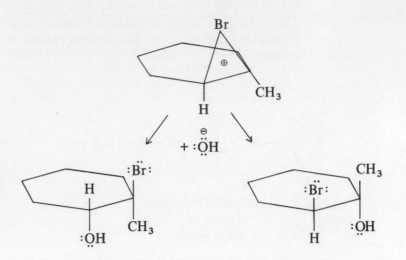

(b) 1. Deuteration will occur preferentially at C1 of the diene as this leads to the resonance stabilised carbocation intermediate

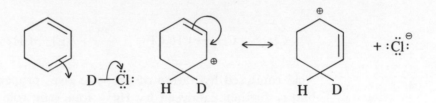

Reaction is completed by addition of :C̈l:$^{\ominus}$ to either C2 (1,2-addition) or C4 (1,4-addition) to give the two products

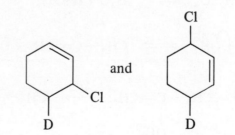

The formation of two products is thus indicative of a mesomeric (resonance) structure for the intermediate.

2. The expected products are $D_2C{=}CHCH_2Br$ and $BrCD_2CH{=}CH_2$. N-bromosuccinimide is used specifically for bromination of alkenes in the allylic position (i.e. adjacent to a double bond). The reagent contains a low concentration of 'free' bromine, and in sunlight these bromine molecules can dissociate into bromine atoms:

$$:\ddot{B}r:_2 \rightleftharpoons 2\ :\dot{B}r:$$

The substitution steps then are

$$:\dot{B}r: \quad H{-}CH_2{-}CH{=}CD_2 \longrightarrow :\ddot{B}r{-}H + \dot{C}H_2{-}CH{=}CD_2$$

followed by

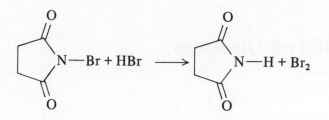

$$CD_2{=}CH{-}\dot{C}H_2 \quad + \quad :\ddot{B}r{-}\ddot{B}r: \quad \longrightarrow \quad CD_2{=}CH{-}CH_2{-}\ddot{B}r: \; + \; :\dot{\ddot{B}}r:, \; etc.$$

The role of the N-bromosuccinimide is to supply and maintain a constant low concentration of bromine by converting the by-product HBr into $Br_2$:

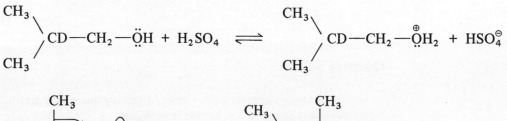

Removal of a hydrogen atom ('hydrogen abstraction') from the allylic position occurs very readily because the resulting (allylic!) free radical is resonance stabilised:

$$\dot{C}H_2{-}CH{=}CD_2 \quad \longleftrightarrow \quad CH_2{=}CH{-}\dot{C}D_2$$

As a consequence, further reaction of this radical with a molecule of bromine can take place at the other terminal carbon atom, giving rise to the alternative product 3-bromo-3,3-dideuteropropene:

$$CH_2{=}CH{-}\dot{C}D_2 \; + \; :\ddot{B}r_2 \quad \longrightarrow \quad CH_2{=}CH{-}CD_2\ddot{B}r: \; + \; :\dot{\ddot{B}}r:$$

Again, then, the formation of two products tells us that there are two reactive centres in the intermediate which must therefore have a mesomeric structure.

3. This question is mechanistically similar to Example 3.4(b) above. The main product will be $CH_3CD{=}CHCH_3$, together with a smaller amount of $CH_2{=}CDCH_2CH_3$.

$$\begin{array}{c} CH_3 \\ \diagdown \\ \diagup \\ CH_3 \end{array} CD{-}CH_2{-}\ddot{O}H \; + \; H_2SO_4 \quad \rightleftharpoons \quad \begin{array}{c} CH_3 \\ \diagdown \\ \diagup \\ CH_3 \end{array} CD{-}CH_2{-}\overset{\oplus}{O}H_2 \; + \; HSO_4^{\ominus}$$

$$\begin{array}{c} CH_3 \\ | \\ CH_3{-}C{-}CH_2{-}\overset{\oplus}{O}H_2 \\ | \\ D \end{array} \qquad \begin{array}{c} CH_3 \quad CH_3 \\ \diagdown \;\;\; | \\ \overset{\oplus}{C}{-}CH_2 \; + \; \ddot{O}H_2 \\ \diagup \\ D \end{array}$$

The intermediate has two different $\beta$-positions from which a proton could be expelled

$$CH_3 \overset{\oplus}{\underset{\beta}{-CD}} \underset{\beta'}{-CH_2} -CH_3 \quad \xrightarrow{-\beta H^\oplus} \quad CH_2 = CDCH_2CH_3 \qquad (1)$$

$$\xrightarrow{-\beta'H^\oplus} \quad CH_3CD = CHCH_3 \qquad (2)$$

The main product is the more alkylated (more stable) alkene (Saytzeff Rule).

## 3.12 Self-Test Questions

### Question 3.1

(a) Explain, giving examples, the terms nucleophile and electrophile. *(8 marks)*

(b) Classify each of the following substituents as +I or −I:
(i) −Cl; (ii) −NO$_2$; (iii) −CH$_3$. *(3 marks)*

(c) Classify each of the following substituents as +M or −M:
(i) −OH; (ii) −NO$_2$; (iii) −COCH$_3$; (iv) −SCH$_3$; (v) −Br. *(5 marks)*

(d) Classify each of the following species as electrophiles or nucleophiles:
(i) NO$_2^\oplus$; (ii) SO$_3$; (iii) CH$_3$MgI; (iv) H$^\oplus$; (v) CH$_3$COCHCOCH$_3^\ominus$. *(5 marks)*

(e) State which type of reaction each of the following would undergo (e.g. electrophilic addition):

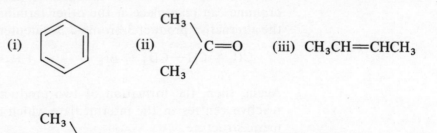

(i)    (ii)  $\overset{CH_3}{\underset{CH_3}{\diagdown}}C=O$    (iii)  CH$_3$CH=CHCH$_3$

(iv)  $\overset{CH_3}{\underset{CH_3}{\diagup}}CH-Br.$

*(4 marks)*
(Trent Polytechnic, 1985)

### Question 3.2

Give structures and systematic names to the organic product(s) of FOUR of the following reactions and suggest a mechanism in each case:

(a)  CH$_2$=CH$_2$ + H$_2$SO$_4$ $\xrightarrow{\text{no solvent}}$

(b)  (CH$_3$)$_2$CH−CH$_3$ $\xrightarrow[CCl_4]{h\nu}$

(c)  CH$_3$(CH$_2$)$_3$CH$_2$OH + H$_2$SO$_4$ + NaBr $\xrightarrow{\text{water as a solvent}}$

(d)  CH$_3$CH$_2$CH$_2$CN + HCl $\xrightarrow{H_2O}$

(e)  CH$_3$CH$_2$CHO + EtMgBr $\xrightarrow{Et_2O}$

(University College of North Wales, 1985)

## Question 3.3

Explain fully the following observations, illustrating your answers by giving structures of intermediates where appropriate.

(a) When t-butyl benzene is first nitrated and the product is then oxidised, the final product is mainly 4-nitrobenzoic acid. If, however, initial oxidation of the t-butyl benzene is followed by nitration, 3-nitrobenzoic acid is produced.

(b) Bromobenzene is hydrolysed by alkali with difficulty, but if a nitro-group is present in the 2- or 4-position the rate of hydrolysis is greatly enhanced.

(c) When 2-bromotoluene reacts with sodium amide in liquid ammonia a mixture of the isomeric 2- and 3-aminotoluenes is produced.

(d) Anthracene undergoes addition reactions predominantly in the 9- and 10-positions.

(University of London, 1984)

## Question 3.4

Give a detailed explanation of FOUR of the following observations:

(a) Under ionic conditions addition of HBr to $CH_3CH{=}CH_2$ gives $CH_3CHBrCH_3$, but addition to $CCl_3CH{=}CH_2$ gives $CCl_3CH_2CH_2Br$.

(b) Treatment of $CH_2{=}CHCH(OH)CH{=}CHCH_3$ with methanol in the presence of dilute sulphuric acid produces $CH_2{=}CHCH{=}CHCH(OCH_3)CH_3$.

(c) Treatment of

with thionyl chloride in ethoxyethane results in the exclusive formation of

(d) Addition of HCl to 3,3-dimethyl-1-butene gives a mixture of two isomeric chloroalkanes.

(e) Elimination of HBr from 3-bromo-2,3-dimethylpentane gives 2,3-dimethyl-2-pentene as the major product.

(Teesside Polytechnic, 1984)

## Question 3.5

Outline the accepted mechanisms of FOUR of the following reactions:

(a) The formation of acetals from aldehydes and ethanol

$$\left( RCHO \xrightarrow[\text{dry HCl}]{\text{EtOH}} RCH(OEt)_2 \right)$$

(b) The formation of semicarbazones from ketones

$$\left( RCOR \xrightarrow[\text{NaOAc}]{H_2NCONHNH_2\,HCl} \begin{array}{c} R \\ \diagdown \\ \diagup \\ R \end{array}\!\!C{=}NNHCONH_2 \right)$$

(c) The anti-Markovnikov hydration of alkenes by hydroboration-oxidation

$$\left( RCH{=}CH_2 \xrightarrow[\text{2. NaOH/H}_2\text{O}_2]{\text{1. B}_2\text{H}_6} RCH_2CH_2OH \right)$$

(d) The acetylation of alcohols by means of acetic anhydride in pyridine

$$\left( RCH_2OH \xrightarrow[\text{pyridine}]{(\text{MeCO})_2O} RCH_2OCOMe \right)$$

(e) The Friedel-Crafts acylation of aromatic hydrocarbons

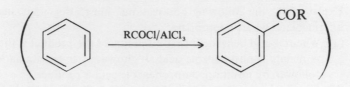

(f) The Baeyer-Villiger oxidation of ketones

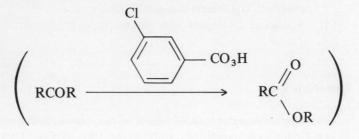

(University of Leeds, 1984)

## Question 3.6

Rationalise the following observations.

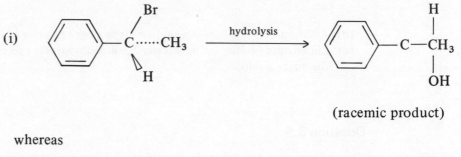

(racemic product)

whereas

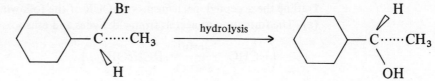

(ii)

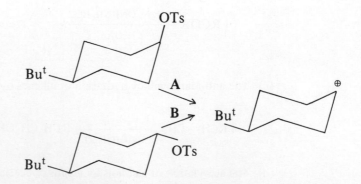

with reaction **A** faster than reaction **B** by a factor of 3.2.

(iii) At room temperature in solution, 1,2-dibromoethane is principally in the *anti* conformation whilst 1,2-ethanediol is almost entirely in the *gauche* conformation.

(iv)

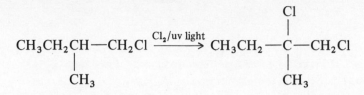

(University of Durham, 1983)

## Question 3.7

Answer ALL parts.
(a) Write a reasonable mechanism for each of the following reactions:

(i)

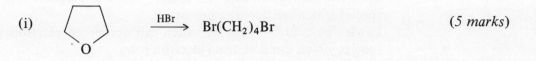

(5 marks)

(ii)

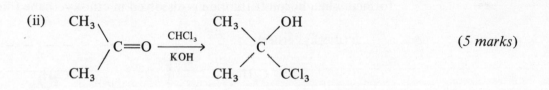

(5 marks)

(iii)

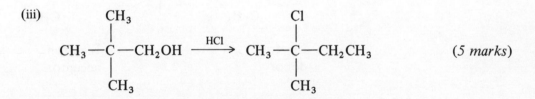

(5 marks)

(b) Predict the major product of each of the following reactions:

(i)

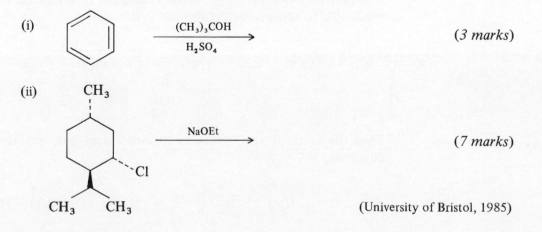

(3 marks)

(ii)

(7 marks)

(University of Bristol, 1985)

# 4 Organic Acids and Bases

## 4.1 Some Definitions

**Bronsted–Lowry**: an acid is a species which can donate a proton, and a base is a species which can accept a proton.

**Lewis**: an acid is a species which can accept an electron pair, and a base is a species which can donate an electron pair.

A typical example of a Lewis acid–base reaction is the coordination complex formed when boron trifluoride is dissolved in ethoxyethane (diethyl ether):

(diethyl ether):

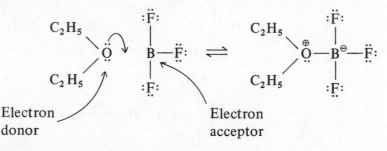

Important Lewis acids are anhydrous metallic salts such as $FeCl_3$ and $AlCl_3$, used as catalysts in reactions such as Friedel–Crafts alkylation and acylation, and halogenation of aromatic compounds. Transition metal atoms have unfilled electron shells able to accept an electron pair:

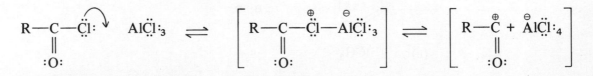

That the concept also embraces Bronsted acid–base reactions is clear from the equation:

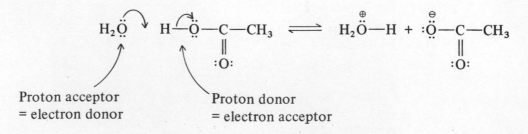

## 4.2 Amphoteric Compounds

Acidity and basicity are relative properties. For example, propanone is amphoteric and may function as an acid or as a base, depending on reaction conditions:

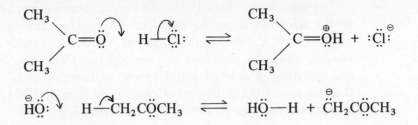

## 4.3 Measurement of Acidity

Acid strength may be expressed in terms of the dissociation constant, $K_a$, for the reaction

$$RC\ddot{O}\ddot{O}H + H_2\ddot{O} \rightleftharpoons RC\ddot{O}\ddot{O}:^{\ominus} + H_3\overset{\oplus}{O}$$

or more conveniently as the acidity constant, $pK_a$, defined by the expression

$$pK_a = -\lg(K_a/\text{mol dm}^{-3})$$

Values are normally in the range 0–14 and a lower value of $pK_a$ indicates a stronger acid. Since it is a logarithmic scale, *one unit change in $pK_a$ signifies a tenfold difference in acidity*.

## 4.4 Measurement of Basicity

The corresponding basicity constant, $pK_b$, for organic bases such as the amines is defined as

$$pK_b = \lg(K_b/\text{mol dm}^{-3})$$

where $K_b$ is the equilibrium constant for the reaction

$$R\ddot{N}H_2 + H_2\ddot{O} \rightleftharpoons R\overset{\oplus}{N}H_3 + :\overset{\ominus}{O}H$$

As with acidity constants, the normal range is 0–14, with a low value corresponding to a strong base. However, it is becoming more common to express basicity in terms of the *acidity constant for the conjugate acid of the base*, $R\overset{\oplus}{N}H_3$. In this respect, we can use the relationship

$$pK_a + pK_b = 14$$

Thus, the $pK_b$ of methylamine ($CH_3\ddot{N}H_2$) is 3.4, from which the $pK_a$ of the conjugate acid ($CH_3N\overset{\oplus}{H}_3$) is 10.6.

## 4.5 Conjugate Acids and Bases

In the general reaction

$$HA + H_2\ddot{O} \rightleftharpoons H_3\overset{\oplus}{O} + \ddot{A}^{\ominus}$$

both the forward and reverse processes involve reaction between an acid and a base. The species $H_3\overset{\oplus}{O}$ is the **conjugate acid** of the base $H_2\ddot{O}$, and $\ddot{A}^{\ominus}$ is the **conjugate base** of the acid HA.

## 4.6 Factors Influencing Acidity

The major factor determining the position of equilibrium — and hence the strength of the acid HA — is the *stability of the conjugate anion* $\overset{..}{A}{}^{\ominus}$ *relative to the acid* HA.

The ease with which the anion can accommodate the negative charge is related to the *size* (the larger the atom the more diffuse the charge) and *electronegativity* of the atom A. For a group of atoms, the *ability to disperse the negative charge* is crucial, and both inductive and mesomeric effects may be important. The *ionising* and *solvating power* of water is also influential, as is shown by the observation that an aqueous solution of ethanoic acid is some 100 000 times more acidic than an ethanolic solution of the same concentration.

## 4.7 Acidity of Alcohols

Ethanol (p$K_a$ 15.9) is an extremely weak acid. The negative charge of the conjugate anion $C_2H_6\overset{..}{O}\overset{..}{\phantom{O}}{}^{\ominus}$ is localised on the oxygen atom and the species is a powerful base. Thus alkoxides can only be prepared in the absence of water.

## 4.8 Acidity of Phenols

Phenol (p$K_a$ 10.0) is an appreciably stronger acid than ethanol, largely because of mesomeric (resonance) stabilisation of the conjugate phenoxide ion, not possible in the alkoxide ion:

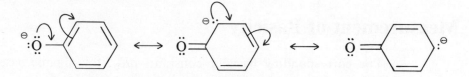

etc.
Of course, the phenol molecule is itself resonance-stabilised, but delocalisation is restricted by charge separation which does not occur in the anion:

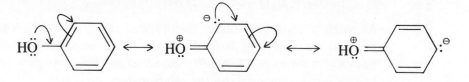

etc.
Part of the 'driving force' for ionisation is the gain in resonance energy on going from the neutral phenol molecule to the negatively charged phenoxide ion. Solvation is also more effective in the charged species.

## 4.9 Acidity of Carboxylic Acids

Although only two limiting forms can be written for the anion of ethanoic acid,

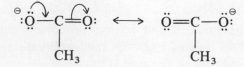

delocalisation is particularly effective as the negative charge is shared equally by the two electronegative oxygen atoms. Thus ethanoic acid ($pK_a$ 4.76) is a much stronger acid than phenol.

Again, it is the *differential* effect of resonance which is important. Because of charge separation, stabilisation of the parent acid is less effective than in the anion:

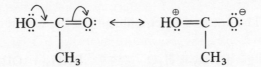

The significance of this is perhaps made more apparent by comparing energy profiles for ionisation of ethanoic acid with and without the operation of resonance stabilisation (see Fig. 4.1).

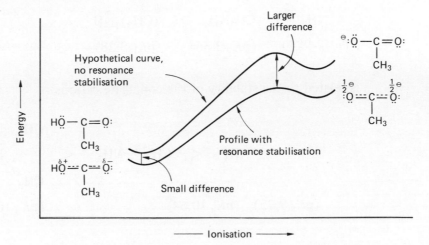

**Fig. 4.1** Energy profiles for ionisation of ethanoic acid

# 4.10 The Effect of Substituents on Acidity

As before, it is the *differential* effect of a substituent on an acid and its conjugate anion which is important. The chlorine atom in chloroethanoic acid, for example, not only increases the stability of the chloroanion, but it also actually has a destabilising effect on the parent acid (see Fig. 4.2.).

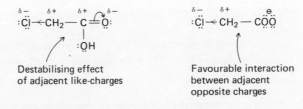

**Fig. 4.2** The effect of a chlorosubstituent on the stability of ethanoic acid and its conjugate anion

Other examples are given in the worked examples which follow. With substituted aromatic acids, 'anomalous' effects may be observed where a substituent can exert a negative inductive effect, but a positive mesomeric effect (from the *ortho* or *para* positions) (see Example 4.3).

75

# 4.11 Acidity in Compounds with an Activated Methylene Group

This concept is considered in Chapter 7.

# 4.12 Basicity of Amines

The position of equilibrium in the reaction

$$R\ddot{N}H_2 + H_2\ddot{O} \rightleftharpoons R\overset{\oplus}{N}H_3 + :\overset{\ominus}{\ddot{O}}H$$

is largely determined by the ability of the conjugate cation $R\overset{\oplus}{N}H_3$ to accommodate the positive charge. The positive inductive effect of alkyl groups is stabilising, and increasing basicity of amines in the series

$$\ddot{N}H_3 \quad < \quad CH_3\ddot{N}H_2 \quad < \quad (CH_3)_2\ddot{N}H$$
$$(pK_b\ 4.75) \quad (pK_b\ 3.36) \quad (pK_b\ 3.28)$$

is related to the stability of the conjugate cations

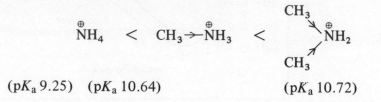

$$(pK_a\ 9.25) \quad (pK_a\ 10.64) \qquad\qquad (pK_a\ 10.72)$$

The role of water as a solvent is apparent in the *decreased* basicity of trimethyl-amine ($pK_b$ 4.20), where the three relatively large methyl groups impede close approach of solvating water molecules to the positively charged nitrogen atom.

# 4.13 Aromatic Amines

The reduced basicity of phenylamine ($pK_b$ 9.38) compared with cyclohexylamine ($pK_b$ 3.36) is attributed to loss of resonance energy on uptake of a proton by the aromatic amine (see Fig. 4.3).

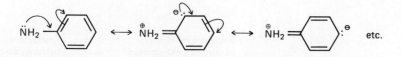

In the phenylamine molecule, resonance involves the amino group

 In the conjugated cation, the protonated amino group is no longer able to take part in resonance

**Fig. 4.3** Resonance in the phenylamine molecule and its conjugate cation

Electron attracting substituents have a destabilising effect on the conjugate cations. Thus 4-nitrophenylamine (p$K_b$ 13.02) is a weaker base than phenylamine (see Fig. 4.4).

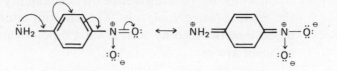

Involvement of the amino group in resonance stabilises the *para* nitrophenylamine molecule

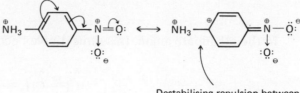

Destabilising repulsion between like charges on adjacent atoms

**Fig. 4.4** The effect of a *para* nitro group on the stability of phenylamine and its conjugate cation

## 4.14 Worked Examples

### Example 4.1

Define the acid dissociation constant $K_a$ using ethanoic acid as an example. Give a detailed account of the effect of substituents on the acidity of aliphatic carboxylic acids.

Under what conditions is the pH of an aqueous solution of an acid equal to the p$K_a$ of the acid?

(Polytechnic of North London, 1984)

*Solution 4.1*

The extent of dissociation of ethanoic acid in aqueous solution,

$$CH_3COOH + H_2O \rightleftharpoons CH_3COO^\ominus + H_3O^\oplus$$

is expressed by the equilibrium constant $K$,

$$K = \frac{(CH_3COO^\ominus]_{aq}\,[H_3O^\oplus]_{aq}}{[CH_3COOH]_{aq}\,[H_2O]_1}$$

(The quantities in brackets are strictly activities, but in dilute aqueous solution can be taken as concentrations.)

Since the concentration of water is large and effectively constant, we can define the dissociation constant for ethanoic acid as

$$K_a = K\,[H_2O]_1 = \frac{[CH_3COO^\ominus]_{aq}\,[H_3O^\oplus]_{aq}}{[CH_3COOH]_{aq}}$$

The question says give a 'detailed account' of the effect of substituents, so you will need to do more than simply state that an electron withdrawing substituent increases acidity and an electron releasing substituent decreases acidity.

You could begin with a statement that substituents effect the position of equilibrium by their influence on the stability of the conjugate anion relative to the undissociated molecule. The chlorine atom in chloroethanoic acid tends to destabilise the acid by setting up a dipole which causes like charges to appear on adjacent atoms:

$$:\overset{\delta-}{\underset{..}{\ddot{C}l}} \leftarrow \overset{\delta+}{CH_2} - \overset{\delta+}{C} = \overset{\delta-}{\ddot{O}:}$$
$$|$$
$$CH_3$$

In the ethanoate anion, this same dipole becomes stabilising:

$$:\overset{\delta-}{\underset{..}{\ddot{C}l}} - \overset{\delta+}{CH_2} - C\overset{\ominus}{\underset{..}{\ddot{O}}\ddot{O}}:$$

(An alternative, simpler, view is that the negative inductive effect of the chlorine atom stabilises the anion by helping to disperse the negative charge.)

The significance of these effects is made clearer by drawing comparative energy diagrams illustrating how the activation energy for ionisation is reduced in the chloroacid:

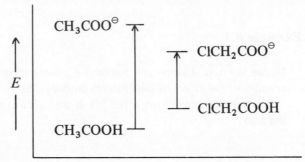

Other strongly electron withdrawing substituents such as $-NO_2$ and $-C\equiv N$ will have similar effects. In longer chain carboxylic acids, the magnitude of these effects decreases as the substituent is moved further from the carboxyl group. Thus, 2-chlorobutanoic acid is quite a strong acid, but the dissociation constant for 4-chlorobutanoic acid is only marginally greater than that for butanoic acid. Electron releasing substituents have the opposite effect, and ethanoic acid is weaker than methanoic acid:

$$CH_3 \rightarrow \overset{\delta+}{C} = \overset{\delta-}{\ddot{O}:}$$
$$|$$
$$OH$$

acid stabilised

$$CH_3 - C\overset{\ominus}{\underset{..}{\ddot{O}}\ddot{O}}:$$

conjugate anion destabilised

The first dissociation constant for propandioic acid is larger than that for ethanoic acid because of the $-I$ effect of the $-COOH$ group. But after ionisation the group becomes electron repelling and the second dissociation constant is correspondingly reduced:

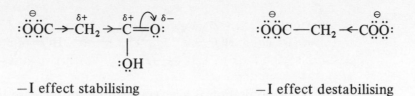

−I effect stabilising                    −I effect destabilising

The final part of the question requires only a concise statement. It is readily shown that

$$pH = pK_a + \lg \frac{[RCOO^{\ominus}]_{aq}}{[RCOOH]_{aq}}$$

Thus, pH will equal the $pK_a$ at the equivalence point when

$$[RCOO^{\ominus}]_{aq} = [RCOOH]_{aq}$$

## Example 4.2

Give qualitative explanations for the following observations:
(a)  4-nitrophenol (*p*-nitrophenol) is a stronger acid than 3-nitrophenol (*m*-nitrophenol):

(b)

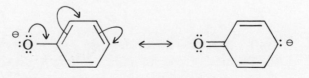

is a stronger base than

(c)

has $pK_a = 15$, whereas cyclopentane has $pK_a = 45$

(University of Leeds, 1983)

*Solution 4.2*

(a) Acidity in phenols is related to the ability of the aromatic ring to share the negative charge in the phenoxide ion:

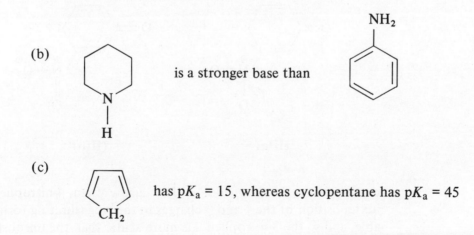

79

The hydroxyl group in the parent molecule is also conjugated with the ring, but resonance is restricted by charge separation:

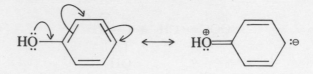

Thus, ionisation is accompanied by an increase in resonance energy. Nitrophenols are more acidic than the parent compound because the electron attracting nitro group increases dispersal of the negative charge in the phenoxide ion. When the nitro group is in the 4-position, delocalisation is particularly effective, as can be seen from the following limiting forms:

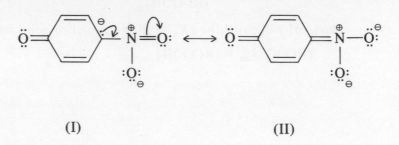

(I)                                    (II)

The additional limiting form (II) cannot be written when the nitro group is in the 3-position:

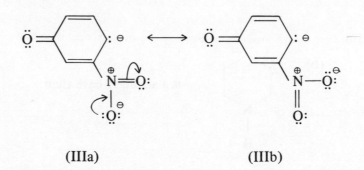

(IIIa)                                 (IIIb)

consequently 3-nitrophenol is less acidic than 4-nitrophenol. However, juxtaposition of the ⊕ and ⊖ charges in the ring (limiting form III) is favourable, and so the 3-nitrophenol is more acidic than the unsubstituted phenol.

(b) Phenylamine (aniline) is a relatively weak base because uptake of a proton results in some loss of resonance energy associated with conjugation of the amino group with the aromatic nucleus:

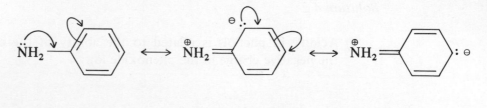

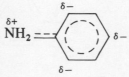

Once the nitrogen lone pair is donated to a proton, resonance is limited to the two Kekulé forms

No such 'energy barrier' is present in piperidine, which has about the same basicity as dimethylamine. The methylene groups of the ring have a similar inductively stabilising effect on the conjugate cation as the methyl groups in the conjugate cation of dimethylamine:

(c) Cyclopentane has a very high $pK_a$ value as loss of a proton would give a carbanion in which the negative charge was localised on one carbon atom:

However, in the corresponding cyclopentadienyl anion, conjugation of the negatively charged carbon atom with the alkene double bonds allows the charge to be shared by all the ring carbon atoms:

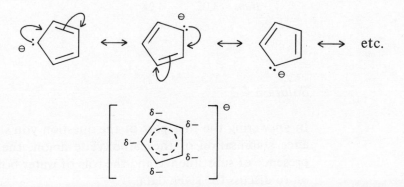

Resonance stabilisation of the conjugate anion lowers the activation energy for ionisation and cyclopentadiene is much more acidic than cyclopropane.

**Example 4.3**

Review the factors influencing the acidity of carboxylic acids in aqueous solution. (*8 marks*)
Account for the variation in $pK_a$ values of the following compounds:

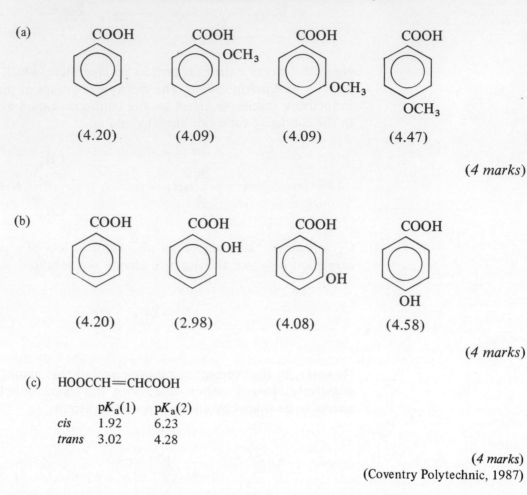

(a)

(4.20)　　　(4.09)　　　(4.09)　　　(4.47)

(*4 marks*)

(b)

(4.20)　　　(2.98)　　　(4.08)　　　(4.58)

(*4 marks*)

(c)　HOOCCH=CHCOOH

|       | $pK_a(1)$ | $pK_a(2)$ |
|-------|-----------|-----------|
| *cis*   | 1.92      | 6.23      |
| *trans* | 3.02      | 4.28      |

(*4 marks*)
(Coventry Polytechnic, 1987)

*Solution 4.3*

In answering the first part of the question you should mention differential resonance stabilisation of the carboxylate anion, the effect (in general terms) of the
presence of substituents, and the role of water both as a base and solvent. Then go
on to discuss the given data.

(a) Oxygen is more electronegative than carbon, but it also has an unshared
electron pair which can be involved in conjugation with the ring. Consequently, the methoxy and hydroxy groups can exert both a negative
inductive effect (all positions) and a positive mesomeric effect (*ortho* and
*para* positions).

The positive mesomeric effect is stabilising in the acid itself, but destabilising in the carboxylate anion:

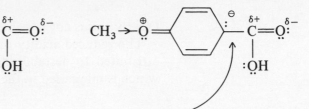

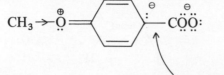

Favourable juxtaposition
of opposite charges in the
carboxylic acid

Repulsion of adjacent like charges
is destabilising in the carboxylate anion

2-Methoxybenzoic acid is marginally stronger than benzoic acid and it would appear that inductive stabilisation of the carboxylate anion just outweighs any destabilisation due to the positive mesomeric effect. No direct mesomeric destabilisation is possible from the *meta* position, and the negative inductive effect is weaker over the greater distance, so acidity of 3-methoxybenzoic acid stays about the same as the *ortho* compound.

From the *para* position the negative inductive effect is very weak, and destabilisation of the carboxylate anion by the positive mesomeric effect results in the acidity of 4-methoxybenzoic acid being less than that of unsubstituted benzoic acid.

(b) Similar considerations apply to the hydroxyacids, but clearly another factor must be involved in the unexpectedly high acidity of the *ortho* compound.

From a study of the infrared spectrum of this compound it is evident that additional stabilisation of the carboxylate anion results from hydrogen bonding between the hydroxyl and carboxylate groups, not possible with the other isomers (or with 2-methoxybenzoic acid):

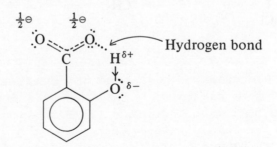

(Some hydrogen bonding may occur in the unionised acid, but bonding is much stronger in the anion.)

(c) Hydrogen bonding is also influential in stabilisation of the conjugate anion of *cis* butenoic acid:

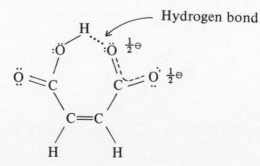

Clearly this is not possible with the *trans* isomer.

The reduced acidity of the second carboxyl group in the *cis* acid can be attributed to destabilising repulsion between adjacent carboxylate groups which is minimised in the *trans* isomer:

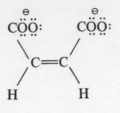

# 4.15  Self-Test Questions

### Question 4.1

Arrange the following molecules into an increasing order of acidity; justify the order you select by a description of the factors which affect relative acidities.

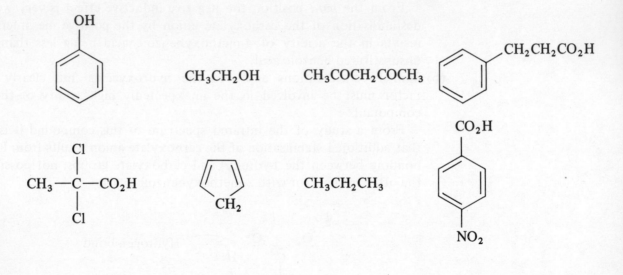

$CH_3CH_2OH$      $CH_3COCH_2COCH_3$

$CH_3CH_2CH_3$

(University of Leeds, 1984)

### Question 4.2

Study the following pairs of compounds and state which is the strong acid of

(i)      $HCO_2H$          and          $CH_3CO_2H$

(ii)

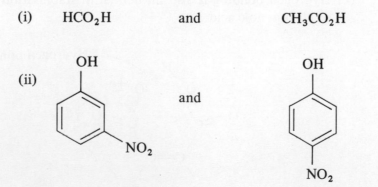

and

84

and the stronger base of

(iii)    $CH_3NH_2$    and    $CH_3OCH_2NH_2$

(iv)

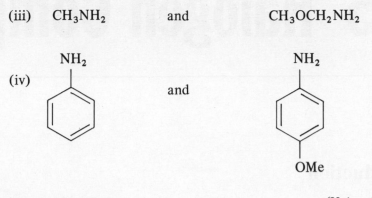

and

Explain your reasoning.

(University of Salford, 1985)
(part question)

## Question 4.3

Place the following compounds in order of decreasing acid strength, giving reasons for your choice:

2-chlorobutanoic acid, phenol, butanoic acid, 2-fluorobutanoic acid, 4-bromobutanoic acid, 3-chlorobutanoic acid.

(Kingston Polytechnic, 1985)
(part question)

## Question 4.4

(i)  Predict the relative acidities of the following phenols.

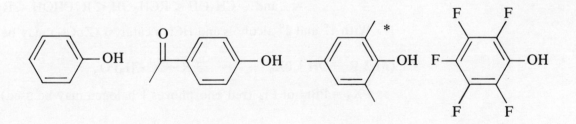

(ii)  The amine (3) (below) is a much weaker base than amine (4). Explain this observation.

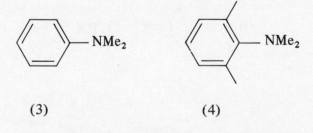

(3)                    (4)

(University of Nottingham, 1985)
(part question)

*This is a conventional shorthand way of representing methyl substituents.

# 5 Halogen Compounds

## 5.1 Introduction

Many first year courses begin with the chemistry of haloalkanes because a study of their substitution ($S_N1$, $S_N2$) and elimination (E1,E2) reactions introduces many of the basic concepts which are later applied to the reactions of other classes of compounds.

## 5.2 Formation of Haloalkanes (Alkyl Halides)

### (a) From Alcohols (Most Useful Method)

(i) $R\!-\!OH + HX \longrightarrow R\!-\!X + H_2O$

HX = HCl, HBr, HI ($NaX + H_2SO_4$ may sometimes be used)

Order of reactivity: $HCl < HBr < HI$

and: $CH_3OH < RCH_2OH < R_2CHOH < R_3COH$

With 1° and 2° alcohols and HCl, a catalyst ($ZnCl_2$) may be required.

(ii) $3\,R\!-\!OH + PX_3 \longrightarrow 3\,R\!-\!X + H_3PO_3$

$PX_3 = PBr_3$ or $PI_3$ (red phosphorus + halogen may be used)

### (b) From Alkenes and Alkynes

(i)

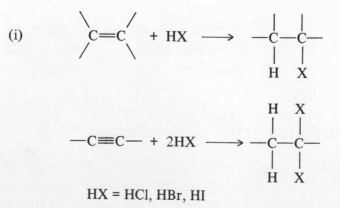

HX = HCl, HBr, HI

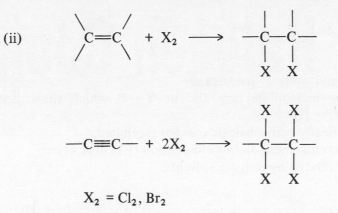

(ii)

$$\ce{C=C} + X_2 \longrightarrow \ce{-C-C-} \text{ (with X, X)}$$

$$\ce{-C#C-} + 2X_2 \longrightarrow \ce{-C-C-} \text{ (with X X, X X)}$$

$$X_2 = Cl_2, Br_2$$

**(c) Halogen Exchange**

$$R\!-\!X + NaI \longrightarrow R\!-\!I + NaX \text{ (propanone solution)}$$

$$X = Cl, Br$$

**(d) Direct Halogenation of Alkanes**

$$R\!-\!H + X_2 \longrightarrow R\!-\!X + HX$$

$$X = Cl_2, Br_2 \ (F_2)$$

Industrial method, of little use in the laboratory.

## 5.3 Formation of Haloarenes (Aryl Halides)

**(a) Direct Halogenation**

$$Ar\!-\!H + X_2 \xrightarrow{\text{LAC}} Ar\!-\!X + HX$$

$$X = Cl, Br$$

LAC = Lewis acid catalyst (e.g. anhydrous $AlCl_3$, $FeCl_3$, etc.)

See Chapter 10.

**(b) Reactions of Diazonium Salts** (Most Useful)

$$Ar\!-\!H \xrightarrow[\text{conc. } H_2SO_4]{\text{conc. } HNO_3} Ar\!-\!NO_2 \xrightarrow[\text{Sn/HCl}]{[H]} Ar\!-\!NH_2 \xrightarrow[\text{5-10°C}]{HNO_2} Ar\!-\!\overset{\oplus}{N}\!\!\equiv\!\!N$$

Then with

(i) CuX $\xrightarrow{\Delta}$ Ar—X

    X = Cl, Br (Sandmeyer reaction)

(ii) HBF$_4$ $\xrightarrow{\Delta}$ Ar—F (Balz–Schieman reaction)

(iii) KI $\longrightarrow$ Ar—I

## 5.4 Structure and Reactivity of Haloalkanes

The characteristic reaction of haloalkanes is nucleophilic substitution. Br, I, and to a lesser extent Cl (but not F) are good leaving groups and may be displaced in *unimolecular* ($S_N1$) or *bimolecular* ($S_N2$) processes:

$S_N1$  $-\overset{|}{\underset{|}{C}}-\ddot{\underset{\cdot\cdot}{X}}:$ $\underset{}{\overset{slow}{\rightleftharpoons}}$ $-\overset{|}{\underset{|}{C}}{}^{\oplus} + :\ddot{\underset{\cdot\cdot}{X}}:^{\ominus}$ $\overset{\ddot{Y}^{\ominus}}{\underset{Fast}{\longrightarrow}}$ $-\overset{|}{\underset{|}{C}}-Y\,(+\,:\ddot{\underset{\cdot\cdot}{X}}:^{\ominus})$

3° and some 2° haloalkanes

The nucleophile may also be Y—H which subsequently loses a proton (e.g. $\ddot{N}H_2$—H)

Optically active haloalkanes are racemised.

Rearrangement and/or elimination (E1) may occur.

Facilitated by ionising solvents.

$S_N2$  $\overset{\ominus}{\ddot{Y}}{\downarrow}$ $-\overset{|}{\underset{|}{C}}\frown\ddot{\underset{\cdot\cdot}{X}}:$ $\rightleftharpoons$ $\left[Y\text{----}\overset{|}{C}\text{----}\ddot{\underset{\cdot\cdot}{X}}:\right]^{\ominus}_{TS}$ $\longrightarrow$ $Y-\overset{|}{\underset{|}{C}}- + :\ddot{\underset{\cdot\cdot}{X}}:^{\ominus}$

Methyl, 1° and most 2° haloalkanes.

Accompanied by *inversion of configuration*.

Some elimination (E2) may occur.

In reaction with nucleophiles which are also strong bases (e.g. $Y^{\ominus} = RO^{\ominus}$), a competing reaction is elimination of HX. With some 3° haloalkanes this may be the predominant reaction.

There are two processes:

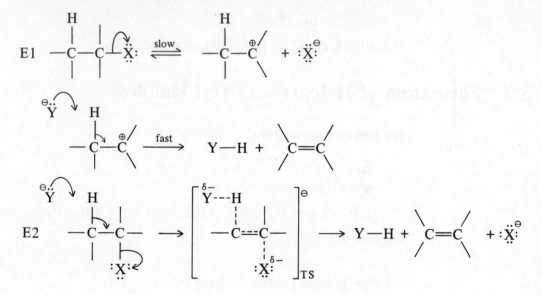

Factors which tend to favour substitution or elimination are summarised in Table 5.1.

**Table 5.1** Factors which tend to favour substitution or elimination in haloalkanes

| Factor | $S_N1$ | $S_N2$ | E1 | E2 |
|---|---|---|---|---|
| Structure of haloalkane | 3° ≫ 2° > 1° | 1° > 2° > 3° | 3° | 3° or 2° |
| Nature of reagent | weak nucleophile | strong nucleophile | strong base | strong base |
| Concentration of reagent | low | high | low | high |
| Nature of solvent | highly polar | lower polarity | polar | low polarity |
| Temperature | moderate | moderate | high | high |

## 5.5 Principal Reactions of Haloalkanes

### (a) Replacement of Halogen

$$R—X + Y^{\ominus} \text{ (or } H—Y) \longrightarrow R—Y + X^{\ominus} \text{ (or } HX)$$

| $Y^{\ominus}$ (H—Y) | Product, R—Y |
|---|---|
| $H\ddot{\underset{..}{O}}\!:^{\ominus}$ | ROH, alcohols |
| $H\ddot{\underset{..}{S}}\!:^{\ominus}$ | RSH, thiols |
| $R\ddot{\underset{..}{O}}\!:^{\ominus}$ | ROR, ethers |
| $R\ddot{\underset{..}{S}}\!:^{\ominus}$ | RSH, thioethers |
| $:\ddot{\underset{..}{X}}\!:^{\ominus}$ | RI, haloalkanes |
| $\ddot{\underset{..}{N}}\ddot{C}^{\ominus}$ | RCN, nitriles |
| $\ddot{N}H_3$ | $RNH_2$, 1° amines |
| $R\ddot{N}H_2$ | $R_2NH$, 2° amines |
| $R_2\ddot{N}H$ | $R_3N$, 3° amines |

Remember also, Friedel Crafts alkylation (Chaper 10), malonic ester and ethyl acetoacetate syntheses (Chapter 8).

### (b) Elimination of Hydrogen Halide

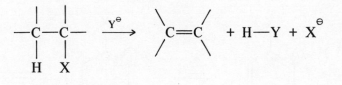

$$Y^{\ominus} = CH_3CH_2\ddot{\underset{..}{O}}\!:^{\ominus}, \text{ etc.}$$
$$X = Cl, Br, I$$

### (c) Formation of Grignard Reagents

$$RX + Mg \xrightarrow{\text{ether}} RMgX$$
$$X = Cl, Br, I$$

### (d) Reduction to an Alkane

$$RX \xrightarrow{\text{metal/acid}} RH + X^{\ominus}$$

89

## 5.6 Structure and Reactivity of Haloarenes

Chlorobenzene has a resonance structure in which the carbon to chlorine bond is shorter and stronger than in chloroalkanes. Consequently, the chlorine atom is difficult to displace unless activated by one or more strongly electron-withdrawing substituents such as nitro groups.

There are two mechanisms for the displacement of halogen:

### (a) Bimolecular Nucleophilic Substitution

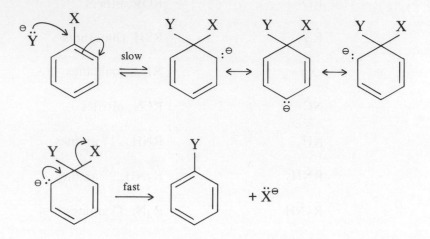

Hydrolysis of chlorobenzene requires 10% NaOH at 350° under pressure. 1-Chloro-4-nitrobenzene is hydrolysed at 160°C and 1-chloro-2,4,6-trinitrobenzene is hydrolysed by boiling water.

### (b) The Aryne (Benzyne) Mechanism (addition—elimination mechanism)

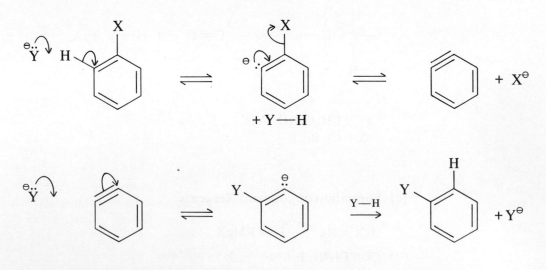

X = F, Cl, Br or I
Y = e.g. $NH_2^{\ominus}$ (NaNH$_2$ in liquid ammonia)

This mechanism may also operate in the hydrolysis of chlorobenzene with aqueous alkali at high temperature and pressure.

When substituents are present, rearranged products may be obtained (**ciné-substitution**), e.g.

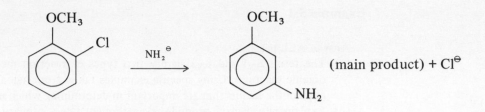

(main product) + $Cl^{\ominus}$

## 5.7 Principal Reactions of Haloarenes

### (a) Replacement of Halogen

#### (i) *Nucleophilic Substitution*

$$Ar—X + Y^{\ominus} \text{ (or H—Y)} \qquad Ar—Y + X^{\ominus} \text{ (or HX)}$$

The molecule must normally be 'activated' by one or more strongly electron-withdrawing substituents *ortho* and/or *para* to X

| $Y^{\ominus}$ (H—Y) | Product, ArY |
|---|---|
| $H\ddot{\underset{..}{O}}\!:^{\ominus}$ | ArOH phenols |
| $R\ddot{\underset{..}{O}}\!:^{\ominus}$ | ArOR ethers |
| $\ddot{N}H_2^{\ominus}$ | $ArNH_2$ 1° amines |
| $R\ddot{N}H^{\ominus}$ | ArNHR 2° amines |

#### (ii) *Via an Aryne*

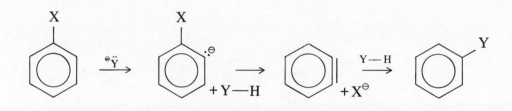

Strongly electron-withdrawing substituents should not be present.

$$X = F, Cl, Br, I. \quad Y = \ddot{N}H_2^{\ominus}, \ddot{R}^{\ominus}$$

The aryne will also react with weak nucleophiles such as ROH, $\ddot{C}\ddot{N}^{\ominus}$, etc.

### (b) Aromatic Electrophilic Substitution

Halogen is weakly deactivating and *ortho/para* directing (see Chapter 10).

## 5.8  Worked Examples

**Example 5.1**

Answer ALL parts

(a) The terms $S_N1$ and $S_N2$ signify two types of reaction mechanisms that can occur in organic reactions. Using specific examples taken from haloalkane or alcohol chemistry, explain the factors that are important in determining which mechanism operates.

(b) Explain why benzyl bromide, even though it is a primary halide, often reacts by an $S_N1$ mechanism.

(c) Arrange the following compounds in order of increasing reactivity towards substitution involving the $S_N1$ mechanism.

(University of Bradford, 1984)

*Solution 5.1*

(a) Nucleophilic substitution is a topic common to nearly all first year courses in organic chemistry, and this question is typical of those relating to the $S_N1$ and $S_N2$ mechanisms. Substitution reactions of alcohols are considered in Chapter 6, so we will take haloalkanes as an example.

Begin by explaining the terms $S_N1$ and $S_N2$, i.e. substitution, nucleophilic, unimolecular or bimolecular. Make it clear that you understand that molecularity defines the number of species involved in the rate determining step, *not* the kinetic order of reaction.

In the unimolecular process, the slow rate determining step is ionisation of the haloalkane substrate. A good example would be hydrolysis of 2-bromo-2-methylpropane (*tert*-butyl bromide) in warm, dilute aqueous alkali:

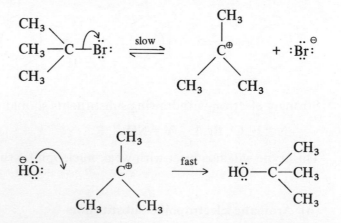

Summarise the energy relationships by means of an energy profile for the process, and if you have covered that sort of detail in your lecture course, include drawings of the transition states for the formation and subsequent reaction of the carbocation intermediate:

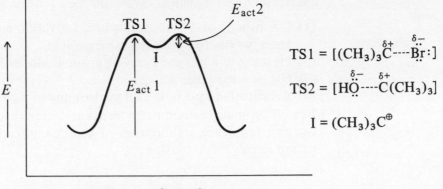

——— Course of reaction ———→

The bimolecular mechanism is a *concerted* process in which expulsion of the leaving group is accompanied by formation of the new bond with the nucleophile. You could illustrate this by reference to hydrolysis of iodoethane with moderately concentrated aqueous alkali. Only one step is involved:

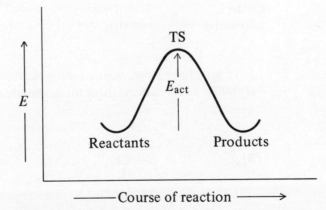

and the energy profile is of the simple form

$$E \qquad \text{TS} \qquad E_{act}$$

Reactants          Products

——— Course of reaction ———→

Factors which favour the $S_N1$ process are:

(1) Use of 3° haloalkane substrates where alkyl substituents on the $\alpha$-carbon atom help to stabilise the carbocation (thus lowering the activation energy for its formation) and hinder the approach of a nucleophile. (More specifically, steric crowding caused by three alkyl groups would destabilise a bimolecular transition state, thus increasing the activation energy for its formation.)

(2) Presence of a good leaving group i.e. bromo or iodoalkanes rather than chloroalkanes.

(3) Use of a highly polar (ionising) solvent.

(4) Substitution by a relatively weak nucleophile, in dilute solution.

Conversely, factors which favour the $S_N2$ process are:

(1) Use of a 1° or 2° substrate where activation energy for ionisation is too great for this to occur as a separate step.
(2) Presence of a less good leaving group (chlorine).
(3) Use of a less polar solvent.
(4) Substitution by a powerful nucleophile in high concentration.

(b) Benzyl bromide (bromomethylbenzene) often reacts by an $S_N1$ mechanism because resonance stabilisation of the benzyl carbocation lowers the activation energy for ionisation:

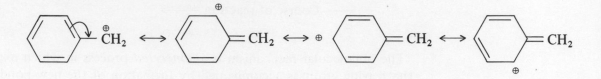

Thus the benzyl carbocation is much more stable than, say, the ethyl carbocation, $CH_3CH_2^{\oplus}$ where the +ve charge is essentially localised on one carbon atom.

(c) Keeping in mind the dual role of alkyl groups in stabilising a carbocation and hindering the necessary close approach of an attacking nucleophile, the order of increasing reactivity by an $S_N1$ process is

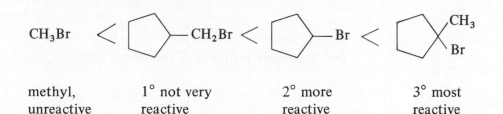

| methyl, unreactive | 1° not very reactive | 2° more reactive | 3° most reactive |

If you have time, you could indicate that this is the order of increasing stability of the carbocation intermediates:

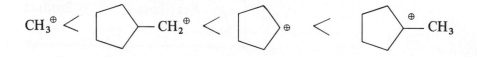

## Example 5.2

Discuss the $S_N1$ and $S_N2$ nucleophilic substitution mechanisms of the alkyl halides including reference to the stereochemistry of the two reactions and the factors (e.g. nature of the alkyl group, nucleophile and choice of solvent) which favour each.

*(14 marks)*

Show how 1-chlorobutane may be converted into
(a) 1-hexyne and
(b) n-butyl methyl ether.

*(6 marks)*
(Kingston Polytechnic, 1985)

*Solution 5.2*

This question differs from Example 5.1 in that you are specifically required to discuss the stereochemistry of the two reactions. You will therefore need an additional section indicating that

(i) an $S_N2$ process is accompanied by inversion of configuration, e.g.

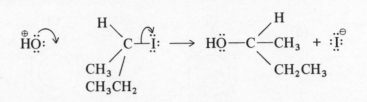

It follows that an optically active substrate should give rise to an optically active product, although actual inversion of configuration can only be demonstrated, for example, by reaction of an enantiomer of 2-iodooctane with radioactive iodide ion, $^{128}:\ddot{I}:^{\ominus}$.*

(ii) in a $S_N1$ reaction of an optically active substrate, the product would be largely racemised because the carbocation intermediate has a planar structure which could be attacked by the nucleophile from either side:

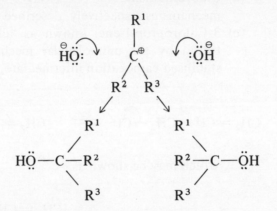

You might mention here that racemisation is often not complete because attack of the nucleophile may occur before the leaving group has moved out of shielding distance, resulting in a slight predominance of product molecules with inverted configuration.

The required syntheses in the second part of the question are

(a) $HC{\equiv}\overset{\ominus}{C}$ ⌒ $CH_2$—$\ddot{C}l:$ → $HC{\equiv}C$—$CH_2CH_2CH_2CH_2CH_3$

$(Na^{\oplus})$ $CH_2CH_2CH_2CH_3$ $(+ Na^{\oplus}:\ddot{C}l:^{\ominus})$

This is a standard method for obtaining higher alkynes from simpler ones.

(b) $CH_3$—$\ddot{O}:$ ⌒ $CH_2$—$\ddot{C}l:$ → $CH_3$—$\ddot{O}$—$CH_2CH_2CH_2CH_3$

$(Na^{\oplus})$ $CH_2CH_2CH_3$ $(+ Na^{\oplus}:\ddot{C}l:^{\ominus})$

*See the author's book *Ionic Organic Mechanisms*, Macmillan Education, 1986.

**Example 5.3**

Discuss the reactivities towards nucleophilic reagents of the following compounds paying particular attention to the mechanisms of the observed reactions.

(a)  $CH_3Cl$     (b)  $(CH_3)_3CCl$     (c)  $CH_2{=}CHCH_2Cl$

(d) 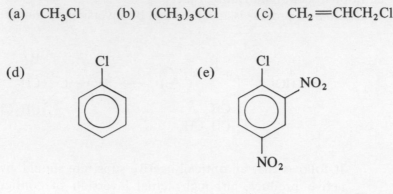     (e)

(Teesside Polytechnic, 1984)

*Solution 5.3*

(a) Chloromethane and (b) 2-chloro-2-methylpropane react by $S_N2$ and $S_N1$ mechanisms respectively, described above.

(c) 3-Chloroprop-1-ene, known as 'allyl chloride', readily undergoes substitution by the unimolecular mechanism. Loss of $:\ddot{C}l^{\ominus}$ gives a resonance stabilised carbocation intermediate,

$$CH_2{=}CH{-}CH_2{-}\ddot{C}l: \rightleftharpoons [CH_2{=}CH{-}\overset{\oplus}{C}H_2 \longleftrightarrow \overset{\oplus}{C}H_2{-}CH{=}CH_2]$$

which may be shown as

$$[\overset{\delta+}{C}H_2{=}{=}CH{=}{=}\overset{\delta+}{C}H_2]^{\oplus}$$

Delocalisation of the positive charge stabilises the ion and lowers the activation for its formation.

(d) The carbon–chlorine bond in chlorobenzene is shorter and stronger than in a chloroalkane, and the chlorine atom is very difficult to displace. Ionisation is not possible as the resulting carbocation would have the positive charge localised on one carbon atom. Furthermore, normal rearside attack by a nucleophile ($S_N2$) is sterically impossible.

To bring about hydrolysis, chlorobenzene has to be heated to a high temperature under pressure with strong aqueous alkali. It is possible to formulate a two-step bimolecular mechanism in which the hydroxide ion attacks the ring α-carbon from one side:

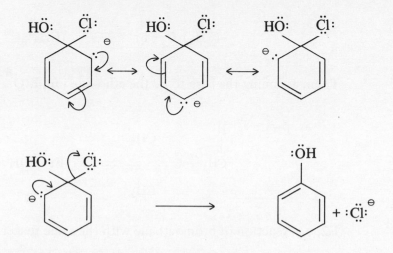

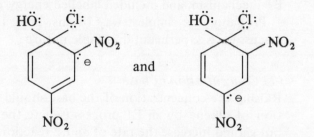

Although a proportion of chlorobenzene may react this way, there is evidence that in the absence of activating substituents (such as *ortho* and/or *para* nitro groups), haloarenes undergo the addition–elimination (aryne) process outlined in Section 5.6.

(e) Chloro-2,4-dinitrobenzene does react via the bimolecular mechanism. The two strongly electron-attracting nitro groups are placed on ring carbon atoms which carry negative charges in the carbanionic intermediate, as shown in the limiting forms

Thus the two nitro groups have a considerable stabilising effect on the intermediate, and chloro-2,4-dinitrobenzene is readily hydrolysed simply on boiling with strong aqueous alkali.

## Example 5.4

Give the E1 and E2 mechanisms by which hydrogen bromide can be eliminated form alkyl bromides under the influence of a base.

*(5 marks)*

Describe the ways which might be used to establish whether it was the E1 or E2 mechanism which was predominating in a given elimination situation.

*(15 marks)*
(University of Keele, 1985)

*Solution 5.4*

Only 5 marks are allocated for the first part of the question, so only a brief outline of the two mechanisms is required.

**E1**, e.g. reaction of 2-bromo-2-methylpropane (*tert*-butyl bromide) with alcoholic KOH solution:

$$(CH_3)_3C-\overset{..}{Br}: \rightleftharpoons (CH_3)_3C^{\oplus} + :\overset{..}{\underset{..}{Br}}:^{\ominus}$$

Then, assuming the base to be the ethoxide ion, $EtO:^{\ominus}$,

E2, e.g. reaction of bromoethane with the same reagent:

Had a higher proportion of marks been allocated to this part of the question, you could have added the structures of the two transition states involved in the E1 mechanism, and included labelled energy profiles for the two processes.

Probably the simplest way of answering the second part of the question is to discuss each experimental variable in turn.

### (1) Concentration of base
Raising the concentration of the base should have no effect on the rate of production of alkene by an E1 process (where the rate determining step is ionisation), but should increase the rate of an E2 reaction.

### (2) Base strength
Changing from a weaker to a stronger base should increase the rate of an E2 reaction, but again will have little effect on an E1 process.

### (3) Nature of the solvent
Changing from a less to a more-ionising solvent should increase the rate of an E1 process, but have minimal effect on an E2 reaction.

### (4) Reaction of deuterated substrates
About $5\ kJ\ mol^{-1}$ more energy is required for heterolysis of a C–D bond than for a C–H bond. Thus, since removal of the $\beta$-proton is already underway in the E2 transition state, replacing the haloalkane substrate with the $\beta$-deuterated analogue should reduce the rate of formation of the alkene. No such kinetic isotope effect will be observed in an E1 process.

### (5) Rearranged products
With certain substrates, the intermediate carbocation in an E1 process may undergo rearrangement before losing the $\beta$-proton. This would not be observed in an E2 reaction. (However, where more than one kind of $\beta$-proton is present, mixed alkenes may result from either E1 or E2 reactions.)

# 5.9 Self-Test Questions

### Question 5.1

Answer ALL parts.

(i) Explain the difference between nucleophilic strength and base strength.

(ii) Indicate the products and predict a mechanism for the reaction, in propanone, of $XCH=CH-CH_2I$ with water, where X is, in turn, (a) $CH_3$, (b) $OCH_3$, and (c) CN. Indicate the order of reactivities for the substituents, X, and explain your reasoning.

(iii) Draw and explain an energy profile for the reaction described in (ii).

(University of Durham, 1984)

### Question 5.2

Comment on the following observations:

(a) The rate of solvolysis of *tert*-butyl bromide is not affected by the addition of sodium azide ($NaN_3$). However, this addition results in the formation of *tert*-butyl azide as well as *tert*-butanol.

(b) Hydrolysis at room temperature of 1-chlorobut-2-ene and 3-chlorobut-1-ene in the presence of silver oxide leads to a mixture of but-2-en-1-ol and but-1-en-3-ol. The hydrolyses of these chlorobutenes are considerably faster than those of 1-chloro- or 2-chlorobutane under comparable conditions.

(c) Hydrolysis of *R*-2-bromopropanoic acid in aqueous alkaline solution leads to a product with the same configuration as the starting material. Hydrolysis of *R*-2-bromopropanoic acid under $S_N2$ conditions leads to inversion of configuration.

(University of Warwick, 1982)

### Question 5.3

Briefly describe the bimolecular elimination mechanism (E2) applied to alkyl halides paying particular attention to any orientation or stereochemical considerations involved.

Assuming each of the following reactions proceeds via the E2 mechanism, draw the structure(s) of the product(s) in each case. If more than one product is formed, state which will be obtained in greatest amount.

(a) $(CH_3)_2CHCHBrCH_3 \xrightarrow{H_2O}$

(b) $(CH_3)_2C=CHCHClCH_3 + \overset{\ominus}{O}H \xrightarrow{EtOH}$

(c) *trans*-1-bromo-2-methylcyclohexane $+ \overset{\ominus}{O}H \xrightarrow{EtOH}$

(d) *cis*-1-bromo-2-methylcyclohexane $+ \overset{\ominus}{O}H \xrightarrow{EtOH}$

(Teesside Polytechnic, 1985)

### Question 5.4

(a) Summarise *briefly* the main methods by which bromoalkanes can be prepared from alkenes and alkanes.

(b) Give reasons why the reaction of 1-iodo-2-phenylethane with methoxide ion leads to the formation of an ether as the major product, whereas similar treatment of 1-iodo-1-phenylethane yields mainly an alkene.

(c) When 2-iodo-2-methylpentane is treated with ethoxide ion, the major product is 2-methylpent-2-ene, whereas the reaction of 2-fluoro-2-methylpentane with ethoxide ion leads predominantly to the formation of 2-methylpent-1-ene. Explain this observation.

(d) The solvolysis of (+)-PhCHClCH$_3$ in ethanol leads to the formation of an ether; explain why the optical activity of the ether isolated from several experiments decreased steadily as the temperature of the reaction was increased.

(e) Identify compounds J, K and L.

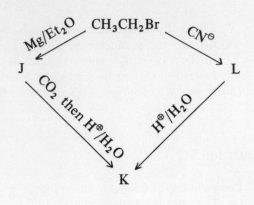

(University of York, 1985)

# 6 Alcohols, Phenols and Ethers

## 6.1 Introduction

Acidic and basic properties of alcohols and phenols are considered in Chapter 4. You might find it worthwhile to compare the behaviour of the hydroxyl group in alcohols and phenols, and also to draw parallels between the reactions of alcohols and haloalkanes (Chapter 5).

## 6.2 Formation of Alcohols

### (a) Alkaline Hydrolysis of Primary Haloalkanes

$$R-X \xrightarrow{HO^{\ominus}} R-OH + X^{\ominus}$$

X = Cl, Br, I    (2° or 3° haloalkanes give mainly elimination products)

### (b) Hydration of Alkenes

(a) Acid catalysed addition of water

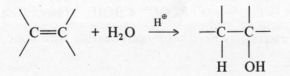

(Markownikov product)

(b) Oxymercuration–demercuration

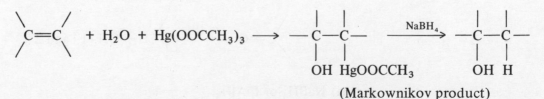

(Markownikov product)

(c) Hydroboration–oxidation

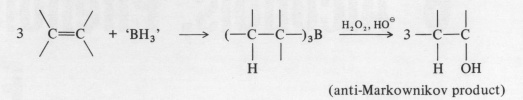

(anti-Markownikov product)

(c) **Use of Grignard Reagents**

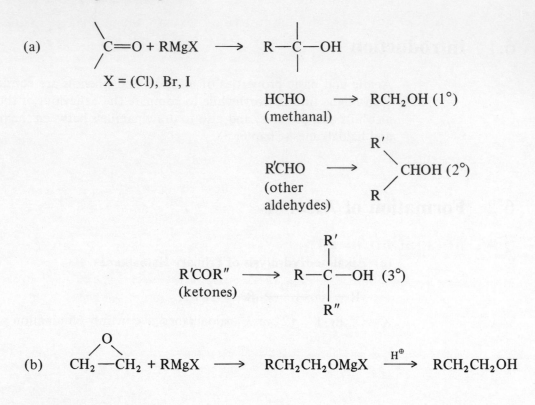

(a) $\quad\ce{C=O} + RMgX \longrightarrow R-\overset{|}{\underset{|}{C}}-OH$

$X = (Cl),\ Br,\ I$

$HCHO \longrightarrow RCH_2OH\ (1°)$
(methanal)

$R'CHO \longrightarrow \underset{R}{\overset{R'}{>}}CHOH\ (2°)$
(other
aldehydes)

$R'COR'' \longrightarrow R-\overset{R'}{\underset{R''}{\overset{|}{\underset{|}{C}}}}-OH\ (3°)$
(ketones)

(b) $\quad\underset{CH_2-CH_2}{\overset{O}{\triangle}} + RMgX \longrightarrow RCH_2CH_2OMgX \xrightarrow{H^{\oplus}} RCH_2CH_2OH$

(d) **Hydrolysis of Esters**

(a) $RCOOR + H_2O \underset{}{\overset{H^{\oplus}}{\rightleftharpoons}} RCOOH + ROH$ (equilibrium)
(b) $RCOOR + HO^{\ominus} \longrightarrow RCOO^{\ominus} + ROH$ (saponification)

Resistant esters may need prolonged heating with conc. alkali.

(e) **Reduction of Carbonyl Compounds**

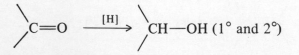

$\ce{C=O} \xrightarrow{[H]} \ce{CH-OH}$ (1° and 2°)

e.g. (i) $NaBH_4$ or $LiAlH_4$
    (ii) $H_2$/catalyst.

## 6.3 Formation of Phenols

**(a) Alkali Fusion of Sulphonates**

$$ArSO_2\overset{\ominus}{O}\ \overset{\oplus}{Na} \xrightarrow[\text{fuse}]{\Delta} Ar\overset{\ominus}{O}\ \overset{\oplus}{Na} \xrightarrow{H^{\oplus}} ArOH$$

**(b) Hydrolysis of Diazonium Salts**

$$Ar\overset{\oplus}{N}\!\!\equiv\!\!N \xrightarrow[\Delta]{H_2O} Ar\!-\!OH\ \ (+H^{\oplus} + N_2\!\uparrow)$$

**(c) Acid-catalysed Rearrangement of Cumene Hydroperoxide**

Most phenol is now made by the 'cumene' process

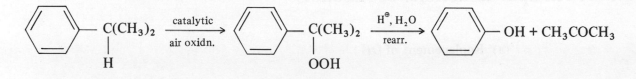

(cumene hydroperoxide)

## 6.4 Formation of Ethers

**(a) Alkoxymercuration–demercuration of Alkenes**

$$\overset{\diagdown}{\underset{\diagup}{C}}\!\!=\!\!\overset{\diagup}{\underset{\diagdown}{C}}\ +\ ROH\ +\ Hg(OOCCF_3)_2 \longrightarrow \ -\overset{|}{\underset{|}{C}}\!-\!\overset{|}{\underset{|}{C}}\!-\ \xrightarrow{NaBH_4}\ -\overset{|}{\underset{|}{C}}\!-\!\overset{|}{\underset{|}{C}}\!-$$

$$\qquad\qquad\qquad\qquad\qquad\qquad\qquad RO\quad HgOOCCF_3 \qquad RO\quad H$$

(Markownikov product)

**(b) Williamson Synthesis**

$$R\!-\!X + R\!-\!ONa \longrightarrow R\!-\!O\!-\!R + NaX\ \text{(see Chapter 5)}$$

$(1° \text{ or } 2°)$

## 6.5 Structure and Reactivity of Alcohols

Alcohols are associated through hydrogen bonding (Chapter 1). They are weak nucleophiles and bases, and are also weakly acidic and the corresponding alkoxide ions are powerful nucleophiles and bases (Section 5.4 and Example 5.4).

Although alcohols have a polar structure, direct displacement of the hydroxyl group does not occur. The carbon–oxygen bond is quite strong, and the strongly basic hydroxide ion is a poor leaving group. However, protonation of the OH group converts it into a good leaving water molecule, and alcohols undergo both substitution and elimination reactions in acid solution.

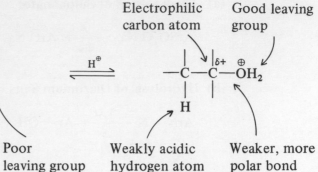

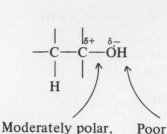

## 6.6  Principal Reactions of Alcohols

### (a)  Replacement of OH

(a)  Reaction with hydrogen halides

$$R—OH + HX \longrightarrow R—X + H_2O$$

order of reactivity: $HCl < HBr < HI$

and  $CH_3OH < RCH_2OH < R_2CHOH < R_3COH < PhCH_2OH$

Most alcohols readily form bromo or iodoalkanes with HBr and HI. Tertiary and phenylmethyl (benzyl) alcohols also form chloroalkanes with HCl, but other alcohols may require the use of a catalyst ($ZnCl_2$). Some reactions may be performed with $NaX + H_2SO_4$.

(b)  Reaction with phosphorus trihalides

$$3 R—OH + PX_3 \longrightarrow 3 R—X + H_3PO_3$$

$PX_3 = PBr_3$ or $PI_3$  (may be preferable if elimination is a problem with HX)

### (b)  Dehydration

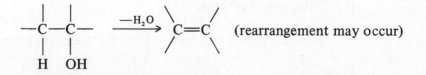

order of reactivity $3° > 2° > 1°$
Also catalytic dehydration, e.g. using $Al_2O_3$.

### (c)  Esterification

(a)  $ROH + RCOOH \xrightleftharpoons{(H^\oplus)} ROOCR + H_2O$  (equilibrium — use Dean and Stark)
(b)  $ROH + RCOCl \longrightarrow ROOCR + HCl$  (may give better yields)

### (d) Alkoxide Formation

$$2\ ROH + 2\ Na \longrightarrow 2\ R\overset{\ominus}{O}\ \overset{\oplus}{Na} + H_2\uparrow$$

Also with K, Ca, etc. (see also Section 4.7).

### (e) Oxidation

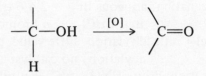

e.g. with $KMnO_4$, $K_2Cr_2O_7$, $CrO_3$, etc.

$1°$ → aldehydes ( → carboxylic acids)
$2°$ → ketones
$3°$   are resistant to oxidation.

## 6.7   Structure and Reactivity of Phenols

Phenols have a resonance structure (Section 1.5) and do not undergo the substitution or elimination reactions of alcohols. Phenols are stronger acids than alcohols (differential resonance stabilisation of the phenoxide ion — Section 4.8). Thus, the phenoxide ion is a weaker nucleophile than the alkoxide ion, but exhibits the same sort of reactivity (e.g. Williamson's synthesis of alkyl–aryl ethers).

The OH group strongly activates the ring towards electrophilic substitution (Chapter 10).

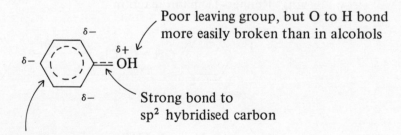

## 6.8   Principal Reactions of Phenols

### (a) Formation of Phenoxides

$$PhOH + HO^{\ominus}\ (Na^{\oplus}) \longrightarrow PhO^{\ominus}\ (Na^{\oplus}) + H_2O$$

### (b) Esterification

$$PhOH + RCOCl \longrightarrow PhOOCR + HCl$$

Direct reaction with carboxylic acids is usually slow and incomplete.

### (c) Electrophilic Substitution in the Ring

HO— and particularly $\overset{\ominus}{O}$ — are very powerfully activating and *ortho/para* directing. Substitution occurs under very mild conditions. e.g.

(i) Nitration: dilute $HNO_3$, 20°C, → *o*- + *p*-nitrophenols
(ii) Sulphonation: mod. c.$H_2SO_4$, 20°C → *o*-sulphonic acid
$\qquad\qquad\qquad\qquad\qquad$ 100°C → *p*-sulphonic acid
$\qquad$ (Reversible reaction: kinetic and thermodynamic control)
(iii) Halogenation: aqueous $Br_2$, → 2,4,6-tribromophenol
$\qquad\qquad\qquad\qquad$ $Br_2$, $CHCl_3$, → *p*-bromophenol
(iv) Friedel–Crafts alkylation: little value, poor yields
$\qquad\qquad\qquad\qquad$ acylation: no catalyst
(v) Nitrosation: dilute $HNO_2$, 5°C, → *p*-nitrosophenol
(vi) Coupling with diazonium salts:

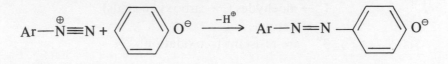

Note also:

(vii) Kolbé reaction:

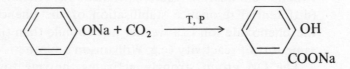

(viii) Reimer–Tiemann reaction:

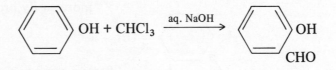

## 6.9 Structure and Reactivity of Ethers

The two carbon–oxygen bonds are less polar than those in alcohols, and the alkoxide ion is a very poor leaving group. Consequently ethers do not readily undergo nucleophilic displacement, and are inert to aqueous alkali or moderate concentrations of acids such as $H_2SO_4$ or HCl. They are, however, rapidly cleaved by concentrated aqueous HI:

$$R—O—R + HI \longrightarrow R—\overset{\oplus}{\underset{H}{O}}—R + I^{\ominus}$$

$$R—\overset{\oplus}{\underset{H}{O}}—R + I^{\ominus} \longrightarrow R—I + R—OH$$

($I^{\ominus}$ is a powerful nucleophile)

106

In electrophilic substitution reactions of alkyl–aryl ethers, the alkoxy substituent is activating and *ortho/para* directing — see Chapter 10.

## 6.10 Worked Examples

### Example 6.1

Outline ONE method in each case by which you would attempt to achieve the following conversions. Comment on the suitability of the methods that you select.

(1) $CH_3CH_2I \longrightarrow CH_3CH_2CH_2CH_2OH$

(2) $CH_3CH=CH_2 \longrightarrow CH_3CH_2CH_2OH$

(3) $(CH_3)_3CCH=CH_2 \longrightarrow (CH_3)_3CCH(OH)CH_3$

(4)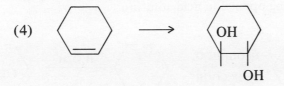

(Coventry Polytechnic, 1981)

*Solution 6.1*

(1) Chain lengthening by two carbon atoms is most simply achieved by reaction of a Grignard reagent with epoxyethane. The steps are:

$$CH_3CH_2I + Mg \xrightarrow{\text{dry ether}} CH_3CH_2MgI, \text{ followed by}$$

$$CH_3CH_2MgI + CH_2\overset{O}{\underset{}{-}}CH_2 \longrightarrow CH_3CH_2CH_2CH_2OMgI \xrightarrow[\text{workup}]{H_2O} CH_3CH_2CH_2CH_2OH$$
$$+ Mg(OH)I$$

(2) The required alcohol is the anti-Markownikov product, so acid catalysed hydration is not suitable. A complementary method giving the right orientation is the reaction with diborane:

$$CH_3CH=CH_2 + \text{'}BH_3\text{'} \longrightarrow CH_3CH_2CH_2BH_2 \quad \text{....} (CH_3CH_2CH_2)_3B$$

The tripropyl borane is then decomposed with hydrogen peroxide in alkaline solution:

$$(CH_3CH_2CH_2)_3B \xrightarrow{H_2O_2, HO^\ominus} 3\ CH_3CH_2CH_2OH + B(OH)_3$$

(3) In principle, acid-catalysed hydration or addition of HBr followed by hydrolysis would give the required orientation, but protonation of this alkene would give a carbocation prone to rearrangement — see Example 6.3.

Diborane would give the anti-Markownikov product, but reaction with mercuric ethanoate (mercuric acetate) followed by demercuration with sodium borohydride gives 3,3-dimethylbutan-2-ol:

$$(CH_3)_3CCH{=}CH_2 + Hg(OOCCH_3)_2 + H_2O \longrightarrow (CH_3)_3CCH{-}CH_2HgOOCCH_3$$

$$\underset{OH}{\mid}$$

$$+ CH_3COOH$$

$$(CH_3)_3CCH{-}CH_2HgOOCCH_3 \xrightarrow{NaBH_4} (CH_3)_3CCH(OH)CH_3$$
$$\underset{OH}{\mid}$$
(free from rearrangement products)

(4) Dilute alkaline $KMnO_4$ and $OsO_4$ both give the *cis* diol. Addition of $Br_2$ followed by hydrolysis is feasible, but some elimination would be unavoidable. The best method involves oxidation with a peracid to give the epoxide, which is ring-opened in acid solution:

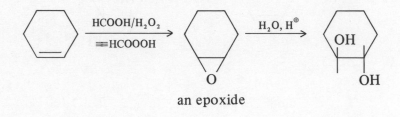

an epoxide

## Example 6.2

(a) Write mechanistic equations to show the reaction of an alcohol, ROH, as
   (1) an acid,
   (2) a base, and
   (3) a nucleophile.

*(6 marks)*

(b) Describe the Lucas test for the classification of primary, secondary and tertiary alcohols and discuss the reactions involved.

     Account for the anomalous behaviour of allyl alcohol ($CH_2{=}CHCH_2OH$) and benzyl alcohol ($PhCH_2OH$) in this test.

*(14 marks)*
(Coventry Polytechnic, 1985)
(Mid-sessional examination)

*Solution 6.2*

(a) (1) Alcohols are only weakly acidic, but form alkoxide salts on treatment with a very strong base. You could use the reaction with alkynide (acetylide) ion, or with a Grignard reagent as an example:

$$Na^{\oplus} CH{\equiv}\overset{\ominus}{\underset{..}{C}} \quad H{-}\overset{..}{\underset{..}{O}}R \longrightarrow CH{\equiv}CH\uparrow + \overset{\oplus}{Na}{:}\overset{\ominus}{\underset{..}{O}}R$$

$$:\overset{\oplus}{\underset{..}{Br}}Mg \ \overset{\ominus}{CH_3} \quad H{-}\overset{..}{\underset{..}{O}}R \longrightarrow CH_4\uparrow + :\overset{\oplus}{\underset{..}{Br}}Mg \ {:}\overset{\ominus}{\underset{..}{O}}R$$

(2) Alcohols are protonated in acid solution,

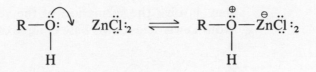

They can also complex with Lewis acids such as zinc chloride:

$$R—\overset{\cdot\cdot}{\underset{H}{O}}: \quad Zn\overset{\cdot\cdot}{\underset{\cdot\cdot}{Cl}}:_2 \quad \rightleftharpoons \quad R—\overset{\oplus}{\underset{H}{\overset{\cdot\cdot}{O}}}—\overset{\ominus}{Zn}\overset{\cdot\cdot}{\underset{\cdot\cdot}{Cl}}:_2$$

(3) Protonation of an alcohol converts the OH group into a good leaving OH$_2$ group. This is the basis of the Williamson continuous etherification process:

$$R—\overset{\cdot\cdot}{\underset{H}{O}}: \quad R—\overset{\oplus}{\overset{\cdot\cdot}{O}}H_2 \longrightarrow R—\overset{\cdot\cdot}{\underset{H}{\overset{\oplus}{O}}}—R + H_2\overset{\cdot\cdot}{O} \longrightarrow R—\overset{\cdot\cdot}{\underset{\cdot\cdot}{O}}—R + H_3\overset{\oplus}{O}:$$

Thus, a neutral alcohol molecule functions as a nucleophile in a bimolecular reaction, with a protonated alcohol molecule as substrate.

(b)  Lucas reagent is a mixture of concentrated hydrochloric acid and zinc chloride. Alcohols with six or less carbon atoms dissolve in the reagent, whereas the corresponding chloroalkanes are insoluble. Formation of a chloroalkane produces a cloudiness in the reagent as the halide separates from solution.

Tertiary alcohols react rapidly, secondary alcohols usually within a few minutes, but primary alcohols react only very slowly at room temperature. Thus the time taken for cloudiness to appear gives an indication of the class of alcohol under investigation.

A tertiary alcohol such as 2-methylpropan-2-ol reacts almost instantaneously with concentrated hydrochloric acid:

$$(CH_3)_3C—\overset{\cdot\cdot}{\underset{H}{O}}: \quad H—\overset{\cdot\cdot}{\underset{\cdot\cdot}{Cl}}: \quad \rightleftharpoons \quad (CH_3)_3C—\overset{\oplus}{\underset{H}{O}}—H + :\overset{\cdot\cdot}{\underset{\cdot\cdot}{Cl}}:^{\ominus}$$

$$(CH_3)_3C—\overset{\oplus}{\overset{\cdot\cdot}{O}}H_2 \quad \overset{slow}{\rightleftharpoons} \quad (CH_3)_3\overset{\oplus}{C} + \overset{\cdot\cdot}{O}H_2$$

$$^{\ominus}:\overset{\cdot\cdot}{\underset{\cdot\cdot}{Cl}}: \quad \overset{\oplus}{C}(CH_3)_3 \quad \xrightarrow{fast} \quad :\overset{\cdot\cdot}{\underset{\cdot\cdot}{Cl}}—C(CH_3)_3$$

(and H$_2$O + HCl $\rightleftharpoons$ H$_3\overset{\oplus}{O}$ + $\overset{\ominus}{Cl}$, of course!)

Secondary alcohols react more slowly, and primary alcohols may take several hours to form the corresponding chloroalkane at room temperature.

Reactivity is related to the stability (i.e. ease of formation) of the corresponding carbocation intermediates:

reactivity of alcohols $\qquad R_3COH > R_2CHOH > RCH_2OH > CH_3OH$

stability of carbocations $\qquad R_3\overset{\oplus}{C} > R_2\overset{\oplus}{C}H > R\overset{\oplus}{C}H_2 > \overset{\oplus}{C}H_3$

Primary and secondary alcohols react much more slowly in the absence of added zinc chloride. The catalyic behaviour of this reagent may involve the formation of the very strong acid $H_2ZnCl_4$, or result from Lewis acid–Lewis base complexing of $ZnCl_2$ with the alcohol:

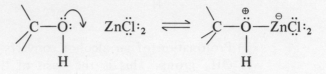

Bimolecular attack of chloride ion on this complex gives the chloro-alkane:

$$:\overset{\ominus}{\underset{..}{\overset{..}{Cl}}}: \quad \overset{\backslash}{\underset{/}{C}}-\overset{\oplus}{\underset{..}{O}}-Zn\overset{..}{\underset{..}{Cl}}:_2 \longrightarrow :\overset{..}{\underset{..}{Cl}}-\overset{/}{\underset{\backslash}{C}} + H\overset{..}{\underset{..}{O}}-Zn\overset{..}{\underset{..}{Cl}}:_2 \xrightarrow{H\overset{..}{\underset{..}{Cl}}:} H_2\overset{..}{O} + Zn\overset{..}{\underset{..}{Cl}}:_2 + :\overset{..}{\underset{..}{Cl}}:^{\ominus}$$

Allyl and benzyl alcohols are anomalous in the sense that although they have a primary structure, they both rapidly react with Lucas reagent. The reason for this enhanced reactivity is that ionisation of the protonated alcohol gives a carbocation intermediate which is resonance stabilised:

$$CH_2{=}CH{-}CH_2{-}\overset{\oplus}{O}H_2 \rightleftharpoons [CH_2{=}CH{-}\overset{\oplus}{C}H_2 \leftrightarrow \overset{\oplus}{C}H_2{-}CH{=}CH_2] + \overset{..}{O}H_2$$

A further complication is that allyl chloride initially remains in solution because of protonation of the alkene double bond.

**Example 6.3**

Answer TWO from (a), (b) and (c).
(a) Suggest a synthesis for 3-methylhexan-3-ol from alcohols of four carbon atoms or less (4 steps or more may be required).
(b) It has been observed that secondary alcohols sometimes undergo rearrangements as well as substitution, when treated with hydrogen halide acids (HX). Most primary alcohols do not. Explain why this pattern of reactivity is observed, illustrate your answer with reference to the mechanisms of reactions of HX with ethanol, propan-2-ol, and 3-methyl-butan-2-ol.

(c) Predict, and explain by reference to mechanistic theory, the major products of dehydration of the following by conc. sulphuric acid.

  (i) Pentan-2-ol.

 (ii) 2,2-Dimethylpropan-1-ol.

(iii) 1-Phenylbutan-2-ol.

(iv) 1-Phenyl-1,2-propanediol (the product gives a positive iodoform reaction).

(*20 marks*)

(University of Bradford, 1983)

*Solution 6.3*

(a) Begin by writing down the structure of the alcohol to be synthesised:

$$CH_3-CH_2-\underset{\underset{\textstyle OH}{|}}{\overset{\overset{\textstyle CH_3}{|}}{C}}-CH_2-CH_2-CH_3$$

This suggests that the starting materials should be butan-2-ol and propan-1-ol

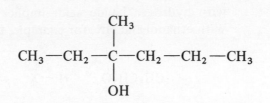

To link the three- and four-carbon fragments together, we can make use of the Grignard reaction. First of all, butan-2-ol must be converted into the ketone, butanone, for example by oxidation with acidified $K_2Cr_2O_7$:

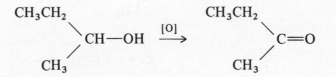

To prepare the Grignard, propan-1-ol is converted into the bromide by treatment with HBr or $PBr_3$, and then reacted with magnesium turnings in dry ether:

$$CH_3CH_2CH_2OH \xrightarrow[\text{or } PBr_3]{HBr} CH_3CH_2CH_2Br \xrightarrow[\text{dry ether}]{Mg} CH_3CH_2CH_2MgBr$$

Finally, reaction between the Grignard and the ketone will give, after the usual workup, the tertiary alcohol, 3-methylhexan-3-ol:

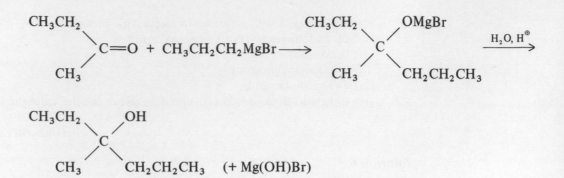

$$CH_3CH_2 \quad OH$$
$$\underset{CH_3}{\overset{\phantom{CH_3CH_2}}{C}}\underset{CH_2CH_2CH_3}{\phantom{C}} \quad (+ Mg(OH)Br)$$

(b) The fact that primary alcohols seldom undergo rearrangement when treated with hydrogen halide acids implies that the reaction is bimolecular, $S_N2$. With ethanol and HX, for example, the steps are

$$CH_3CH_2\overset{..}{\underset{|}{O}} \quad H\overset{\frown}{-}\overset{..}{\underset{..}{X}}: \rightleftharpoons CH_3CH_2\overset{\oplus}{O}H_2 + :\overset{..}{\underset{..}{X}}:^{\ominus}$$
$$\underset{H}{\phantom{CH_3CH_2O}}$$

$$^{\ominus}:\overset{..}{\underset{..}{X}}: \quad \underset{CH_2}{\overset{CH_3}{|}}\overset{\frown}{-}\overset{\oplus}{\underset{}{O}}H_2 \longrightarrow :\overset{..}{\underset{..}{X}}-\underset{CH_2}{\overset{CH_3}{|}} + \overset{..}{O}H_2$$

Since no carbocation intermediate is involved, there is no possibility of rearrangement occurring during substitution.

Secondary alcohols may react by $S_N1$ and/or $S_N2$ mechanisms. With propan-2-ol, unimolecular reaction would not be accompanied by rearrangement since the secondary propyl carbocation is the most stable possible with three carbon atoms.

$$CH_3CH\overset{\frown}{-}\overset{\oplus}{O}H_2 \rightleftharpoons CH_3\overset{\oplus}{C}H + \overset{..}{O}H_2$$
$$\underset{CH_3}{|} \qquad \qquad \underset{CH_3}{|}$$

$$^{\ominus}:\overset{..}{\underset{..}{X}}: \quad \overset{\oplus}{C}HCH_3 \longrightarrow :\overset{..}{\underset{..}{X}}-CHCH_3$$
$$\underset{CH_3}{|} \qquad \qquad \underset{CH_3}{|}$$

However, with 3-methylbutan-2-ol, the structure is such that in unimolecular reaction the first-formed secondary carbocation *can* undergo rearrangement to a more stable tertiary carbocation:

$$CH_3-CH-CH-\overset{\oplus}{O}H_2 \rightleftharpoons CH_3-CH-\overset{\oplus}{C}H + \overset{..}{O}H_2$$
$$\phantom{CH_3-}\underset{CH_3}{|}\ \underset{CH_3}{|} \qquad \qquad \phantom{CH_3-}\underset{CH_3}{|}\ \underset{CH_3}{|}$$

$$\overset{H}{\overset{|}{\phantom{C}}}$$
$$CH_3-\underset{\underset{CH_3}{|}}{\overset{|}{C}}-\overset{\frown}{C}H-CH_3 \xrightarrow{\text{'hydride shift'}} CH_3-\underset{\underset{CH_3}{|}}{\overset{\oplus}{C}}-CH_2-CH_3$$
$$\qquad\qquad \overset{\oplus}{\phantom{CH}}$$

112 $\qquad\qquad\qquad$ (2°) $\qquad\qquad\qquad\qquad$ (3°)

followed by

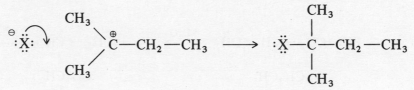

(c) There are two factors to keep in mind here.
1. When more than one alkene is possible, the major product will be the most alkylated (most stable) alkene — the Saytzeff rule.
2. In unimolecular reactions, rearrangement to a more stable carbocation may occur before loss of $H^\oplus$ to give an alkene.

(i) Pentan-2-ol

$$CH_3CH_2CH_2-\underset{\underset{\oplus OH_2}{|}}{CH}-CH_3 \xrightarrow{-H_2O} CH_3CH_2CH_2-\underset{\beta}{\overset{\oplus}{CH}}-\underset{\beta'}{CH_3} \xrightarrow{-\beta-H^\oplus} CH_3CH_2CH{=}CHCH_3$$

pent-2-ene

In a more detailed answer, you might point out that alkylation of the incipient (forming) double bond in the transition state is important, and that activation energy for removal of the $\beta$-H atom is less than that for removal of the $\beta'$-H atom:

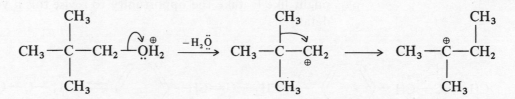

(ii) 2,2-Dimethylpropan-1-ol
With a 'longish' systematic name, check carefully that you have the correct structure! The steps are

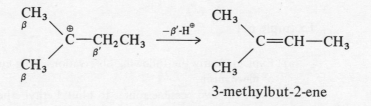

followed by preferential loss of the $\beta'$-H atom to give the most alkylated alkene:

$$\underset{\underset{\beta}{CH_3}}{\overset{\overset{\beta}{CH_3}}{\underset{|}{\overset{|}{C}}}}{\overset{\oplus}{C}}-\underset{\beta'}{CH_2CH_3} \xrightarrow{-\beta'-H^\oplus} \underset{CH_3}{\overset{CH_3}{C}}{=}CH-CH_3$$

3-methylbut-2-ene

Since this alcohol very readily undergoes dehydration, it may be that migration of the methyl group accompanies (i.e. facilitates) loss of the water molecule (neighbouring group participation — anchimeric assistance).

(iii) 1-Phenylbutan-2-ol

$$Ph-CH_2-CH-OH_2 \xrightarrow{-H_2O} Ph-\overset{\beta}{CH_2}-\overset{\oplus}{CH} \xrightarrow{-\beta-H^{\oplus}} Ph-CH=CH-CH_2CH_3$$
$$\underset{CH_3CH_2}{|} \qquad \underset{\underset{\beta'}{CH_3CH_2}}{|}$$

Loss of the β-H atom adjacent to the benzene ring gives the most stable alkene, in which the double bond is conjugated with the aromatic nucleus:

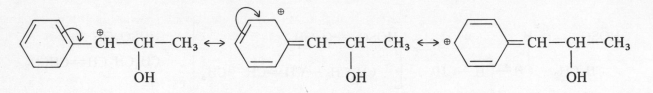

(iv) 1-Phenylpropane-1,2-diol

A positive iodoform test indicates a methyl ketone, $CH_3CO-$ (or $CH_3CH(OH)-$, which can oxidise to $CH_3CO-$). This implies retention of the OH group next to the $CH_3$ group. Furthermore, loss of OH adjacent to the phenyl group will lead to a resonance stabilised (benzylic) carbocation:

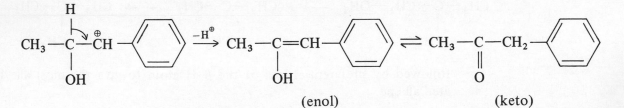

etc.

The next step, loss of a proton, gives an alkene which in fact is the *enol tautomer* of the methyl ketone product. The question does not require you to give the mechanism for the tautomeric change, but you might like to take the opportunity to revise this if you are unsure of the detail.

$$CH_3-\overset{H}{\underset{OH}{C}}-\overset{\oplus}{CH}\text{—}\bigcirc \xrightarrow{-H^{\oplus}} CH_3-\underset{OH}{C}=CH\text{—}\bigcirc \rightleftharpoons CH_3-\underset{O}{\overset{}{C}}-CH_2\text{—}\bigcirc$$
$$\text{(enol)} \qquad\qquad \text{(keto)}$$

### Example 6.4

(a) Explain clearly the following observation, using curly arrows to indicate clearly points of mechanism.

Of the two possible routes to t-butyl ethyl ether using either $Me_3CO^{\ominus}$ and EtBr, or $Me_3CBr$ and $EtO^{\ominus}$, only one is successful.

(University of Southampton, 1983)

(b) List the advantages and disadvantages of ethoxyethane (diethyl ether) as a medium for solvent extraction.

Account for the observation that whereas ethoxyethane is resistant to attack by moderately concentrated hydrochloric acid, it is rapidly cleaved on treatment with concentrated hydriodic acid, even at room temperature.

(Coventry Polytechnic, 1985)

(c) The methoxy group of methyl phenyl ether (methoxybenzene) is not easily displaced by nucleophiles. Nitration of methyl phenyl ether occurs readily to give a product $C_7H_6N_2O_5$ which becomes radioactively labelled when treated with $^{14}C$-labelled methoxide ion. Explain.

(University of Warwick, 1981)

*Solution 6.4*

(a) Formation of ethers by this method involves the use of alkoxide ion, which is not only nucleophilic, but also powerfully basic. Hence, competition from elimination is always a problem.

When the primary haloalkane EtBr is the substrate, reaction is bimolecular. This favours substitution, since activation energy for formation of the transition state is usually lower for substitution than for elimination. The mechanistic steps are

If the tertiary haloalkane, $Me_3CBr$, is employed, reaction is unimolecular, and the first step is ionisation to give the tertiary butyl carbocation, $Me_3C^{\oplus}$:

$$Me_3C{-}\ddot{B}\ddot{r}: \; \underset{}{\overset{slow}{\rightleftharpoons}} \; Me_3C^{\oplus} + :\ddot{B}\ddot{r}:^{\ominus}$$

The positively charged carbon atom is strongly electron attracting ($-I$ effect), and consequently the methyl hydrogen atoms are appreciably more acidic than the $\beta - H$ atoms in bromoethane:

$$\overset{\delta\delta+}{H}{\rightarrow}\overset{\delta+}{CH_2}{\rightarrow}\overset{\oplus}{C(CH_3)_2}$$

Since the carbocation has a planar structure, these methyl hydrogen atoms are easily accessible to the basic ethoxide ion, and so elimination is now the major reaction:

2-methylpropene

(b) Advantages
   (1) Dissolves a wide range of compounds.
   (2) Chemically inert to moderate concentrations of acids and alkalis.
   (3) Low boiling point makes it easy to remove.
   Disadvantages
   (1) The combination of flammability, low flash point and volatility is potentially hazardous.
   (2) Toxic (anaesthetic!)
   (3) Some losses result from slight solubility of, and in, water.
   Ethoxyethane is protonated by acids, HX, which converts the poor leaving ethoxy group into the better leaving ethanol molecule:

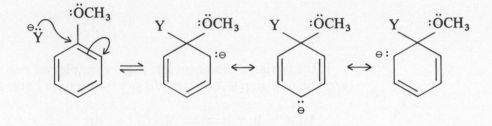

Hydrogen iodide is a stronger acid than hydrochloric acid, and equilibrium lies further to the right. The key factor is that iodide ion is a stronger nucleophile than chloride ion, and is thus able to displace ethanol in a bimolecular reaction:

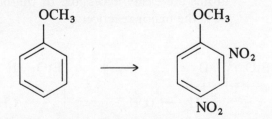

(c) Aromatic nucleophilic substitution involves a resonance stabilised carbanion intermediate,

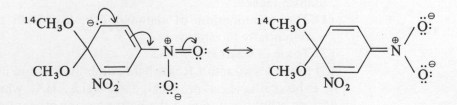

However, it is unlikely that this intermediate is formed under moderate conditions unless the substrate is activated by electron attracting substituents.

The methoxy group is activating and *ortho/para* directing towards aromatic electrophilic substitution, and nitration of methyl phenyl ether gives 2,4-dinitromethoxybenzene:

These strongly electron-attracting substituents are able to stabilise the carbanion intermediate formed by attack of the $^{14}C$ labelled methoxide ion. Illustrate this by drawing at least one of the additional limiting forms which contribute to the resonance hybrid, e.g.

Reaction is completed by expulsion of a methoxide ion, which clearly can be either the labelled or the unlabelled ion. When a large excess of labelled methoxide ion is used, most of the methoxybenzene will quickly become labelled in this way.

# 6.11 Self-Test Questions

### Question 6.1

Give FOUR methods available for the preparation of alcohols and describe which methods would be most suitable for the preparation of the alcohols (1) and (2) from benzoic acid.

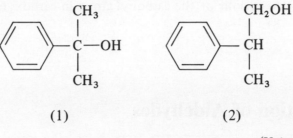

(1)                    (2)

(University of Leeds, 1983)

### Question 6.2

Compare and contrast the chemical properties of alcohols and amines. Within this discussion you should consider the reactivity of the lone pair of electrons, the relative ease of cleavage of the O—H and N—H bonds, the relative ease of cleavage of the C—O and C—N bonds and any other reactions which clearly illustrate their differing chemistry.

(University of Southampton, 1981)

### Question 6.3

Account for the reactions of alcohols, amines and Grignard reagents with aldehydes, acid chlorides and carboxylic acids.

(University of Exeter, 1985)

### Question 6.4

(a) Discuss the effect of hydrogen bonding upon the physical properties of alcohols and phenols.

(Trent Polytechnic, 1985)

(b) How would you prepare pure samples of both *ortho*- and *para*-nitrophenol starting from phenol?

(Trent Polytechnic, 1985)

(c) Cyclohexanol is insoluble in sodium hydroxide solution, whereas phenol readily dissolves. However, phenol is insoluble in sodium hydrogen carbonate solution, whereas 2,4,6-trinitrophenol dissolves with the evolution of carbon dioxide. Explain.

(University of York, 1984)
(part questions)

# 7 Aldehydes and Ketones

## 7.1 Introduction

Two useful comparisons you could make in connection with this chapter are in the chemistry of aldehydes and ketones, and in the effect of substituents on the behaviour of the carbonyl group in carboxylic acids and their derivatives (Chapter 8).

## 7.2 Formation of Aldehydes

### (a) Oxidation Methods

(i) 1° alcohols

$$RCH_2OH \xrightarrow[\text{or } CrO_3/\text{pyridine}]{K_2Cr_2O_7/H_2SO_4} RCHO$$

(ii) Methylbenzenes

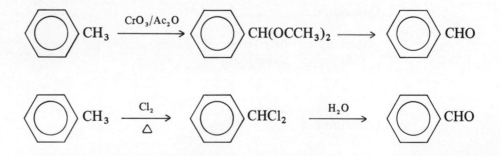

### (b) Reduction of Alkanoyl Chlorides (Acid Chlorides)

$$RCOCl \xrightarrow{LiAlH[(CH_3)_3CO]_3} RCHO$$

### (c) 'Hydroformylation' of Alkenes ('Oxo' Process)

$$RCH{=}CH_2 + CO + H_2 \xrightarrow{Co_2(CO)_8} RCH_2CH_2CHO \text{ (main product)}$$

### (d) Reimer–Tiemann Reaction (Phenolic Aldehydes)

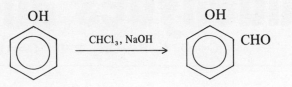

## 7.3 Formation of Ketones

### (a) Oxidation Methods

2° alcohols

$$\underset{R'}{\overset{R}{\diagdown}}CHOH \xrightarrow[\text{or } CrO_3/\text{pyridine}]{K_2Cr_2O_7/H_2SO_4} RCOR'$$

### (b) Alkylation of Alkanoyl Chlorides (Acid Chlorides)

$$RCOCl \xrightarrow[\text{or } R'_2CuLi]{R'_2Cd} RCOR'$$

### (c) Friedel–Crafts Acylation

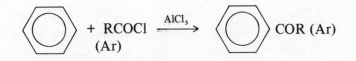

### (d) 'Acetoacetic Ester' Synthesis

$$CH_3COCH_2COOEt \xrightarrow[\text{(2) RI}]{\text{(1) NaOEt}} \underset{\text{(I)}}{CH_3COCHRCOOEt} \xrightarrow[\text{(2) RI}]{\text{(1) NaOEt}} \underset{\text{(II)}}{CH_3COCRRCOOEt}$$

$$I \text{ (or II)} \xrightarrow{H^{\oplus}H_2O} CH_3COCHR \text{ (or } CH_3COCRR) + CO_2 + EtOH$$

## 7.4 Structure and Reactivity of Aldehydes and Ketones

The carbonyl group has a planar polar structure, $\overset{\delta+}{\diagup}\overset{\delta-}{C}=\overset{\delta-}{O}$, described as a mesomer or resonance hybrid of two limiting forms, $\diagup C=\underset{\cdot\cdot}{O}: \longleftrightarrow \diagup\overset{\oplus}{C}—\overset{\ominus}{\underset{\cdot\cdot}{O}}:$ (see Chapter 4).

Thus the carbonyl group can be attacked at the oxygen atom by *electrophiles* (commonly $H^{\oplus}$) and at the carbon atom by *nucleophiles* (commonly compounds containing oxygen or nitrogen). Many reactions are acid-catalysed. Protonation of

119

the oxygen atom gives a resonance-stabilised cation which is readily attacked by weak nucleophiles:

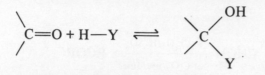

## 7.5  Principal Reactions of Aldehydes and Ketones

### (a)  Nucleophilic Addition, Reversible

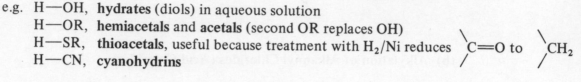

(proton may be supplied by acid catalyst)

e.g.  H—OH,  **hydrates** (diols) in aqueous solution
H—OR,  **hemiacetals** and **acetals** (second OR replaces OH)
H—SR,  **thioacetals**, useful because treatment with $H_2$/Ni reduces $\diagdown\!\!C\!\!=\!\!O$ to $\diagdown\!\!CH_2$
H—CN,  **cyanohydrins**

note also the formation of 'bisulphite' compounds,

### (b)  Nucleophilic Addition, Non-reversible

(i)  *Addition of Hydride Ion ($\ddot{H}^{\ominus}$)*

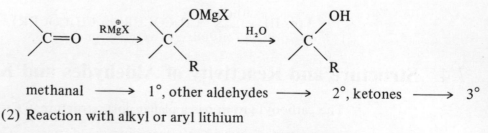

(ii)  *Addition of Carbanions*
(1)  Reaction with Grignard reagents

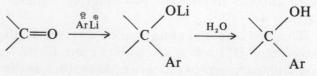

methanal $\longrightarrow$ 1°, other aldehydes $\longrightarrow$ 2°, ketones $\longrightarrow$ 3°

(2)  Reaction with alkyl or aryl lithium

120

## (3) Reaction with alkynides (acetylides)

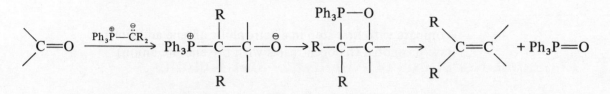

## (4) Reaction with phosphorus ylides (Wittig)

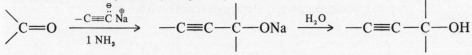

### (c) Nucleophilic Addition with Elimination

With certain derivatives of ammonia ($NH_2$ — G), nucleophilic addition is followed by elimination (which may be acid or base catalysed):

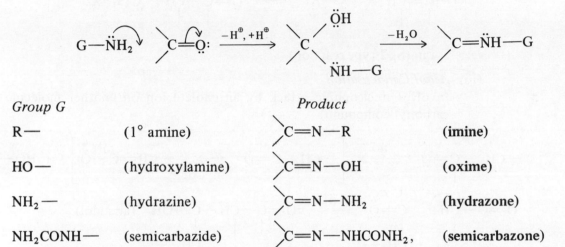

e.g.

| Group G | | Product | |
|---------|---|---------|---|
| R— | (1° amine) | $\,>$C=N—R | **(imine)** |
| HO— | (hydroxylamine) | $\,>$C=N—OH | **(oxime)** |
| $NH_2$— | (hydrazine) | $\,>$C=N—$NH_2$ | **(hydrazone)** |
| $NH_2CONH$— | (semicarbazide) | $\,>$C=N—$NHCONH_2$, | **(semicarbazone)** |

### (d) Reactions Involving Acidic α-Hydrogen

Hydrogen adjacent (α) to a carbonyl group is weakly acidic and can be removed by a strong base (e.g. $HO^\ominus$, $RO^\ominus$) to give a resonance stabilised enolate anion:

$$\overset{\ominus}{HO}: \quad H—CH—C=O: \underset{-H^\oplus, +H^\oplus}{\rightleftharpoons} HO—H + \left[ \overset{\ominus}{C}H—C=O: \longleftrightarrow CH=C—\overset{\ominus}{O}: \right]$$

Reactions consequent upon this include:

(i) *Formation of Tautomers (Tautomerism)*

Protonation of the enolate ion at the carbonyl oxygen atom gives rise to the tautomeric enol:

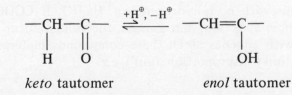

*keto* tautomer        *enol* tautomer

(equilibrium well to left unless enol stabilised)

121

(ii) α-*Halogenation*

1. Acid catalysed (halogen reacts with enol)

$$-CH=C-\overset{..}{O}H \rightleftharpoons^{:\overset{..}{X}_2} \left[ -\overset{|}{\underset{:\overset{..}{X}:}{CH}}-\overset{\oplus}{C}-\overset{..}{O}H \leftrightarrow -\overset{|}{\underset{:\overset{..}{X}:}{CH}}-C=\overset{\oplus}{O}H \right] + :\overset{..}{\underset{..}{X}}:^{\ominus} \xrightarrow{-H^{\oplus}} -\overset{|}{\underset{X}{CH}}-C=\overset{..}{O}$$

(compare with first step in electrophilic alkene addition)

2. Base promoted (halogen reacts with the enolate anion)

$$-\overset{|}{CH_2}-C=\overset{..}{O} \rightleftharpoons^{\overset{..}{B}^{\ominus}} -\overset{|}{\overset{..}{C}H}-C=\overset{..}{O} \leftrightarrow -CH=\overset{|}{C}-\overset{..}{\underset{..}{O}}:^{\ominus} + HB$$

(resonance stabilised)

$$^{\ominus}:\overset{..}{\underset{..}{O}}\overset{\frown}{\frown}\overset{|}{C}=\overset{|}{CH} \quad :\overset{..}{X}\overset{\frown}{\frown}\overset{..}{X}: \longrightarrow \overset{..}{O}=\overset{|}{C}-\overset{|}{CH}-\overset{..}{X}: + :\overset{..}{\underset{..}{X}}:^{\ominus}$$

(an ($S_N2$ type reaction)

(iii) **Aldol Condensation**

Involves nucleophilic attack by an enolate ion on another molecule of carbonyl compound:

$$-\overset{|}{CH_2}-C=\overset{..}{O}: \rightleftharpoons^{H\overset{..}{O}:^{\ominus}} \left[ -\overset{|}{\overset{..}{C}H}-C=\overset{..}{O}: \leftrightarrow -CH=\overset{|}{C}-\overset{..}{\underset{..}{O}}:^{\ominus} \right] + H\overset{..}{O}-H$$

$$:\overset{..}{O}=\overset{|}{C}-\overset{|}{\overset{..}{C}H} \quad \overset{|}{\underset{CH_2-}{C}}=\overset{..}{O}: \xrightarrow{+H^{\oplus}} :\overset{..}{O}=\overset{|}{C}-\overset{|}{CH}-\overset{|}{\underset{CH_2-}{C}}-\overset{..}{O}H \quad \text{(an aldol)}$$

The resulting hydroxycompounds (aldols) are readily dehydrated, loss of water often occurring spontaneously, e.g.

$$2\ CH_3CHO \longrightarrow CH_3CH(OH)CH_2CHO \xrightarrow{-H_2O} CH_2CH=CHCHO$$
$$\qquad\qquad\qquad\qquad \text{3-hydroxybutanal} \qquad\qquad \text{but-2-enal}$$

**Crossed condensation** may occur with two different molecules (aldehydes and/or ketones), one or both of which have an α-hydrogen atom. Note also condensation with esters, etc.

## (e) Cannizzaro Reaction

Compounds with no α-hydrogen (e.g. HCHO, $R_3$CCHO, PhCHO, PhCOPh) can react with enolate ion, but cannot themselves form such ions. Instead, when warmed with aqueous alkali these compounds undergo simultaneous oxidation and reduction ('disproportionation'), e.g.

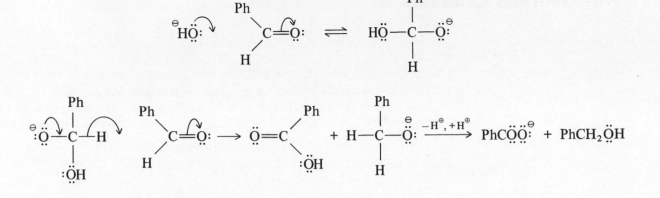

### (f) Oxidation and Reduction

#### (i) *Oxidation*

$$\text{Aldehydes} \quad \text{RCHO} \quad \xrightarrow[\text{or } K_2Cr_2O_7]{\text{KMnO}_4} \quad \text{RCOOH} \quad (\text{R} = \text{alkyl or aryl})$$

Ketones are resistant and do not have the reducing properties of aldehydes.

#### (ii) *Reduction* (see also Section 7.5(b)(i))
1. Aldehydes and ketones are reduced to the corresponding alcohol with Na/EtOH, or by hydrogenation using Pt or Pd catalysts.
2. The carbonyl group can be reduced to $\diagdown$CH$_2$ by
   (a) **Clemmensen**, amalgamated Zn/conc. HCl.
   (b) **Wolff–Kishner**, hydrazone heated with KOH in non-aqueous solution.
   (c) **Mozingo**, reduction of thioacetal or thioketal

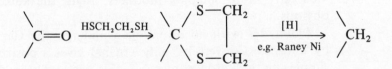

(This reaction is also useful for the 'protection' of the C=O group which can be regenerated by hydrolysis of the thioacetal or thioketal with dilute acid)

### Additional Notes

# 7.6 Worked Examples

**Example 7.1**

What are the chief differences between the properties of aldehydes and ketones?

How does acetone (propanone) react with (a) methylamine, (b) nitrous acid, and (c) sodium hypochlorite?

(Brunel University, 1983)

*Solution 7.1*

Aldehydes and ketones are characterised more by their similarities than differences, and there are as many points of difference between ethanal and benzaldehyde as between ethanal and propanone. Methanal also differs from the other aldehydes in a number of ways. Questions involving such comparisons are quite common and you should summarise the main points in your notes.

The main difference between aldehydes and ketones is in their ease of oxidation. Ketones are very resistant, as oxidation requires cleavage of a carbon–carbon bond. Aldehydes, on the other hand, are readily oxidised to the corresponding carboxylic acid:

$$RCHO \xrightarrow[\text{or KMnO}_4]{\text{K}_2\text{Cr}_2\text{O}_7} RCOOH$$

Consequently, aldehydes have reducing properties not shown by ketones. They react with Fehlings solution and Tollens reagent, while ketones do not.

As a class, aldehydes are much more prone to polymerisation both in acidic and basic solution. Ketones either do not react, or undergo more simple condensation reactions. Only certain aldehydes give a simple addition product with ammonia. Most give more complex products, as do the ketones, so the distinction is less obvious here.

Aldehydes with no α-hydrogen undergo the Cannizzaro reaction, not shown by ketones. Conversely, only ethanal gives a positive iodoform test, whereas all simple methylketones respond this way. These differences between the two classes of compound are clearly related to structure, ketones having a second alkyl or aryl group in place of the aldehydic hydrogen atom. As far as the present answer is concerned, there is no requirement to discuss this point. But should the wording ask you to 'Account for (or discuss the reasons for) the differences. . .' then you could expand your answer along the following lines. There are two main factors involved.

(1) *Loss of inductive stabilisation* on going from reactant to transition state, where the positive inductive effect (Chapter 1) of alkyl groups may even become destabilising:

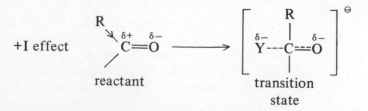

Two groups are more effective than one, so there is a bigger barrier to the formation of the transition state for a ketone.

(2) *Steric compression* in the transition state. On passing from reactant to transition state, groups move from a planar arrangement towards the tetrahedral one of the intermediate:

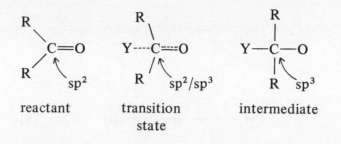

reactant       transition       intermediate
               state

Crowding together of the groups attached to the carbonyl carbon atom ('**steric crowding**') destabilises the transition state, and again the effect will be greatest in ketones. Steric crowding increases with the size of the attacking nucleophile, and it is with larger nucleophiles that the difference in reactivity between aldehydes and ketones becomes most apparent. For example, ketal formation with alcohols and dry hydrogen chloride is often slow and incomplete.

In answering Sections (a) to (c) it is not sufficient to state the reaction and give an equation. The examples have been chosen to illustrate different aspects of the chemistry of carbonyl compounds, and these should be emphasised in your answers.

(a) This is an example of nucleophilic addition with elimination. The first formed compound is unstable and rapidly eliminates a molecule of water to give an *imine*:

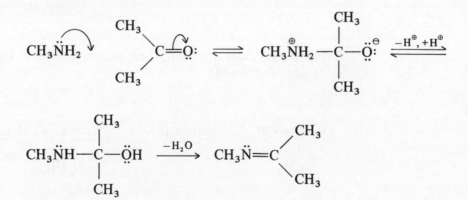

If your treatment of the first part of the question was relatively brief, you might have time here to include a short discussion of the pH dependence of the reaction. Loss of water in the second stage is acid catalysed (protonation converts the poor leaving —OH group into the good leaving —$\overset{\oplus}{O}H_2$ group). However, if the solution is too acidic, protonation of the amine will effectively reduce its concentration. Thus the optimum pH is often about 4–5.

Finally, you could mention that simple aliphatic imines are relatively unstable and readily polymerise.

(b) This reaction is not always mentioned in first year courses, so do not be surprised if it is new to you! The process being illustrated is an electrophilic attack on the *enol* form of the ketone.

A variety of electrophilic species may be present in nitrous acid (see Example 11.3), but for simplicity the attacking electrophile is usually assumed to be the nitrosyl cation, $NO^{\oplus}$. The steps are then formulated as

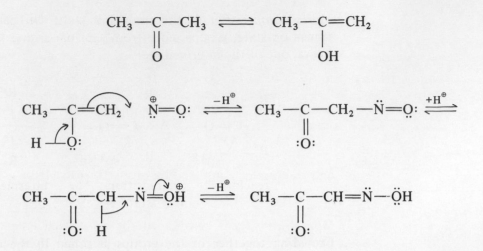

The product is the monoxime of the dicarbonyl compound $CH_3COCHO$.

(c) This is an example of a reaction which results from the activating effect of the carbonyl group on the adjacent ($\alpha$) hydrogen atoms. The reagent behaves as though it were simply a solution of 'free' chlorine in an excess of aqueous sodium hydroxide, and the rate determining step is conversion of propanone into its (resonance stabilised) conjugate anion:

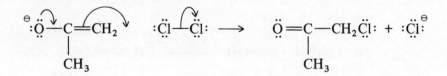

(enolate ion)

Rapid bimolecular reaction then occurs between the enolate ion and a molecule of chlorine:

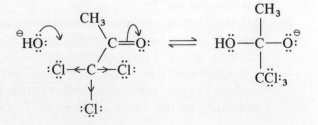

As a result of the $-I$ effect of the chlorine atom, the remaining $\alpha$-hydrogen atoms are more acidic than the first, and so are quickly replaced by repetition of the above steps to give the trichloroketone, $CH_3COCCl_3$.

Nucleophilic attack by hydroxide ion on the carbonyl carbon atom then follows, facilitated by the combined effect of the three electron attracting chlorine atoms:

The process continues with the expulsion of the trichloromethyl carbanion by the intermediate:

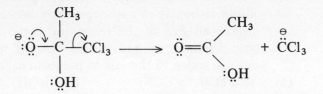

and reaction is completed by proton transfer from acid to base:

$$CH_3COOH + \overset{\ominus}{\ddot{C}}Cl_3 \longrightarrow CH_3C\overset{\ominus}{\ddot{O}}O + HCCl_3$$

More time should be allowed for this rather longer answer, which would probably carry a higher proportion of marks than Sections (a) and (b). You could round off your answer by mentioning that the corresponding reaction with iodine is used as a diagnostic test (formation of 'iodoform') for methylketones.

## Example 7.2

Suggest TWO methods suitable for the preparation of benzaldehyde from benzene or toluene on an industrial scale. *(4, 4 marks)*

Discuss the reaction of benzaldehyde with the following substances:
(a) hydrogen cyanide,
(b) hydroxylamine,
(c) methanol and anhydrous hydrogen chloride,
(d) sodium borohydride,
(e) aniline. *(5 × 5 = 25, total 33 marks)*

(Teesside Polytechnic, 1984)

## Solution 7.2

The first part of this question is pure bookwork. Methylbenzene (toluene) is available in large quantities from coal tar and by catalytic dehydrogenation ('hydroforming') of methylcyclohexane, itself obtained from certain types of petroleum. Oxidation of methylbenzene to benzaldehyde is accomplished on an industrial scale by:

(1) catalytic air oxidation

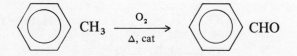

A mixture of methylbenzene and air is passed over the heated catalyst, e.g. vanadium(V) oxide at 350°C. (As an alternative, you could quote liquid phase oxidation using manganese(IV) oxide ($MnO_2$) in 65% sulphuric acid at 40°C.)

(2) Side chain chlorination of methylbenzene to dichloromethylbenzene (benzal chloride) and hydrolysis of the product:

127

Controlled side chain chlorination is achieved by passing chlorine into boiling methylbenzene until the calculated increase in weight is observed. Hydrolysis occurs simply on boiling the product with water containing sodium carbonate or a suspension of calcium carbonate.

Since all five of the remaining parts are to be answered — giving you about 5 min for each reaction — you will have to be fairly concise and limit your discussion to the salient points in each case.

(a)     PhCHO + HCN $\longrightarrow$ PhCH(OH)CN
benzaldehyde cyanohydrin
(2-hydroxy-2-phenylethanonitrile)

This nucleophilic addition reaction may be summarised as:

$$H_2O + HCN \rightleftharpoons H_3\overset{\oplus}{O} + \overset{\ominus}{C}N$$

(remember that it is the *carbon* atom which carries the negative charge)

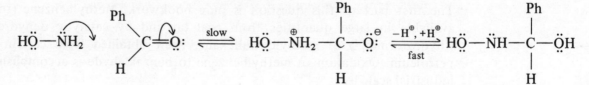

Give as supporting evidence the observation that the rate depends on the concentration of both benzaldehyde and hydrogen cyanide, and the reaction is base catalysed. At low pH, ionisation of HCN is suppressed, but the solution must not be too alkaline because of competition from the stronger nucleophile, $HO^\ominus$ ion. Optimum conditions are met by using a solution of sodium cyanide to which is added a few drops of strong acid.*

(b)     PhCHO + NH$_2$OH $\longrightarrow$ PhCH=NOH
(benzaldehyde oxime)

In this reaction, nucleophilic addition of hydroxylamine is followed by elimination of water. The steps are:

followed by dehydration:

Hydroxide ion is a poor leaving group and dehydration is acid catalysed. Strongly acidic conditions must be avoided, however, because protonation of hydroxylamine would compete with the addition step. (If more time was available for this question, you could give a more detailed discussion of pH dependence, and mention the use of oxime derivatives.)

*With alcoholic KOH, benzaldehyde also undergoes condensation to form *benzoin*, PhCH(OH)COPh.

(c)

$$PhCHO + CH_3OH \rightleftharpoons PhCH(OH)OCH_3 \xrightarrow{CH_3OH} PhCH(OCH_3)_2$$

methoxyphenylmethanol      dimethoxyphenylmethane

(a hemiacetal)          (an acetal)

Methanol is only weakly nucleophilic, and the role of the hydrogen chloride is catalytic. The mechanism of this reaction takes a little longer to write out, so allow extra time for this.

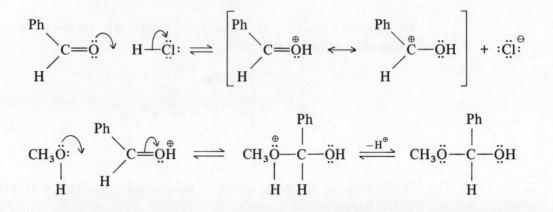

Protonation of hydroxyl converts it into a better leaving group, making ionisation possible. This is also facilitated by resonance stabilisation of the resulting cation:

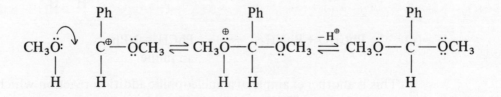

The cation is then attacked by another molecule of methanol:

Equilibrium is driven to the right by reaction of anhydrous hydrogen chloride with water,

$$HCl + H_2O \rightleftharpoons H_3\overset{\oplus}{O} + \overset{\ominus}{Cl}$$

but for good yields the water should be distilled off as it is formed. (In a longer answer, you could mention 'protection' of the carbonyl group by reaction with ethan-1,2-diol, and the importance of acetal formation in carbohydrate chemistry.)

(d)      $PhCHO \xrightarrow{NaBH_4} PhCH_2OH$

Reduction of carbonyl compounds with sodium tetrahydridoborate (boro-hydride) involves transfer of hydride ion ($\overset{\ominus}{\ddot{H}}$) from $BH_4^{\ominus}$ to the carbonyl

129

carbon atom, the residual $BH_3$ then complexing with the carbonyl oxygen atom:

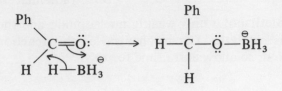

Repetition transfers the other three hydrogen atoms to benzaldehyde molecules, giving the structure

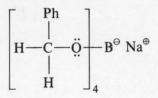

In aqueous or alcoholic solution, decomposition occurs with protonation of the oxygen atom:

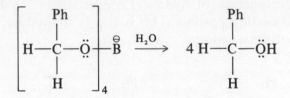

Overall, then, hydrogen is added across the original carbonyl double bond as a hydride ion from the reagent, and a proton from the solvent:

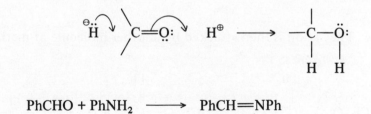

(e) $\qquad$ $PhCHO + PhNH_2 \longrightarrow PhCH{=}NPh$
an imine

This is another example of a nucleophilic addition reaction which is followed by elimination. The steps may be shown as

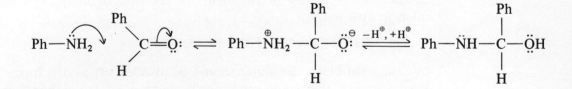

130

$$Ph-\overset{\cdot\cdot}{N}H-\underset{H}{\overset{Ph}{\underset{|}{\overset{|}{C}}}}-\overset{\cdot\cdot}{\overset{\cdot\cdot}{O}}H \underset{}{\overset{H^{\oplus}}{\rightleftharpoons}} Ph-\underset{H}{\overset{H}{\underset{|}{\overset{|}{N}}}}-\underset{H}{\overset{Ph}{\underset{|}{\overset{|}{C}}}}-\overset{\oplus}{\overset{\cdot\cdot}{O}}H_2 \overset{-H^{\oplus}}{\longrightarrow} Ph-\overset{\cdot\cdot}{N}=C\overset{Ph}{\underset{H}{}} + \overset{\cdot\cdot}{\overset{\cdot\cdot}{O}}H_2$$

The dehydration step is acid catalysed, but again the pH must not be too low or concentration of phenylamine (aniline) will be reduced by protonation.

Unlike aliphatic imines, which tend to polymerise, aromatic imines are stable compounds known as Schiff bases.

## Example 7.3

By considering the mechanisms of the reactions concerned explain the following observations:

(a) Chloral (trichloroethanal) most commonly exists as a crystalline hydrate while acetone forms very little hydrate even in aqueous solution. (*5 marks*)

(b) One mole of the aldehyde, RCHO, reacts with $LiAlH_4$ (0.25 mole) to give a complex which is hydrolysed to the alcohol, $RCH_2OH$, while 1 mole of the ester, RCOOR, requires 0.5 mole of $LiAlH_4$ to give, after work-up, the same alcohol. (*3 marks*)

(c) The base catalysed addition of HCN to methyl ethyl ketone (butan-2-one) gives the cyanohydrin while similar reaction with methyl vinyl ketone (but-3-en-2-one) gives 4-oxopentanenitrile ($CH_3COCH_2CH_2CN$) (*4 marks*)

(d) Acetone is brominated in acid solution to form a monobromo product while in alkaline solution the product is bromoform (tribromomethane) (*8 marks*)

## Solution 7.3

For discussion of this question, see Solution 3.3, pp. 57–61.

## Example 7.4

Define the term tautomerism and discuss the extent of its presence in propanone (A) and 2,4-pentanedione (C). A synthesis of (C) is as follows:

$$\underset{\text{(A)}}{CH_3COCH_3} + \underset{\text{(B)}}{CH_3CO_2Et} \overset{\overset{\ominus}{O}Et}{\longrightarrow} \underset{\text{(C)}}{CH_3COCH_2COCH_3}$$

(a) Write a mechanism for the synthesis.
(b) Suggest the course of the reaction if (B) were omitted.
(c) Suggest the course of the reaction if (B) were replaced by $HCO_2Et$.

(Loughborough University of Technology, 1983)

*Solution 7.4*

Try to avoid beginning your answer with 'Tautomerism is when. . .': It is a good idea to learn concise definitions of all the common phenomena and terms that you meet with, so that you do not have to think about the wording in an exam. Tautomerism is a particular type of functional group isomerism where the two isomers (tautomers) are in dynamic equilibrium with one another.

Interconversion of tautomers may be acid or base catalysed, and involves a very rapid proton exchange:

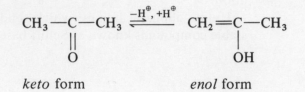

keto form                    enol form

Simple carbonyl compounds such as propanone exist almost entirely in the more stable keto form. Stabilisation of the enol tautomer does occur in some compounds, and pentan-2,4-dione is 80% enol:

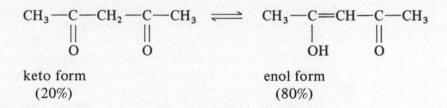

keto form                    enol form
(20%)                        (80%)

Position of equilibrium does vary with solvent when one is employed, but that need not be discussed here. The important point is that the open chain representation of the enol form is an oversimplification. It has been shown that the actual structure is a six membered ring, formed by intramolecular hydrogen bonding:

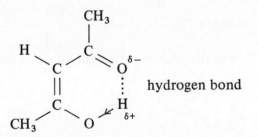

Furthermore, the structure is stabilised by resonance. The alkene double bond of the enol is conjugated with the remaining carbonyl double bond, and the oxygen atom of the hydroxyl group is also involved:

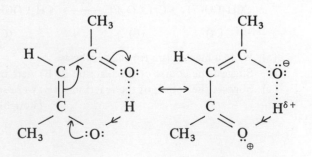

Another factor is that the larger charges on the oxygen and hydrogen atoms will increase the strength of hydrogen bonding.

(a) Propanone is more acidic than the ester (B), and it is the enolate ion of this compound which attacks the ester molecule:

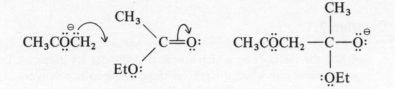

(resonance stabilised)

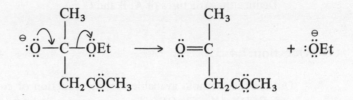

Expulsion of the ethoxide ion gives the dione product:

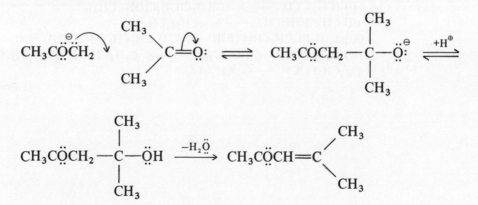

(b) In the absence of ester (B), the propanone enolate ion will simply attack another molecule of propanone to give an aldol:

(c) Replacing ester (B) with $HCO_2Et$ will give 4-oxobutan-2-one, $CH_3COCH_2CHO$, by the mechanism outlined in (a).

## 7.7 Self-Test Questions

### Question 7.1

Give the products of the reaction between acetophenone (phenylmethylketone, $C_6H_5COCH_3$) and FIVE of the following reagents: (1) $NaOD/D_2O$, (2) $C_6H_5NHNH_2$, (3) $LiAlH_4$ followed by $H^{\oplus}/H_2O$, (4) $C_6H_5Li$ followed by $H^{\oplus}/H_2O$, (5) $NaOH/C_6H_5CHO$, (6) $Ph_3\overset{\oplus}{P}\overset{\ominus}{C}H_3I$/strong base. (5 × 4 marks)

(Huddersfield Polytechnic, 1985)

133

## Question 7.2

(a) Write *one* general mechanism for the reaction between a ketone and a nucleophile, and *one* general mechanism for the reaction of an acid chloride and a nucleophile.

(*3 marks each*)

(b) Suggest *two* syntheses of hexan-2-one. (*3 marks each*)

(c) Give the organic products when ethyl magnesium bromide reacts with (1) hexan-2-one and (2) methyl benzoate. (*2 marks each*)

(d) Give the organic products of reaction of lithium aluminium hydride with (1) hexan-2-one and (2) butanamide ($CH_3CH_2CH_2CONH_2$). (*2 marks each*)

(University of Keele, 1984)

## Question 7.3

Describe the methods by which aromatic aldehydes are prepared. Discuss those reactions of aromatic aldehydes which differ from those of aliphatic aldehydes.

An aromatic compound A ($C_8H_8O$) reacts with ethyl magnesium bromide in ether to give B ($C_{10}H_{14}O$). Under the appropriate acid conditions B loses the elements of water to form C ($C_{10}H_{12}$). If treated with a strong oxidising agent, A, B and C each give phthalic acid.

Deduce the structures of A, B and C. (University of Bradford, 1982)

## Question 7.4

Outline the methods available for the reduction of compounds of the type RCHO, R.CO.R and RCOX (X = Cl, OR). In your answer indicate any limitations or advantages of the methods you described.

What reagents would you use to achieve the following conversions?

(i) $(CH_3)_2CO \longrightarrow (CH_3)_2C(OH)C(OH)(CH_3)_2$

(ii) $CH_3CHOHCH_3 \longrightarrow (CH_3)_2CO$

(iii) $C_6H_5COCH_2CH_2COOH \longrightarrow C_6H_5CH_2CH_2CH_2COOH$

(iv) $C_6H_5COCH_2CH_2COOC_2H_5 \longrightarrow C_6H_5CHOHCH_2CH_2CH_2OH$

(v) $CH_3COCl \longrightarrow CH_3CHO$

(University of Bradford, 1983)

# 8 Carboxylic Acids and their Derivatives

## 8.1 Introduction

The acidity of carboxylic acids is considered in Chapter 4. This chapter is concerned with the synthesis, reactivity and inter-relationships of carboxylic acids and their derivatives.

## 8.2 Formation of Carboxylic Acids

### (a) Oxidation methods

(i) $RCH_2OH \xrightarrow{[O]} RCOOH$

(ii) $RCHO \xrightarrow{[O]} RCOOH$

(iii) $RCH = CR \xrightarrow{[O]} RCOOH + R_2CO$

(iv) $Ar - R \xrightarrow{[O]} ArCOOH$

e.g. $KMnO_4$ (acid or alkaline solution)

$K_2Cr_2O_7$ (acid solution)

For (i), (ii) and (iii), R can be alkyl or aryl.

### (b) Carboxylation of Grignard Reagents

$$R - X \xrightarrow{Mg/ether} RMgX \xrightarrow[\text{2. } H_2O/H]{\text{1. solid } CO_2} RCOOH$$

$X = Cl, Br, I$
$R = $ alkyl or aryl

### (c) Hydrolysis of Acid Derivatives

(a) $RCOOOCR + H_2O \longrightarrow 2\ RCOOH$
(b) $RCOX + H_2O \longrightarrow RCOOH + HX$
(c) $RCONH_2 + H_2O \longrightarrow RCOOH + NH_3$
(d) $RCOOR + H_2O \longrightarrow RCOOH + ROH$
(e) $RCN + 2H_2O \longrightarrow RCOOH + NH_3$

Most reactions can be carried out in acidic or alkaline conditions, but acid hydrolysis of an ester will give an equilibrium mixture.

### (d) Malonic Ester Synthesis

Preferred to the 'acetoacetic ester' synthesis for carboxylic acids. There are two stages.

#### (i) *Alkylation of the Ester*

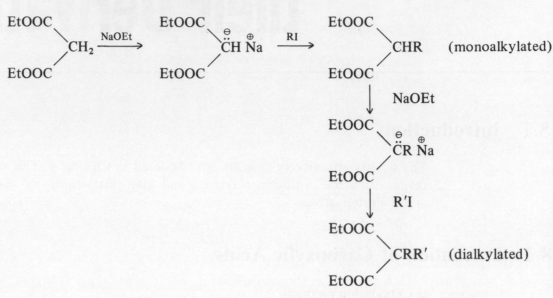

R and R' can be the same or different

#### (ii) *Hydrolysis and Decarboxylation of the Alkylated Ester*

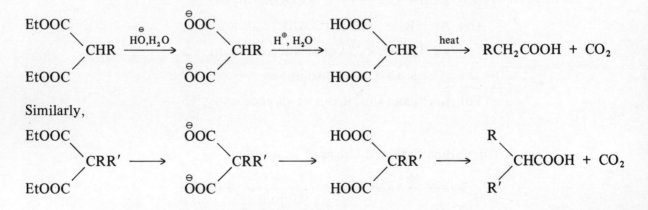

Thus the products are monoalkyl or dialkyl ethanoic acids. Note that because of their lower reactivity, aryl halides, ArI cannot be used in place of alkyl halides, RI. (A mechanism for decarboxylation is given in Example 8.2.)

## 8.3 Formation of Carboxylic Acid Derivatives

### (a) Anhydrides

(i) $RCOO^{\ominus} + RCOCl \longrightarrow (RCO)_2O + Cl^{\ominus}$

(ii) $2 ArCOOH + (CH_3CO)_2O \longrightarrow (ArCO)_2O + 2 CH_3COOH$

136

**(b) Acid Halides (e.g. chlorides)**

$$RCOOH \longrightarrow RCOCl$$

e.g. with $SOCl_2$ ($\rightarrow SO_2 + HCl$)

or $PCl_3$ ($\rightarrow H_3PO_3$)

**(c) Amides**

    (i) $(RCO)_2O + 2\,NH_3 \longrightarrow 2\,RCONH_2 + H_2O$
   (ii) $RCOCl + NH_3 \longrightarrow RCONH_2 + HCl$
  (iii) $RCOOR' + NH_3 \longrightarrow RCONH_2 + R'OH$

Replacing $NH_3$ by $RNH_2$ etc., gives N-alkyl (or aryl) amides.

**(d) Esters**

    (i) $RCOOH + R'OH \rightleftharpoons RCOOR' + H_2O$ (distil off – Dean–Stark)
   (ii) $(RCO)_2O + ROH \longrightarrow RCOOR + RCOOH$
  (iii) $RCOCl + ROH \longrightarrow RCOOR + HCl$
  (iv) $RCOO^\ominus + R'X \longrightarrow RCOOR' + X^\ominus$

Method (i) gives an equilibrium mixture. Methods (ii) and (iii) can be used with phenols. Method (iv) is only of use with reactive haloalkanes (3°, benzyl, etc.).

## 8.4 Structure and Reactivity of Carboxylic Acids and their Derivatives

(Acidity of carboxylic acids is covered in Chapter 4.)

Carboxylic acids and their derivatives have a resonance structure, summarised by the generalised limiting forms:

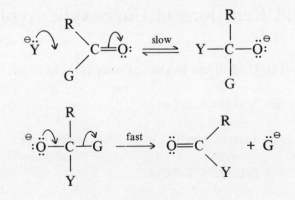

where G = OH, OOCR, Cl, $NH_2$, or OR.

The characteristic reaction is *nucleophilic addition with elimination* of group G, a process which occurs in two steps with the formation of a tetrahedral anionic intermediate:

137

The ease which these reactions occur is determined by a number of factors:

(a) loss of resonance energy on formation of the intermediate;
(b) stabilisation of the anionic intermediate by G;
(c) leaving ability of G; and
(d) the nucleophilic strength of $\ddot{Y}^{\ominus}$.

With acid chlorides, resonance energy is minimised because of relatively poor overlap of the chlorine 3p orbital with the carbonyl carbon 2p orbital, and inductive stabilisation of the anionic intermediate by the chlorine atom (strong $-I$ effect) is appreciable. Furthermore, the chloride ion is a good leaving group, and so acid chlorides are readily attacked by even weak nucleophiles such as $H_2O$.

On the other hand, resonance energy in amides is higher because of better (2p-2p) orbital overlap, inductive stabilisation of the intermediate is minimised and the amide ion ($NH_2^{\ominus}$) is a very poor leaving group. Consequently, amides are only attacked by strong nucleophiles such as hydroxide ion (or weaker water molecules in acid solution, where protonation of the $NH_2$ group inhibits resonance with the carbonyl group and provides a good leaving ammonia molecule).

Anhydrides and esters are intermediate in their reactivity.

## 8.5 Principal Reactions of Carboxylic Acids

### (a) Acidity: Formation of Salts

See Chapter 4.

### (b) Formation of Various Derivatives

See Section 8.3 above.

### (c) Reduction

$$RCOOH \xrightarrow[\text{(2) } H_2O/H^{\oplus}]{\text{(1) LiAlH}_4} RCH_2OH$$

### (d) Decarboxylation

$$RCOOH \longrightarrow RH + CO_2$$

For example, heat the sodium salt with sodalime (Example 8.2).

## 8.6 Principal Reactions of Carboxylic Acid Derivatives

### (a) Hydrolysis to the Carboxylic Acid

See Section 8.2 above.

### (b) Interconversion with other Derivatives

See Section 8.3 above.

### (c) Reduction with $LiAlH_4$

(i) $RCOCl \longrightarrow RCH_2OH$

(ii) $RCONH_2 \longrightarrow RCH_2NH_2$

(iii) $RCOOR' \longrightarrow RCH_2OH + R'OH$

# 8.7 Worked Examples

### Example 8.1

Give TWO general methods of preparation for aliphatic monocarboxylic acids. (*4 marks*)
Suggest how propanoic acid may be converted into

(a) propanoyl chloride,

(b) ethyl propanoate,

(c) propanamide,

(d) 2-bromopropanoic acid. (*10 marks*)

Place the following compounds in order of decreasing acid strength, giving reasons for your choice: 2-chlorobutanoic acid, phenol, butanoic acid, 2-fluorobutanoic acid, 4-bromobutanoic acid, 3-chlorobutanoic acid. (*8 marks*)

(Kingston Polytechnic, 1985)

### *Solution 8.1*

Of the general methods available, you might like to quote oxidation of primary alcohols,

$$RCH_2OH \xrightarrow{\text{[O]}} RCOOH$$

e.g. by heating with acidified $K_2Cr_2O_7$, and carboxylation of a Grignard reagent

$$RX \xrightarrow[\text{dry ether}]{\text{Mg}} RMgX \xrightarrow{\text{solid } CO_2} RCOOMgX \xrightarrow{H_2O} RCOOH \ (+ \ Mg(OH)X)$$

Alternatively, you could use the malonic ester synthesis outlined in Self-Test Question 8.2.

(a) The usual process is to treat the carboxylic acid with either a chloride of phosphorus ($PCl_3$ or $PCl_5$) or with sulphur dichloride oxide (thionyl chloride, $SOCl_2$). The latter reagent has the advantage that the by-products ($SO_2$ and HCl) are gaseous and therefore easily removed.

(b) Esters are readily prepared by reaction of the acid chloride with an alcohol. Thus,

$$CH_3CH_2COOH \xrightarrow{SOCl_2} CH_3CH_2COCl \xrightarrow{C_2H_5OH} CH_3CH_2COOC_2H_5$$

(c) Likewise, the acid amide is easily made by reaction of the acid chloride with ammonia (concentrated aqueous $NH_3$).

$$CH_3CH_2COCl \xrightarrow{NH_3} CH_3CH_2CONH_2$$

Reaction is extremely vigorous and should be carried out with caution in a fumecupboard.

(d) Specific $\alpha$-halogenation is carried out by the **Hell–Volhard–Zelensky** reaction. In the presence of a small amount of phosphorus, aliphatic carboxylic acids react smoothly with chlorine or bromine to yield the $\alpha$-haloacid:

$$CH_3CH_2COOH \xrightarrow[Br_2]{P} CH_3CH(Br)COOH$$

Order of decreasing acid strength: 2-fluorobutanoic acid > 2-chloro-butanoic acid > 3-chlorobutanoic acid > 4-bromobutanoic acid > butanoic acid > phenol.

The strength of carboxylic acids is related to the position of the equilibrium

$$RCOOH + H_2O \rightleftharpoons RCOO^\ominus + H_3O^\oplus$$

This in turn is largely determined by the relative stabilities of the acid $RCOOH$ and its conjugate anion, $RCOO^\ominus$. Electron-attracting substituents (halogen, $NO_2$, etc.) increase dispersal of charge in the anion and so shift the equilibrium to the right. Conversely, electron-releasing substituents tend to increase stability of the acid and decrease stability of the conjugate anion.

Phenol is less acidic than a carboxylic acid because dispersal of negative charge in the phenoxide ion,

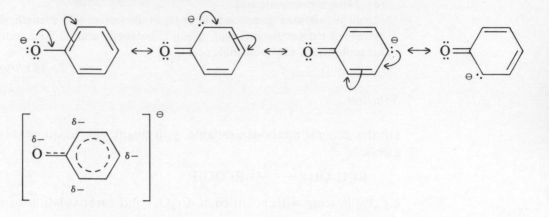

is less effective than dispersal over the two oxygen atoms in the carboxylate anion:

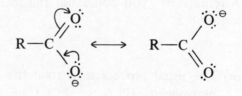

i.e.

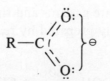

Halogen is able to stabilise a carboxylate anion by a negative inductive effect, and this decreases in the series

$$F > Cl > Br$$

Thus 2-fluorobutanoic acid is stronger than 2-chlorobutanoic acid.

The ability of halogen to stabilise the anion decreases with distance from the carboxylate group. Thus 3-chlorobutanoic acid is weaker than the 2-chloroacid, and 4-bromobutanoic acid is only slightly more acidic than butanoic acid itself.

This is probably as much detail as you would be expected to give for 8 marks, but for a fuller discussion see Chapter 4 and Example 4.1.

**Example 8.2**

Discuss the following statements.
(a) The acidity of carboxylic acids, $RCH_2CO_2H$, varies with the nature of **R**.
(b) Some carboxylic acids decarboxylate at temperatures below 200°C.
(c) Some carboxylic acids are stable as such, but readily decarboxylate on conversion into the carboxylate anions. (University of Leicester, 1985)

*Solution 8.2*

(a) You can assume that 'R' here means not just an alkyl group, but also substituted alkyl groups such as chloroalkyl, nitroalkyl, etc. See previous question, and Example 4.1 for details.

(b) Most simple aliphatic and aromatic carboxylic acids are resistant to decarboxylation, but certain substituted and β-ω unsaturated acids decarboxylate when heated to about 150°C. The ease with which a particular acid loses carbon dioxide is related to the relative stability of the first formed carbanion. The less stable this is, the higher the activation energy for decarboxylation.

With β-ketoacids, the role of the carbonyl group is to accept a proton from the carboxyl group, so converting it into a good leaving CO group. Reaction in this case probably involves a cyclic transition state:

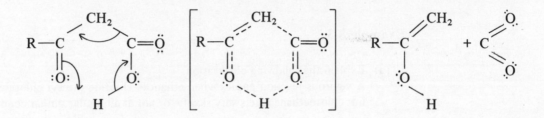

The first formed product is an *enol* which rapidly tautomerises to the more stable *keto* form:

$$R-\underset{\underset{:OH}{|}}{C}=CH_2 \rightleftharpoons R-\underset{\underset{:O:}{||}}{C}-CH_3$$

Similar reaction does not occur with unsubstituted carboxylic acids because COOH as such is a poor leaving group, and the carbanion $\overset{..}{R}{}^{\ominus}$ is highly unstable. α-nitro and α-cyanoacids are more prone to decarboxylation, as in the corresponding anions $R\overset{..}{C}H-NO_2$ and $R\overset{..}{C}H-CN$, the electron attracting nitro and cyano groups are stabilising.

Some aromatic carboxylic acids decarboxylate on heating, and it is thought that in these cases the normal aromatic electrophilic substitution mechanism operates, where the attacking electrophile is $H^{\oplus}$ and the displaced group is COOH. Reaction has been shown to be second order,

$$Rate \propto [H^{\oplus}] [ArCOOH]$$

and no kinetic isotope effect is observed with $Ar^{13}COOH$.

(c) Carboxylate anions tend to decarboxylate more readily than the parent acids because the stable molecule $CO_2$ is a much better leaving group than

COOH. The mechanism can be formulated as an 'internal $S_N2$' process in which the displaced group is the carbanion:

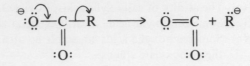

The more stable the anion $\overset{..}{R}{}^{\ominus}$, the more readily decarboxylation occurs. Electron-attracting substituents help to disperse the negative charge, and for example, 2,4,6-trinitrobenzoic acid loses $CO_2$ simply on heating in aqueous solution, and reaction is very fast in alkaline solution.

Similarly, the monanion of malonic acid readily decarboxylates on heating:

$$HOOC-CH_2-\overset{\ominus}{C}O_2 \longrightarrow HOOC-\overset{\ominus}{C}H_2 + CO_2$$

Interestingly, the dianion resists decarboxylation because $-COO^{\ominus}$ destabilises the carbanion:

$$HOOC \leftarrow \overset{\ominus}{C}H_2 \qquad\qquad \overset{\ominus}{O}OC-\overset{\ominus}{C}H_2$$

| | |
|---|---|
| $-I$ effect | similar charges on adjacent |
| is stabilising | groups is destabilising |

## Example 8.3

(a) Explain the following observation.
A vigorous reaction ensues when ethanoyl chloride (acetyl chloride) is added to water, but chloroethane reacts very slowly, or not at all, under similar conditions.

(Wolverhampton Polytechnic, 1985)

(b) Indicate how you would attempt to separate the components of a mixture of phenol and benzoic acid. Comment on the theoretical basis for the proposed procedure.

(Coventry Polytechnic, 1984)

(c) Discuss how the following conversion could be achieved indicating the steps involved and any necessary reagents. Comment, where appropriate, on the reaction mechanisms concerned.

$$C_6H_5\overset{\|}{\underset{O}{C}}-\overset{\|}{\underset{O}{C}}C_6H_5 \longrightarrow (C_6H_5)_2\overset{|}{\underset{OH}{C}}-COOH$$

Describe briefly the important features of the IR spectrum of the product.

(Loughborough University of Technology, 1985)

(part questions)

*Solution 8.3*

(a) The vigorous reaction is hydrolysis of the acid chloride to give the carboxylic acid plus hydrogen chloride:

$$RCOCl + H_2O \longrightarrow RCOOH + HCl$$

There are two distinct steps in this process, addition to the carbonyl carbon atom, followed by elimination of the chloride ion.

## 1. Nucleophilic addition

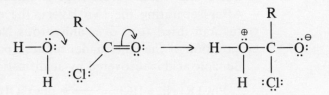

Water is only weakly nucleophilic, but the alkoxide ion is a relatively stable intermediate (stability is helped by the −I effect of the Cl and $O^\oplus$ atoms), and activation energy for its formation is therefore quite low.

## 2. Elimination

In the second step, which follows very quickly, expulsion of the chloride ion is facilitated by an internal 'nucleophilic push' from the negatively charged oxygen atom:

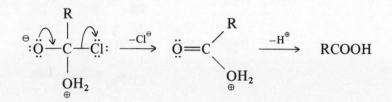

Chloroethane is relatively inert under similar conditions because activation energy for ionisation (in an $S_N1$ reaction) or the formation of a bimolecular transition state (in an $S_N2$ reaction) is too great for either of these processes to occur.

Water is a highly polar medium, which favours ionisation, but a primary carbocation is very unstable

$$CH_3CH_2\text{—}Cl \not\rightleftharpoons CH_3\overset{\oplus}{C}H_2 + Cl^\ominus$$

On the other hand, water is only weakly nucleophilic, and unable to form the $S_N2$ transition state:

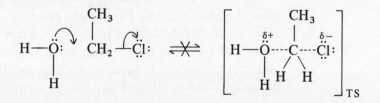

(Bimolecular reaction *is* possible with a more powerful nucleophile, and chloroethane is readily hydrolysed on heating with concentrated aqueous alkali.)

(b) Benzoic acid is a much stronger acid than phenol, and unlike phenol, is able to displace carbonic acid (carbon dioxide) from its salts.

$$NaHCO_3 + PhCOOH \longrightarrow PhCO\overset{\ominus}{O}\,\overset{\oplus}{Na} + CO_2 + H_2O$$

Thus if an ethereal solution of the two compounds is extracted with aqueous $NaHCO_3$, benzoic acid will react and dissolve in the aqueous layer as sodium benzoate. Phenol (which does not react) is only slightly soluble in water, and will therefore largely remain in the ether layer.

After separating the two layers, the ethereal solution is 'washed' with water and dried (e.g. over anhydrous $Na_2CO_3$) and phenol recovered by evaporation of the ether under reduced pressure.

Benzoic acid is recovered by acidification of the aqueous solution,

$$PhCO\overset{\ominus}{O}\ \overset{\oplus}{Na} + HCl \longrightarrow PhCOOH + \overset{\oplus}{Na}\ Cl^{\ominus}$$

followed by suction filtration at the pump.

(c) This reaction is the **benzil–benzylic acid rearrangement**. Benzil is heated on a waterbath with a strong solution of KOH containing a little ethanol to confer some solubility on the diketone. On completion of the reaction, benzylic acid is precipitated by addition of hydrochloric acid.

Reaction involves two main steps, the first of which is rate determining.

1. Addition of hydroxide ion to a carbonyl group

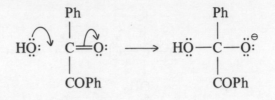

followed by

2. Migration of the phenyl group

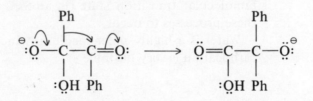

This second step is mechanistically quite interesting. First of all, movement of the phenyl group can be thought of as an internal (intramolecular) $S_N2$ reaction in which $O^{\ominus}$ displaces $Ph^{\ominus}$ (sometimes referred to as a 'nucleophilic push' by the $O^{\ominus}$ atom).

Secondly, since the phenyl group migrates with the electron pair, i.e. as a phenyl carbanion, this part of the reaction can be regarded as (intramolecular) nucleophilic addition to the right-hand carbonyl group. And like many other carbonyl additions, reaction is completed by protonation of the basic alkoxide group ($O^{\ominus}$). Since in alkaline solution the newly formed carboxyl group will ionise, this last step can be viewed as an acid–base reaction.

The main features of the IR spectrum are:
(a) A broad absorption 2500–3500 $cm^{-1}$, due to $O-H_{str}$ of two hydroxyl groups.
(b) Strong absorption near 1700 $cm^{-1}$, due to $C=O_{str}$ of the carbonyl group.
(c) Two strong peaks, near 1500 and 1600 $cm^{-1}$, due to $C=C_{str}$ of the benzene ring.

**Example 8.4**

Show how you would attempt the following transformations. Mechanisms are not required, but where appropriate you should indicate the reasons for your methodology.

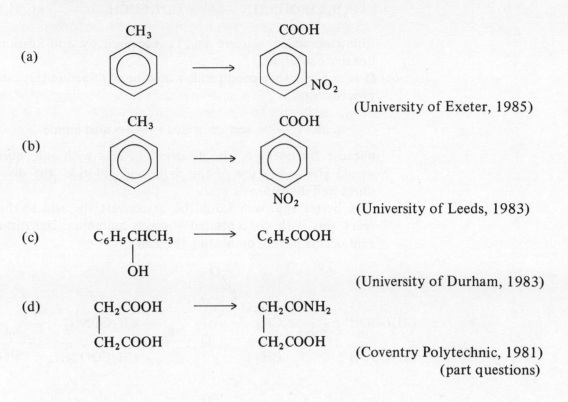

(a)

(University of Exeter, 1985)

(b)

(University of Leeds, 1983)

(c) C₆H₅CHCH₃ ⟶ C₆H₅COOH
    |
    OH

(University of Durham, 1983)

(d) CH₂COOH ⟶ CH₂CONH₂
    |              |
    CH₂COOH       CH₂COOH

(Coventry Polytechnic, 1981)
(part questions)

*Solution 8.4*

(a) The methyl group is *ortho/para* directing, and the carboxyl group is *meta* directing. Therefore toluene is first oxidised, e.g. by heating with acidified $K_2Cr_2O_7$, and the resulting benzoic acid nitrated with the usual mixture of concentrated nitric and sulphuric acids.

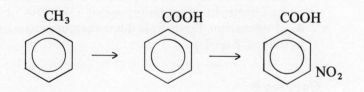

(b) To obtain 4-nitrobenzoic acid, the order of the steps has to be reversed. Nitration of toluene gives a mixture of *para-* together with some *ortho-*nitrotoluenes, from which the *para*-isomer can be isolated by fractional crystallisation, or column chromatography.

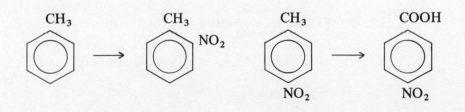

145

(c) Oxidation of 1-phenylethanol gives acetophenone, and the methyl ketone can then be converted into benzoic acid by warming with iodine in alkaline solution (**haloform reaction**). The required sequence is

$$C_6H_5CH(OH)CH_3 \xrightarrow{[O]} C_6H_5COCH_3 \xrightarrow[\text{warm}]{I_2,\ NaOH} C_6H_5COO\ Na + CHI_3$$

following which benzoic acid is recovered by acidification of the sodium benzoate solution.

(d) It is unlikely that good yields would be obtained in this case by the normal procedure of

$$\text{acid} \xrightarrow{SOCl_2} \text{acid chloride} \xrightarrow{NH_3} \text{acid amide}$$

because treatment of the dicarboxylic acid with one equivalent of $SOCl_2$ would give a mixture of the monoacid chloride, the diacid chloride and unreacted dicarboxylic acid.

A better approach would be to convert the acid to the anhydride and react this with concentrated aqueous ammonia. Dehydration of succinic acid occurs merely on heating the acid.

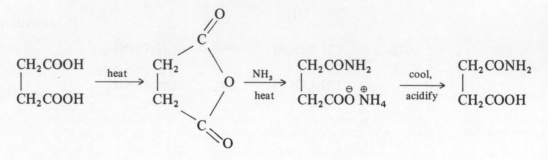

# 8.8 Self-Test Questions

### Question 8.1

Describe TWO methods for preparing a methyl ester from a carboxylic acid.

1,1-Dimethylethyl trifluoroethanoate ($Me_3COOCCF_3$) is hydrolysed in pure water by a $B_{AL}1$ mechanism. However, in dilute aqueous sodium hydroxide the mechanism of hydrolysis is $B_{AC}2$. Explain. (University of Warwick, 1981)

### Question 8.2

Outline the accepted mechanism for:
(a) Hofmann conversion of amides into amines

$$(RCONH_2 \xrightarrow[H_2O]{Br_2/NaOH} RNH_2 + NaBr + Na_2CO_3)$$

(University of Leeds, 1983)

(b) Fischer–Speier esterification of carboxylic acids

$$(RCO_2H \xrightarrow[\text{dry HCl}]{EtOH} RCO_2Et + H_2O)$$

(University of Leeds, 1983)

(c) Malonic ester synthesis of cyclopentane carboxylic acid.

(University of Bradford, 1983)
(part questions)

## Question 8.3

1. Suggest a synthesis of butanoic acid from ethanol as the only source of carbon.

(University of Bradford, 1983)

2. What organic product would you expect from the following reaction?

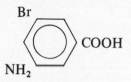

$$PhCOCl \xrightarrow[\text{2. } H_2O]{\text{1. PhMgBr (excess)/ether}}$$

(Sheffield City Polytechnic, 1983)

3. Indicate how the following transformations might be accomplished, specifying the necessary reagents.

$$(CH_3)_2CHCH_2Br \xrightarrow{?} (CH_3)_2CHCH_2CH(OH)CH_3 \xrightarrow{?} (CH_3)_2CHCH_2COOH \quad (A)$$

Make a rough sketch of the expected $^1H$ NMR spectrum of compound A.

(University of Bristol, 1985)

4. Without giving experimental details, outline a synthesis of 3-amino-5-bromobenzene-carboxylic acid,

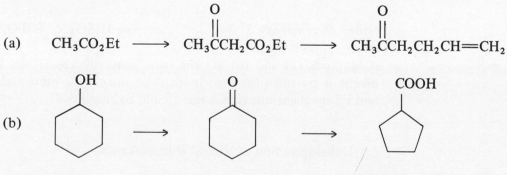

starting with a suitable monosubstituted benzene.

Indicate why it is necessary to carry out the various steps in a particular order.

(Coventry Polytechnic, 1985)

## Question 8.4

1. How may the following transformations be achieved? Each stage may involve more than one step.

(a)   $CH_3CO_2Et \longrightarrow CH_3\overset{O}{\overset{\|}{C}}CH_2CO_2Et \longrightarrow CH_3\overset{O}{\overset{\|}{C}}CH_2CH_2CH=CH_2$

(b)

(University of Leeds, 1984)

2. How would you carry out the following transformations? Outline the reagents and conditions, but detailed experimental conditions and mechanisms are NOT required.

(c) $RCO_2H \longrightarrow RCH_2NHR$

(d) $RCOR \longrightarrow RCONHR$

(e) $PhCH_2Br \longrightarrow PhCH_2CH_2COOH$

(f) $RCOCH_3 \longrightarrow RCOOH$

(Wolverhampton Polytechnic, 1985)

(University of Bristol, 1985)

(part questions)

# 9 Alkenes and Alkynes

## 9.1 Introduction

There are many parallels in the chemistry of alkenes and alkynes, and some of the concepts developed here will be useful in your later study of aromatic compounds.

## 9.2 Formation of Alkenes

### (a) Dehydrohalogenation of Haloalkanes

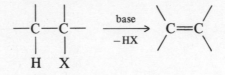

Order of reactivity $3° > 2° > 1°$, e.g. base = $EtO^{\ominus}$

### (b) Dehydration of Alcohols

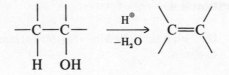

Order of reactivity $3° > 2° > 1°$, e.g. use $H_2SO_4$ or $H_3PO_4$, or pass vapour over $Al_2O_3$ at high temp.

Methods (a) and (b) are the most generally applicable. Where more than one alkene is possible, the predominant product is the **more stable** alkene. For the E1 and E2 mechanisms, see Chapter 5 and Example 5.4.

### (c) Dehalogenation of Vicinal Dihaloalkanes

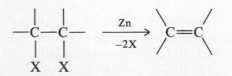

Of limited application. Used to regenerate a double bond previously 'protected' by halogenation.

**(d) Reaction of Carbonyl Compounds with Phosphonium Ylides (Wittig Reaction)**

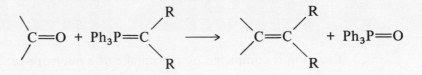

(aldehyde or ketone)
Gives good yields of pure alkene of known structure.

**(e) Reduction of Alkynes**

$$R—C≡C—R' \xrightarrow{[H]} RCH=CHR'$$

(i) catalytic hydrogenation (Lindlar catalyst) ⟶ *cis* isomer
(ii) $NaNH_2$/liquid $NH_3$ ⟶ *trans* isomer

## 9.3 Formation of Alkynes

### (a) Dehydrohalogenation of Vicinal Dihaloalkanes

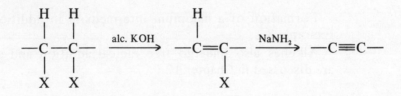

### (b) Synthesis from Lower Alkynes

$$HC≡\ddot{C}\ Li^{\oplus} + RX \longrightarrow HC≡CR + Li^{\oplus}X^{\ominus}$$

## 9.4 Structure and Reactivity of Alkenes

The alkene 'double bond' consists of a sigma bond plus a pi bond. Pi electrons are bound less strongly than sigma electrons and are available for donation to an attacking electrophile. Consequently, alkenes are **nucleophilic reagents**.

Although the pi bond is essentially non-polar, it is readily *polarised* on close approach of an electrophile, $E^{\oplus}$:

Electrons are drawn towards the carbon atom under attack, and eventually form a sigma bond from that carbon atom to the electrophile:

149

Reaction is completed by the uptake of a nucleophile, $\ddot{Y}^{\ominus}$:

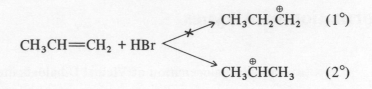

The product is one in which the species $E^{\oplus}$ and $\ddot{Y}^{\ominus}$ have 'added' across the double bond.

Since the first (slow, rate-determining) step involves attack by an electrophile, the reaction is classified as **electrophilic addition**.

In the reaction of a non-symmetrical alkene such as propene, the initial addition occurs so as to give the **more stable** of the two possible carbocations:

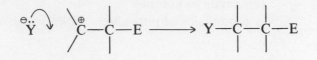

Formation of a halonium intermediate in addition of halogen is discussed in Example 3.2.

Alkenes also undergo free radical addition and substitution reactions, these are discussed in Chapter 12.

# 9.5 Principal Reactions of Alkenes

**(a) Addition of Hydrogen Halides**

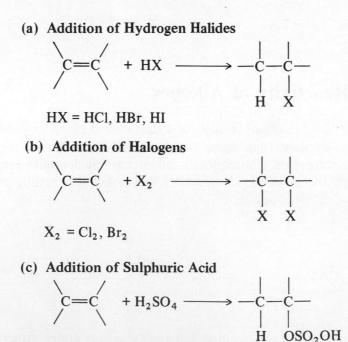

HX = HCl, HBr, HI

**(b) Addition of Halogens**

$X_2$ = $Cl_2$, $Br_2$

**(c) Addition of Sulphuric Acid**

(hydrolysis gives the alcohol).

### (d) Acid Catalysed Hydration

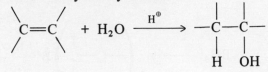

### (e) Addition of HO—X

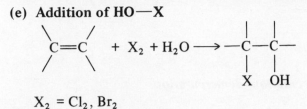

$$X_2 = Cl_2, Br_2$$

### (f) Catalytic Hydrogenation

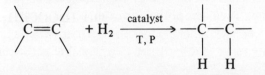

e.g. Pt, Pd, Ni, Raney Ni.

### (g) Oxidation

#### (i) *Hydroxylation*

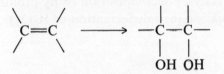

(1) Dilute alkaline $KMnO_4$
(2) Osmium tetroxide, $OsO_4$, then $Na_2SO_3$

#### (ii) *Epoxide Formation*

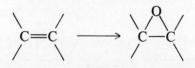

(1) Catalytic air oxidation, Ag catalyst
(2) Peroxides or peracids

Epoxides are readily ring-opened

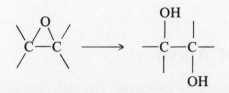

Acid or base catalysed.

### (iii) *Hydroboration–oxidation*

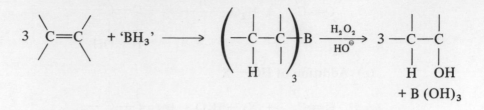

### (iv) *Oxymercuration–demercuration*

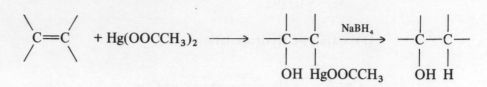

### (v) *Ozonolysis*

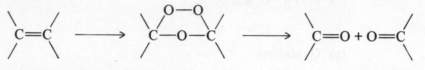

The ozonide is decomposed

(1) with $H_2O$/Zn dust (catalyses decomposition of by-product $H_2O_2$), or
(2) by catalytic hydrogenation (avoids formation of $H_2O_2$).

### (h) Polymerisation

### (i) *Ionic*

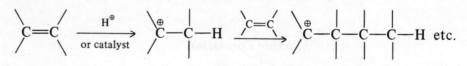

### (ii) *Free Radical*

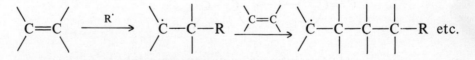

### (i) Free Radical Addition and Substitution

Discussed in Chapter 12.

## 9.6 Structure and Reactivity of Alkynes

The alkyne triple bond consists of one sigma and two pi bonds. Alkynes undergo electrophilic addition reactions, but they sometimes react much more slowly than alkenes with a particular reagent. For example, addition of chlorine or hydrogen chloride is very slow unless a catalyst is employed, and alkynes do not react at all with cold concentrated sulphuric acid.

The reason for this difference in reactivity is not clear, but alkyne pi electrons are contained in a shorter more compact molecular orbital between two sp hybridised carbon atoms, and are perhaps less accessible to an attacking electrophile. Furthermore, activation energy for the formation of an alkenic carbocation $R—CH{=}\overset{\oplus}{C}H$ is quite high, and formation of the cyclic halonium ion in halogen addition might be more difficult than with alkenes.

As with alkenes, addition to a non-symmetrical alkyne occurs so as to give the more stable of the possible carbocations. For example,

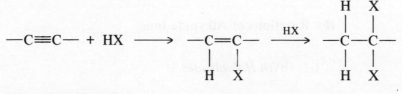

Further addition gives mainly 2,2-dibromopropane, again via the more stable carbocation intermediate, $CH_3—\overset{\oplus}{C}(Br)—CH_3$.

Hydrogen linked to sp hybridised carbon is weakly acidic, and unlike alkenes, alkynes form salts with a strong base such as Li or Na metal, and also react with the carbanion of Grignard reagents.

# 9.7   Principal Reactions of Alkynes

### (a)  Addition of Hydrogen Halides

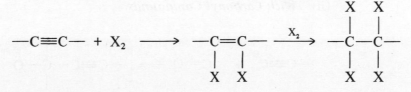

HX = HCl, HBr, HI

### (b)  Addition of Halogens

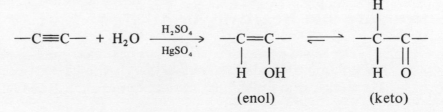

X = Cl, Br
Slow unless catalysed, e.g. by $Hg^{2+}$.

### (c)  Acid Catalysed Hydration

$$—C{\equiv}C— + H_2O \xrightarrow[HgSO_4]{H_2SO_4} \underset{\substack{| \; |\\ H \;\; OH}}{—C{=}C—} \rightleftharpoons \underset{\substack{| \;\; \|\\ H \;\; O}}{\overset{\substack{H\\|}}{—C—C—}}$$

(enol)                          (keto)

153

See Section 7.4 and Example 7.4.

**(d) Hydrogenation**

(i) *Catalytic*

$$-C\equiv C- + 2\,H_2 \xrightarrow{\text{Pt or Ni etc.}} -CH_2CH_2-$$

(ii) *Lindlar Catalyst (deactivated Pd)*

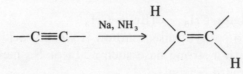

$$-C\equiv C- + H_2 \longrightarrow$$

(*cis* addition)

(iii) *Na or Li in Liquid Ammonia*

$$-C\equiv C- \xrightarrow{\text{Na, NH}_3}$$

(*trans* addition)

**(e) Formation of Metal Alkynides ('Acetylides')**

$$-C\equiv C- \xrightarrow{\text{M}} -C\equiv \ddot{C}^{\ominus}\,M^{\oplus} + \tfrac{1}{2}H_2$$

M = Na, Li (also $\ddot{R}^{\ominus}MgX^{\oplus}$)

**(f) Reactions of Alkynide Ions**

(i) *With Haloalkanes*

$$-C\equiv\ddot{C}^{\ominus} + R\ddot{X}\!: \longrightarrow\; :\ddot{X}\!:^{\ominus} + -C\equiv C-R$$

(ii) *With Carbonyl Compounds*

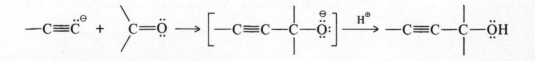

154

## 9.8  Worked Examples

### Example 9.1

Give a detailed explanation of FOUR of the following observations.
(a) Under ionic conditions addition of HBr to $CH_3CH{=}CH_2$ gives $CH_3CHBrCH_3$ but addition to $CCl_3CH{=}CH_2$ gives $CCl_3CH_2CH_2Br$.
(b) Treatment of $CH_2{=}CHCH(OH)CH{=}CHCH_3$ with methanol in the presence of dilute sulphuric acid produces $CH_2{=}CHCH{=}CHCH(OCH_3)CH_3$.

(c) Treatment of 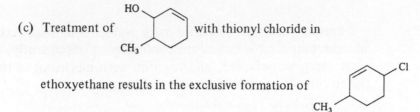 with thionyl chloride in

ethoxyethane results in the exclusive formation of

(d) Addition of HCl to 3,3-dimethylbut-1-ene gives a mixture of two isomeric chloro-alkanes.
(e) Elimination of HBr from 3-bromo-2,3-dimethylpentane gives 2,3-dimethylpent-2-ene as the major product.                                     (Teesside Polytechnic, 1984)

### Solution 9.1

(a) Ionic addition of HBr to alkenes occurs in two steps, the first of which is the slow, rate-determining uptake of a proton to form an intermediate carbocation. With a molecule such as propene, the proton adds to that carbon atom which would result in the formation of the more stable of the two possible carbocations:

$$CH_3CH{=}CH_2 \longrightarrow CH_3\overset{\oplus}{C}HCH_2 \ (2°) \text{ rather than } CH_3CH_2CH_2^{\oplus} \ (1°)$$

Reaction is completed by (fast) uptake of $Br^{\ominus}$ to give $CH_3CHBrCH_3$.

Substitution by three strongly electron attracting chlorine atoms destabilises the intermediate carbocation, and consequently it is now the 1° ion which is more stable as the carbon atom carrying the positive charge is further removed from the influence of the chlorine atoms:

$$Cl_3C{-}CH{=}CH_2 \longrightarrow Cl_3C{-}CH_2{-}\overset{\oplus}{C}H_2 \text{ rather than } Cl_3C{-}\overset{\oplus}{C}H{-}CH_3$$

Subsequent addition of $Br^{\ominus}$ gives $Cl_3CCH_2CH_2Br$.
(b) In the presence of dilute sulphuric acid, some alcohol molecules will be protonated:

$$CH_2{=}CH{-}\underset{\underset{\displaystyle OH}{|}}{CH}{-}CH{=}CH{-}CH \overset{H^{\oplus}}{\rightleftharpoons} CH_2{=}CH{-}\underset{\underset{\displaystyle \oplus OH_2}{|}}{CH}{-}CH{=}CH{-}CH$$

The protonated molecules can then readily lose a molecule of water because the resulting carbocation is resonance stabilised:

155

$$CH_2{=}CH{-}CH{-}CH{=}CHCH_3 \xrightarrow{-H_2O} CH_2{=}CH{-}\overset{\oplus}{CH}{-}CH{=}CHCH_3$$

with $\overset{\oplus}{O}H_2$ group on the middle carbon

$$CH_2{=}CH{-}\overset{\oplus}{CH}{-}CH{=}CHCH_3 \longleftrightarrow CH_2{=}CH{-}CH{=}CH{-}\overset{\oplus}{CH}{-}CH_3$$
$$\text{(I)} \qquad\qquad\qquad \underset{1\quad 2\quad 3\quad 4\quad 5\quad 6}{\text{(II)}}$$

$$\longleftrightarrow \overset{\oplus}{CH_2}{-}CH{=}CH{-}CH{=}CHCH_3$$
$$\text{(III)}$$

Limiting form (II) is particularly stable as it has a secondary structure and the two double bonds are conjugated. Consequently, carbon atom 5 is most electron deficient, and reaction with methanol in this position gives rise to the observed product:

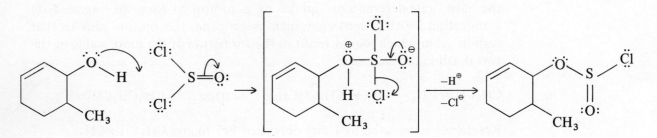

$$\overset{:\ddot{O}CH_3}{\underset{|}{\phantom{.}}}$$
$$CH_2{=}CH{-}CH{=}CH{-}CH{-}CH_3$$

(c) The methylcyclohexenol reacts with sulphur dioxide chloride (thionyl chloride) in this way:

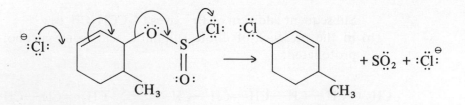

'$SO_2Cl$' is a very good leaving group as it splits up into the stable species $SO_2$ and $Cl^{\ominus}$. Nevertheless, only a rearranged product is observed, so it seems likely that expulsion of $SO_2Cl$ is facilitated by bimolecular attack of the chloride ion on the alkene carbon atom 1:

(d) Addition of a proton occurs so as to give the more stable of the two theoretically possible carbocations:

$$CH_2{=}CH{-}C(CH_3)_3 \underset{}{\overset{H^{\oplus}}{\rightleftharpoons}} CH_3{-}\overset{\oplus}{CH}{-}C(CH_3)_3$$

156

The structure of this secondary carbocation is such that it can readily rearrange to a more stable tertiary ion by migration of a methyl group:

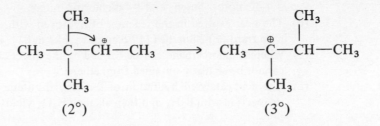

Hence the formation of the two products, 2-chloro-3,3-dimethylbutane and 2-chloro-2,3-dimethylbutane:

$$(CH_3)_3C—CH(Cl)CH_3 \quad \text{and} \quad (CH_3)_2C(Cl)CH(CH_3)_2$$

(e) The tertiary haloalkane 3-bromo-2,3-dimethylpentane can readily ionise, and elimination follows a unimolecular pathway:

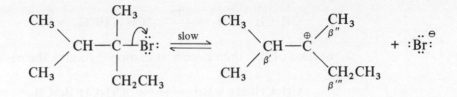

There are three $\beta$ positions in the carbocation, but expulsion of the proton occurs from the $\beta'$ carbon atom as this leads to the most alkylated, i.e. the most stable, of the possible alkenes (Saytzeff rule):

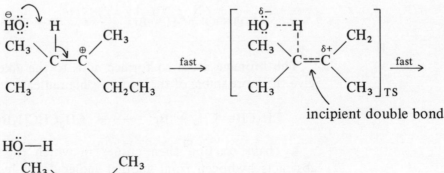

incipient double bond

The most stable alkene is, of course, formed via the most stable transition state (Hammond postulate), which has the same degree of alkylation of the (incipient) double bond as the resulting alkene.


## Example 9.2

Explain FOUR of the following, giving the mechanisms of the reactions where appropriate:
(a) The reaction of propene with hydrogen bromide gives 2-bromopropane when oxygen is absent, but yields 1-bromopropane when oxygen is present.

(b) The reaction of propene with chlorine gives 1,2-dichloropropane at ordinary tempera-tures, but 3-chloropropene (allyl chloride) is the principal product at high temperatures.

(c) The reaction of 1,3-butadiene with one molar proportion of chlorine gives two products, 3,4-dichloro-1-butene and 1,4-dichloro-2-butene.

(d) The heat evolved in the reaction of hydrogen with *cis*-(Z)-2-butene is greater than that with *trans*-(E)-2-butene (119.6 and 115.4 kJ mol$^{-1}$, respectively).

(e) Whereas chloroethane can be formed by the direct chlorination of ethane, chloroethene cannot be similarly obtained from ethene.

(f) The acid-catalysed hydration of 2-methyl-2-butene yields 2-methyl-2-butanol whereas its reaction with $B_2H_6$ and then alkaline $H_2O_2$ yields 3-methyl-2-butanol.

(University of Bradford, 1982)

*Solution 9.2*

With four sections to be covered in 25–30 min, individual answers must necessarily be fairly concise.

(a) In the absence of oxygen, reaction is ionic. The more stable of the two possible carbocations is formed preferentially by addition of $H^{\oplus}$ to carbon 1 of the alkene,

$$CH_3CH{=}CH_2 \xrightarrow{\text{HBr}} CH_3\overset{\oplus}{C}HCH_3 + Br^{\ominus}$$

uptake of $Br^{\ominus}$ then occurs at carbon 2 to give the observed product:

$$CH_3\overset{\oplus}{C}HCH_3 + Br^{\ominus} \longrightarrow CH_3CH(Br)CH_3$$

The change in product when oxygen is present is explained by a change in mechanism. Oxygen can react as a diradical, able to abstract a hydrogen atom from a molecule of HBr:

$$:\!\overset{..}{\underset{..}{O}}{-}\overset{..}{\underset{..}{O}}: \,\,H{-}\overset{..}{\underset{..}{B}}r: \longrightarrow :\!\overset{..}{\underset{..}{O}}{-}\overset{..}{\underset{..}{O}}{-}H + :\overset{..}{\underset{.}{B}}r:$$

The bromine atom so formed adds to the alkene double bond so as to give the more stable of the two possible radicals:

$$CH_3CH{=}CH_2 + :\overset{.}{\underset{..}{B}}r: \longrightarrow CH_3\overset{.}{C}HCH_2Br$$

A chain reaction then ensues in which the bromoalkyl free radical abstracts hydrogen from another molecule of hydrogen bromide, and so on.

(b) At ordinary temperatures, ionic addition of chlorine gives 1,2-dichloro-propane via an intermediate chloronium ion analogous to the bromonium ion

$$CH_3CH{=}CH_2 \quad :\overset{..}{\underset{..}{C}}l{-}\overset{..}{\underset{..}{C}}l: \longrightarrow CH_3CH\overset{\overset{\displaystyle \overset{..}{\underset{.}{C}l}}{\diagup \oplus \diagdown}}{{-}}CH_2 + :\overset{..}{\underset{..}{C}}l:^{\ominus}$$

$$CH_2CH\overset{\overset{\displaystyle \overset{..}{\underset{.}{C}l}}{\diagup \oplus \diagdown}}{{-}}CH + :\overset{..}{\underset{..}{C}}l:^{\ominus} \longrightarrow CH_3CH(Cl)CH_2Cl$$

(Under different conditions, radical addition would give the same product.)

158

At high temperatures, chlorine molecules dissociate into chlorine atoms

$$:\ddot{C}l_2 \rightleftharpoons 2 :\dot{\ddot{C}}l:$$

which then abstract hydrogen from the methyl group of propene:

$$:\dot{\ddot{C}}l: \quad H-CH_2-CH=CH_2 \longrightarrow :\ddot{C}l-H + \dot{C}H_2CH=CH_2$$

Activation energy for this step is low because of resonance stabilisation of the intermediate free radical:

$$\dot{C}H_2-CH=CH_2 \longleftrightarrow CH_2=CH-\dot{C}H_2$$

A chain reaction is set up in which chlorine is abstracted by this free radical from a chlorine molecule, which thus generates another chlorine atom which can repeat the first step:

$$CH_2=CH-\dot{C}H_2 \quad :\ddot{C}l-\ddot{C}l: \longrightarrow CH_2=CH-CH-\ddot{C}l: + :\dot{\ddot{C}}l:$$

Some radical addition to the double bond may also occur, but at high temperature this is readily reversible, so that the predominant product is 3-chloropropene.

(c) See Example 9.3.

(d) Addition of hydrogen to an alkene occurs on the surface of a metal catalyst. Adsorption of hydrogen weakens or breaks the sigma bond. Adsorption of the alkene molecule then results in disruption of the pi bond and formation of two new sigma bonds with hydrogen atoms.

The heat evolved on hydrogenation gives a measure of the thermodynamic stability of an alkene. The *trans* isomer gives out less energy and is therefore more stable than the *cis* isomer by about 4.2 kJ mol$^{-1}$. Presumably this is due to unfavourable interaction between adjacent bulky methyl groups in the *cis* isomer which is not present in the *trans* alkene.

(e) Direct chlorination of ethane involves the formation of an intermediate ethyl free radical:

$$:\ddot{C}l_2 \rightleftharpoons 2 :\dot{\ddot{C}}l:$$

$$:\dot{\ddot{C}}l: \quad H-CH_2CH_3 \longrightarrow :\ddot{C}l-H + \dot{C}H_2CH_3$$

As before, a chain reaction follows in which formation of the product is accompanied by generation of another chlorine atom which can continue the process:

$$CH_3\dot{C}H_2 \quad :\ddot{C}l-\ddot{C}l: \longrightarrow CH_3CH_2-\ddot{C}l: + :\dot{\ddot{C}}l:$$

The failure of ethene to react in a similar way can be attributed to the relative instability of the ethenyl (vinyl) free radical, $CH_2=\dot{C}H$, in which there is destabilising interaction between the unpaired electron and the two pi electrons of the double bond:

159

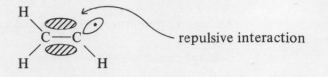

— repulsive interaction

(f) The first step in acid-catalysed hydration involves uptake of a proton so as to give the more stable of the two possible carbocation intermediates:

$$CH_3CH_2C{=}CH_2 \xrightarrow{H^\oplus} CH_3CH_2\overset{\oplus}{C}{-}CH_3$$
(with $CH_3$ substituents on the central carbons)

Subsequent addition of a molecule of water, followed by deprotonation gives the observed product.

In its reactions, $B_2H_6$ behaves as if it had the structure $BH_3$. The boron atom is electron deficient and is able to receive electrons from the alkene double bond. However, reaction is thought to involve a cyclic transition state rather than an ionic intermediate as in the first reaction:

$$CH_3{-}\underset{\underset{H-BH_2}{|}}{C}{=}CH{-}CH_3 \rightarrow \left[ CH_3{-}\underset{\underset{H\text{---}BH_2}{||}}{C}{=\!\!=}\underset{}{C}{-}CH_3 \right] \rightarrow CH_3{-}\underset{}{C}H{-}\underset{\underset{BH_2}{|}}{C}H{-}CH_3$$

Both steric and electronic factors favour addition of the boron atom to carbon 3, and repetition of this step gives rise to the trialkylborane.

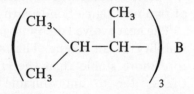

Oxidative decomposition with $H_2O_2$ gives the 'anti Markownikov' product, 3-methylbutan-2-ol:

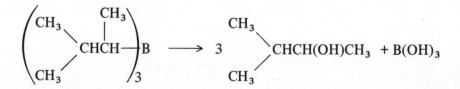

**Example 9.3**

(1) Semihydrogenation of but-2-yne gives an alkene which on addition of bromine gives two enantiomeric products, in equal amounts.
(a) Give conditions under which the semihydrogenation can be accomplished and explain why simple reduction with hydrogen and a platinum catalyst is not successful.
(b) Discuss the stereochemistry of both the semihydrogenation and the bromination steps.

(2) Provide a rationalisation of the fact that addition of bromine to but-1,3-diene in hexane at $-15°C$ gives a mixture of 1,2-dibromobut-3-ene and 1,4-dibromobut-2-ene, which on heating to $+60°C$ is transformed almost exclusively into the 1,4-dibromide.

(University of Sheffield, 1985)

*Solution 9.3*

(a) Semihydrogenation (i.e. reduction to an alkene) of an alkyne can be achieved in a variety of ways. The use of a partially deactivated palladium catalyst gives the *cis* alkene from a non-terminal alkyne (**'syn addition'**). The palladium may be 'poisoned' by addition of quinoline, or deposited in a fine form on a calcium carbonate support (Lindlar catalyst). A relatively new reagent, nickel boride (Ni₂B) can also be used to bring about *syn* addition.

Reduction with lithium or sodium in liquid ammonia gives (by a mechanism which is not yet understood) the *trans* alkene.

Simple reduction with hydrogen and a metal catalyst such as Ni or Pt is unsuitable as it cannot be stopped at the alkene stage.

(b) Hydrogenation occurs on the surface of the metal catalyst. Adsorption of hydrogen molecules disrupts the sigma bonds and new bonds form between hydrogen atoms and metal atoms. Adsorption of the alkyne then leads to the addition of both hydrogen atoms to the same side of the alkyne, leading to the *cis* alkene. Thus but-2-yne gives *cis* but-2-ene:

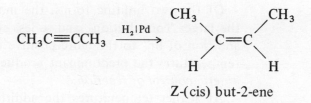

Z-(cis) but-2-ene

The first step in the addition of bromine to this alkene involves uptake of an electrophilic bromine atom to form a cyclic bromonium ion intermediate:

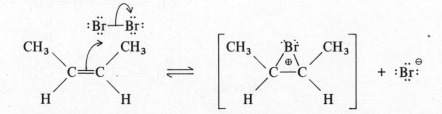

Subsequent addition of the bromide ion can only occur on the opposite side of this intermediate, and since attack on carbon atoms 2 and 3 is random, equal amounts of two enantiomeric 2,3-dibromobutanes will result:

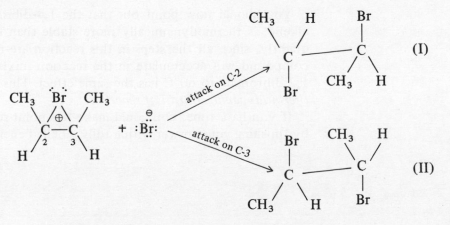

161

That the two products are indeed enantiomeric can be made clearer by drawing the corresponding Fischer projections:

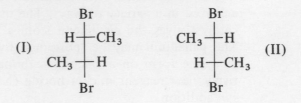

(2) Even at low temperatures, addition of Br occurs very readily because of the formation of a resonance-stabilised bromonium ion intermediate:

$$:\ddot{Br}-\ddot{Br}:$$

$$CH_2=CH-CH=CH_2 \rightleftharpoons \left[ CH_2=CH-\overset{\oplus}{\underset{|}{\overset{:\ddot{Br}:}{CH}}}-CH_2 \longleftrightarrow \overset{\oplus}{C}H_2-CH=CH-\overset{:\ddot{Br}:}{\underset{|}{CH_2}} \right] + :\ddot{Br}:^{\ominus}$$

$$(I) \qquad\qquad\qquad (II)$$

Of the two limiting forms, the more stable secondary structure makes the larger 'contribution', and so less energy of activation is required for the addition of $Br^{\ominus}$ to the more positive carbon atom 2. Consequently, at low temperatures the predominant product is the 1,2-dibromide. This is called *kinetic control of reaction*.

At higher temperatures the additional activation energy required for reaction at carbon 4 is more readily available, and so there is competition from this alternative reaction pathway.

These points may be summarised using the one-formula (composite) representation of the resonance stabilised intermediate:

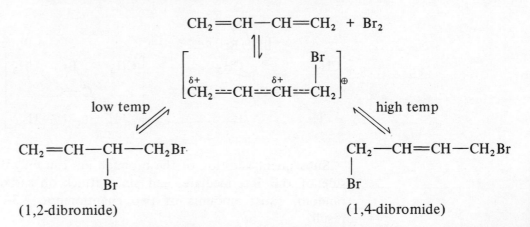

You should now point out that the 1,4-dibromide, being a dialkylated alkene, is thermodynamically more stable than the 1,2-dibromide. Consequently, since all the steps in this reaction are reversible, the more stable compound will accumulate in the reaction mixture. Likewise, heating the 1,2-dibromide to 60°C has the same effect. This behaviour is referred to as *thermodynamic control of reaction*.

If you have time, you could make this point really clear to the examiner by finishing your answer with a fully labelled energy profile:

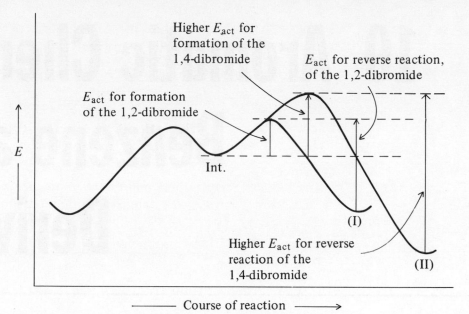

Key:  Int. = $[\overset{\delta+}{C}H_2 \text{====} CH \text{====} \overset{\delta+}{C}H \text{====} CH_2 Br]^{\oplus}$

I = $CH_2 \text{=} CH \text{—} CH(Br)CH_2 Br$

II = $CH_2(Br)CH \text{=} CHCH_2 Br$

## 9.9  Self-Test Questions

### Question 9.1

Answer both parts.

(1) Describe procedures, and discuss the mechanisms, for *syn*- and *anti*-addition of two hydroxyl groups to carbon–carbon double bonds. State whether *anti*-addition to a symmetrical Z-alkene will give a racemic mixture or a *meso* compound. Account for your answer with the aid of clear diagrams.

(2) How is 1,2-ethanediol obtained industrially, and what are its uses?

(University of Durham, 1984)

### Question 9.2

Discuss the mechanism of electrophilic addition of HBr to alkenes, both in the presence of, and in the absence of added peroxides.

You should include an explanation of (i) Markownikov's Rule, (ii) the relative rates of reaction of different alkenes, (iii) the occurrence of 'rearranged' products, and you should briefly list the other reagents which react with alkenes via this mechanism.

(Sunderland Polytechnic, 1985)

### Question 9.3

(a)  Give an account of methods of preparation of alkenes.  (60%)

(b)  Suggest structures for alkenes that give the following reaction products:

(i) ? $\xrightarrow{\text{H}_2/\text{Pd}}$ 2-methylhexane  (10%)

(ii) ? $\xrightarrow{\text{Br}_2/\text{CCl}}$ 2,3-dibromo-5-methylhexane  (10%)

(iii) ? $\xrightarrow{\text{HBr/peroxides}}$ 2-bromo-3-methylheptane  (10%)

(iv) ? $\xrightarrow{\text{O}_3} \xrightarrow{\text{NaHSO}_3}$ acetaldehyde + butan-2-one  (10%)

(Brunel University, 1985)

# 10 Aromatic Chemistry: Benzene and its Derivatives

## 10.1 Introduction

Benzene and other aromatic hydrocarbons are obtained from coal tar, and from petroleum by catalytic cracking/reforming.

The features associated with the concept of 'aromaticity' or 'aromatic character' may be summarised as follows:

1. The unique stability of an aromatic ring. Thus, benzene is unaffected by heating with aqueous $KMnO_4$, or by treatment with $Sn + HCl$. Substituents may be oxidised (e.g. $-CH_3 \rightarrow -COOH$) and reduced (e.g. $-NO_2 \rightarrow -NH_2$) with these reagents without affecting the ring.

2. Although benzene ($C_6H_6$) is highly unsaturated, it does not react with many of the reagents (HBr, aqueous and alkaline halogen, $H_3O^\oplus$, etc.) which readily add to an alkene double bond. Benzene does react additively with chlorine and bromine, but only under conditions for free radical reaction (see Chapter 12) and it is quite resistant to catalytic hydrogenation, which only occurs at high temperature and pressure. Somewhat surprisingly, benzene is easily ozonised to give ethandial, but otherwise the characteristic reaction is *electrophilic substitution*.

3. Substituents are able to influence both the *rate* and *position* of further substitution. Most substituents are either *activating* and *ortho/para-directing*, or *deactivating* and *meta-directing*.

4. Substituents on an aromatic ring may have their properties considerably modified as compared with aliphatic analogues. Thus the OH group is appreciably more acidic in phenol than in ethanol; the amino group is less basic in phenylamine (aniline) than in ethylamine; halogen attached to an aromatic nucleus is far less reactive than in a haloalkane, etc.

## 10.2 The Structure of Benzene and other Aromatic Compounds

In molecular orbital treatment of benzene, the carbon atoms are described as $sp^2$ hybridised, able to form three coplanar sigma bonds at 120° to one another. Two such bonds are with adjacent carbon atoms to form the hexagon ring, and the third is with hydrogen (Fig. 10.1). Sideways interaction (overlap) of the 'spare' 2p atomic orbitals on each carbon atom leads to the formation of additional molecular orbitals in two lobes above and below the plane of the ring. Six atomic orbitals give rise to six molecular orbitals, three high energy unoccupied *anti-bonding* orbitals and three lower energy *bonding* orbitals ($\pi$ orbitals) each with a pair of $\pi$ electrons. Since these orbitals overlap to some extent, the net effect is of a continuous 'streamer-like' orbital in two lobes above and below the plane of the ring and encompassing all six carbon atoms.

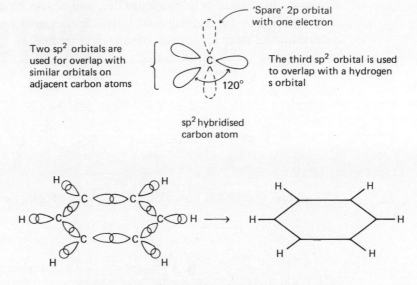

Orbital overlap to form the sigma bonded skeletal structure the six carbon-hydrogen sigma bonds (2p orbitals not shown)

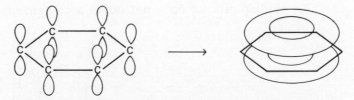

Sideways overlap of the six 2p orbitals leads to the overlapping system of pi orbitals (carbon-hydrogen bonds not shown)

**Fig. 10.1** Structure of benzene

Benzene is accordingly described as having a *delocalised system of $\pi$ electrons*, and as a result is more stable than (hypothetical) cyclohexatriene by about 150 kJ mol$^{-1}$ ('**resonance energy**' – see Chapter 1).

The valence bond approach regards benzene as a **mesomer** or **resonance hybrid** of two Kekulé (cyclohexatriene) structures, with minor contributions from other limiting (canonical, or contributing) forms.

According to the Hückel theory, any planar monocyclic ring containing $(4n + 2)$ $\pi$ electrons, where $n = 0, 1, 2, 3$, etc., will have a delocalised system of $\pi$ electrons, and will exhibit aromatic character. Thus benzene ($6\pi$), and polycyclic benzenoid hydrocarbons such as naphthalene ($10\pi$) and anthracene ($14\pi$) are all aromatic, as are the non-benzenoid heterocyclic compounds pyridine, pyrrole, furan and thiophen, all with $6\pi$ electrons.

Larger monocyclic rings (annulenes) whose formal structure consists of alternating double and single bonds may also have $(4n + 2)$ $\pi$ electrons and exhibit aromatic behaviour, but you are not likely to discuss their chemistry in the first year of your course. The same applies to modern physical evidence for aromaticity such as the strong deshielding of aromatic protons which is observed in the $^1$H NMR spectrum.

## 10.3 Reactivity of Benzene

Benzene, like alkenes, is nucleophilic, and reacts by donation of electrons to an electrophile. In the reaction of both alkenes and benzene, the first, slow, rate-determining, step is uptake of an electrophile ($E^\oplus$) to form an intermediate carbocation:

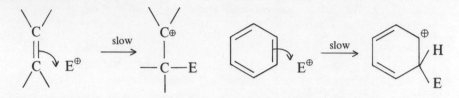

The alkene reaction is completed by uptake of a nucleophile, $\ddot{B}^\ominus$

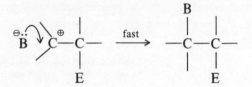

(e.g. where $E^\oplus = H^\oplus$ and $\ddot{B}^\ominus = :\ddot{B}\ddot{r}:^\ominus$ )
A similar process does not occur with benzene because the resulting cyclohexadiene

has only 17 kJ mol$^{-1}$ resonance energy, and its formation would therefore be accompanied by an overall 'loss' of 133 kJ mol$^{-1}$ of resonance energy. Instead, the carbocation (called variously a benzenium ion, benzenonium ion or **sigma complex intermediate**) attains stability by rapid expulsion of a proton (the nucleophile $\ddot{B}^\ominus$ then functioning as a base):

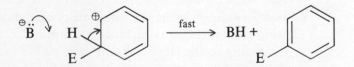

The energy profile (Fig. 10.2) should help to make this clearer. Uptake of an electrophile by benzene can only occur at the expense of loss of most of the resonance energy, and the fact that this is possible under relatively mild condi-

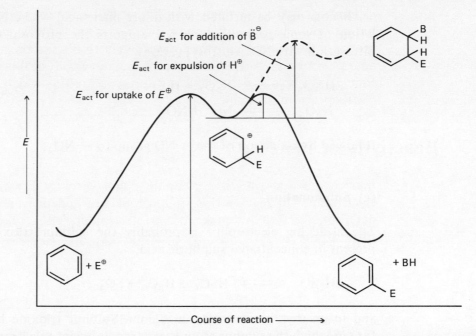

$E_{\text{act}}$ for addition of $\overset{\cdot\cdot}{B}{}^{\ominus}$

$E_{\text{act}}$ for expulsion of $H^{\oplus}$

$E_{\text{act}}$ for uptake of $E^{\oplus}$

$+ E^{\oplus}$

$+ BH$

Course of reaction

**Fig. 10.2** Energy profile for electrophilic substitution and addition to benzene (energy differences not to scale)

tions is attributed to **resonance stabilisation** of the sigma complex intermediate, which lowers the activation energy for its formation:

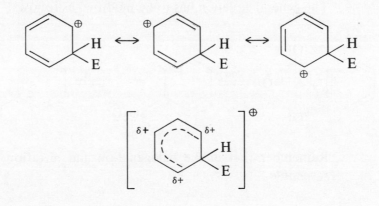

# 10.4   Nature of the Electrophile, $E^{\oplus}$

### (a) Halogenation

E.g. $Br{-}Br + FeBr_3 \rightleftharpoons [\overset{\delta+}{Br}{----}\overset{\delta-}{Br}{----}FeBr_3] \rightleftharpoons [\overset{\oplus}{Br} \ \overset{\ominus}{FeBr_4}]$

For simplicity in writing the mechanistic equation, replace $E^{\oplus}$ by $Br^{\oplus}$.

### (b) Nitration

$$HNO_3 + 2H_2SO_4 \rightleftharpoons 2HSO_4 + H_3O^{\oplus} + NO_2^{\oplus} \ (\textit{nitryl cation})$$

Replace $E^{\oplus}$ by $NO_2^{\oplus}$.

167

Phenols may be nitrated with dilute nitric acid where $NO_2^{\oplus}$ cannot exist. Isolation of some *para*-nitrosophenol supports the proposal that reaction involves nitrosation with $NO^{\oplus}$ (nitrosyl cation)

$$2HNO_3 \rightleftharpoons N_2O_3 + H_2O$$
$$N_2O_3 \rightleftharpoons NO^{\oplus} + NO_2^{\ominus}$$

followed by oxidation of the —NO group to —$NO_2$.

### (c) Sulphonation

The attacking electrophile is probably the sulphur trioxide molecule. This is present in concentrated sulphuric acid,

$$2H_2SO_4 \rightleftharpoons HSO_4^{\ominus} + H_3O^{\oplus} + SO_3$$

and in greater concentration in oleum. Sulphur trioxide has a resonance structure in which the sulphur atom is electron deficient (electrophilic):

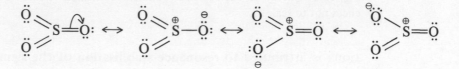

The general equation has to be modified as follows:

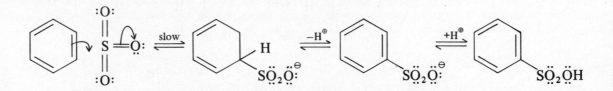

Remember that unlike halogenation and nitration, *all steps in sulphonation are reversible*.

### (d) Friedel-Crafts Reaction

#### (i) *Alkylation*

$$\text{E.g. } R—Cl + AlCl_3 \rightleftharpoons [\overset{\delta+}{R}---Cl---\overset{\delta-}{AlCl_3}] \rightleftharpoons [\overset{\oplus}{R}\ \overset{\ominus}{AlCl_4}]$$

Again it is acceptable to replace $E^{\oplus}$ by $R^{\oplus}$ in the general mechanistic equation.

#### (ii) *Acylation*

$$\text{E.g. } RCO—Cl + AlCl_3 \rightleftharpoons [\overset{\oplus}{RCO}\ \overset{\ominus}{AlCl_4}]$$

Replace $E^{\oplus}$ in the general equation by $RCO^{\oplus}$ ($R—\overset{\oplus}{C}=\overset{..}{O}$, an acylinium ion)

168

## 10.5  **Behaviour of Substituents**

In the reaction of a monosubstituted benzene, the substituent already present influences both the *rate* and *position* of further substitution. In general, *electron releasing* substituents are *activating*: they are able to stabilise the sigma complex intermediate and therefore lower the activation energy for its formation. Conversely, *electron withdrawing* substituents destabilise the sigma complex and are therefore *deactivating*.

Since in the sigma complex the positive charge is distributed over the carbon atoms *ortho* and *para* to the position attacked by the electrophile, substituents will have maximum effect when they are in those positions.

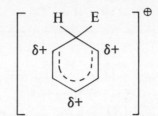

Electron releasing substituents are *ortho*- and *para*-directing because activation energy is lower for the formation of *ortho* and *para* sigma complexes than for the *meta* sigma complex. This means that collision of an electrophile with a ring carbon atom *ortho* or *para* to an electron releasing substituent is more likely to lead to the formation of a sigma complex than when collision occurs at the *meta* position.

It follows that since electron withdrawing substituents have a greater deactivating effect when *ortho* or *para* to the position attacked, collision of an attacking electrophile is more likely to be successful when it occurs at the *meta* position.

These electronic effects of substituents may be described in terms of *inductive* and *mesomeric* displacements of electrons.

### (a)  Inductive Stabilisation

+I effect is stabilising.

(similarly with *ortho* substitution, but
less effective when *meta*)

### (b)  Mesomeric (Resonance) Stabilisation

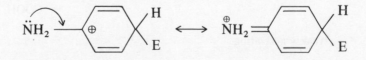

+M effect is stabilsing.

(similarly with *ortho* substitution,
but not possible when *meta*)

### (c) Destabilisation

Usually results from unfavourable juxtaposition of positive charges on the substituent and the ring carbon atom. Again, inductive or mesomeric effects may be involved, e.g.

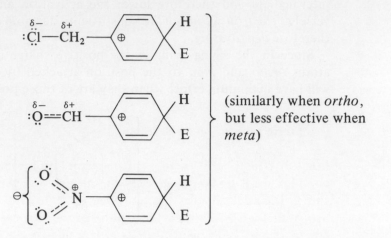

(similarly when *ortho*, but less effective when *meta*)

Halogen is unique in that although it is electron withdrawing by a powerful negative inductive effect, it can also exert a weak positive mesomeric effect when *ortho* or *para* to the position of attack:

weak +M effect

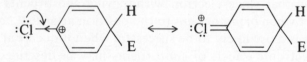

(similarly for *ortho* substitution, but not possible when *meta*)

strong −I effect

Chlorine is thus weakly deactivating but *ortho/para* directing. The orienting effects of common substituents are summarised in Table 10.1.

**Table 10.1** Orienting effects of some common substituents

| *Groups which are ortho/para directing* | *Groups which are meta directing* |
|---|---|
| **Strongly activating** | **Strongly deactivating** |
| —OH, —O$^\ominus$ | —NO$_2$, —$\overset{\oplus}{N}$R$_3$ |
| —NH$_2$, —NHR, —NR$_2$ | —CCl$_3$, —CF$_3$ |
| **Moderately activating** | **Moderately deactivating** |
| —OR, —NHCOCH$_3$ | —CHO, —COR |
|  | —COOH, —SO$_2$OH, —CN |
| **Weakly activating** | |
| -alkyl | |
| **Weakly deactivating** | |
| —F, —Cl, —Br, —I | |

## 10.6 Steric (Size) Effects of Substituents

### (a) The Ortho/Para Ratio

Large substituents may hinder close approach of the electrophile to adjacent *ortho* positions, and destabilise the transition state for formation of the *ortho* sigma complex by *steric compression* with the (partially) bonded electrophile. The net result is to reduce the ratio of *ortho* to *para* isomers ('*ortho-para ratio*') in the product. Even moderate sized substituents may have an effect when the electrophile is very large.

### (b) Steric Limitation of Resonance

For full effect, substituents which activate by resonance stabilisation of the sigma complex must be *co-planar with the aromatic ring*. Molecular orbital theory sees this as the necessity for the p atomic orbital of the substituent atom to be lined up parallel with the 'spare' 2p orbital of the ring carbon atom for maximum overlap to occur. Bulky substituents either side of (i.e. *ortho* to) the activating substituent may prevent this and so limit or even inhibit mesomeric (resonance) interaction with the ring.

For example, N,N-dimethylphenylamine readily couples with benzenediazonium cation, but 2,6-dimethyl-N,N-dimethylphenylamine does not react:

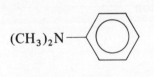

Readily couples with
benzenediazonium cation

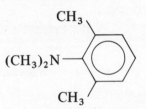

Does not couple with
benzenediazonium cation

## 10.7 Principal Reactions of Benzene

### (a) Halogenation

$$\bigcirc + X_2 \xrightarrow[\Delta]{Fe} \bigcirc\!\!-X \; + HX$$

X = Cl or Br, catalysed by Fe or $FeX_3$.

Amines and phenols are chlorinated and brominated in aqueous solution without the aid of a catalyst. Fluoro and iodo compounds are best prepared via diazonium salts (see Chapter 11).

### (b) Nitration

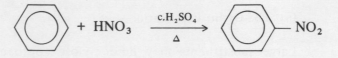

Heat with mixed c.$HNO_3$ + c.$H_2SO_4$ ('mixed acid') at about 45–50°C. Further nitration requires higher temperature, preferably with fuming $HNO_3$ and c.$H_2SO_4$.

An amino group when present must be acylated (e.g. acetylated). (This limits electron release by the nitrogen atom and prevents the formation of $-\overset{\oplus}{N}H_3$ which would be strongly *deactivating* and *meta-directing*.)

Phenols are nitrated with dilute nitric acid, but some oxidation occurs.

### (c) Sulphonation

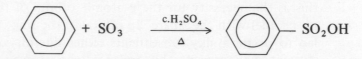

Heat with c.$H_2SO_4$ or oleum. Reversible.

Because of steric effects, sulphonation of alkylbenzenes gives mainly the *para* isomer.

### (d) Friedel–Crafts Reaction

#### (i) *Alkylation*

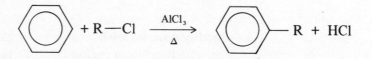

Other Lewis acid catalysts may be used (e.g. $SnCl_4$, $BF_3$, etc.), also alcohols or alkenes + c.$H_2SO_4$.

Limitations:

1. R is activating – difficult to stop at monosubstitution.
2. Primary R groups may rearrange, e.g. with 1-chloropropane and benzene the main product is methylethyl (isopropyl) benzene.
3. Haloarenes (aryl halides) are less reactive and cannot be used.
4. Strongly deactivated substrates (e.g. nitrobenzene) do not react.
5. Amines and phenols cannot normally be used because of complexing of the substituent with the Lewis acid catalyst.

#### (ii) *Acylation*

$$\bigcirc + RCOCl \xrightarrow{AlCl_3} \bigcirc-COR + HCl$$

Both aliphatic and aromatic acyl halides may be used. Multiple substitution is not a problem as the acyl group is deactivating. Reduction of the keto group, e.g.

Clemmensen (amalgamated Zn + c.HCl) or Wolff–Kishner (hydrazine + base) gives an alkylbenzene which might be difficult to synthesise by direct alkylation, e.g.

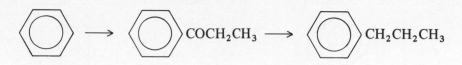

## 10.8  Aromatic Synthesis

The compound to be synthesised is often called the **target molecule**, and the usual approach is to work backwards step by step from the target molecule to stated or readily available starting materials. The problem is to decide the *correct order in which substituents should be introduced*, and if they are to be further modified or replaced, at which stage this is to be carried out. Generally speaking, minute experimental details are not expected, but you do need to be familiar with:

(i) outline conditions for the introduction of various substituents and their conversion into other groups by oxidation, reduction or further reaction; and

(ii) the directive effect of all the common substituents.

When devising a synthesis for a given target molecule, keep the following points in mind:

1. *Ortho/para* directing groups tend to give mixtures of the two isomers. Hence a suitable separatory technique must be available, such as column chromatography or fractional crystallisation. For a pair of *ortho* and *para* isomers, the *para* isomer is usually less soluble in a given solvent.

2. Good yields of an *ortho* isomer may often be obtained by first temporarily blocking the *para* position. The synthesis of *ortho*-bromomethylbenzene illustrates the principle:

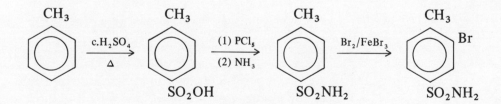

Hydrolysis of the sulphonamide regenerates the acid and the sulphonic acid group is readily removed by heating under reflux with dilute sulphuric acid:

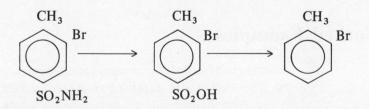

3. *Meta*-directing groups usually give good yields of that isomer with very little of the other isomers.

4. Finally, remember that the Friedel–Crafts reaction will not work when a

strongly deactivating group is present. Synthesis of 3-nitrophenylethanone (*meta*-nitroacetophenone) illustrates this:

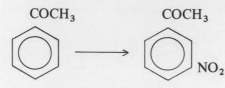

acetylate and then nitrate

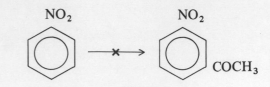

this will not work

Additional Notes

## 10.9 Worked Examples

The chemistry of aromatic compounds features in virtually all first year degree courses. The extent and depth of treatment varies, and may or may not include polycyclic, non-benzenoid or heterocyclic compounds. Three areas are commonly examined; mechanisms of electrophilic substitution, theory of substituent effects and synthesis of di- and tri-substituted derivatives of benzene. Questions may, of course, embrace one, two or all three of these topics as the following examples will show.

**Example 10.1**

1. Discuss in detail the mechanism for nitration of benzene using mixed acid (conc. $HNO_3/H_2SO_4$) and give all the evidence available in support of the proposed mechanism.
*(12 marks)*

2. Offer rational explanations for the fact that toluene nitrates in the *ortho* and *para* positions at a faster rate than benzene, whereas nitrobenzene nitrates at a slower rate than benzene and gives *meta*-dinitrobenzene.
*(8 marks)*

(Portsmouth Polytechnic, 1985)

*Solution 10.1*

1. You could begin with a statement that in nitration with 'mixed acid' the nitrating species is believed to be the **nitryl cation** (nitronium ion), $NO_2^{\oplus}$. Give equations to show how it is formed:

$$NO_2{-}OH + H_2SO_4 \rightleftharpoons NO_2{-}\overset{\oplus}{O}H_2 + HSO_4^{\ominus}$$

$$NO_2{-}\overset{\oplus}{O}H \rightleftharpoons NO_2^{\oplus} + OH_2$$

$$H_2O + H_2SO_4 \rightleftharpoons H_3O^{\oplus} + HSO_4^{\ominus}$$

Overall, $\quad HNO_3 + 2H_2SO_4 \rightleftharpoons NO_2^{\oplus} + H_3O^{\oplus} + 2HSO_4^{\ominus}$

and quote some supporting evidence, e.g.

   (i) electrolysis of mixed acids evolves oxides of nitrogen at the cathode, consistent with migration of $NO_2^{\oplus}$ to the negative electrode;

   (ii) addition of c.$H_2SO_4$ to c.$HNO_3$ results in a *fourfold* depression of freezing point, as expected if equilibrium in the above equation is well to the right;

   (iii) salts such as nitryl borofluoride, $\overset{\oplus}{N}O_2\overset{\ominus}{B}F_4$ have been prepared and characterised;

   (iv) the absorption peak attributed to $NO_2^{\oplus}$ in the IR spectrum of its salts is also present in the IR spectrum of mixed acid.

Demonstrating the existence of the nitryl cation does not of itself prove that it actually is the attacking species. Support for this comes from observations that benzene can be nitrated with nitryl borofluoride, and that there is good agreement between measured rates of reaction and those predicted from the calculated concentration of $NO_2^{\oplus}$ in mixed acid.

$$\text{Rate} = k\,[NO_2^{\oplus}]\,[\text{benzene}]$$

Now outline the accepted two-step mechanism which is consistent with this rate equation.

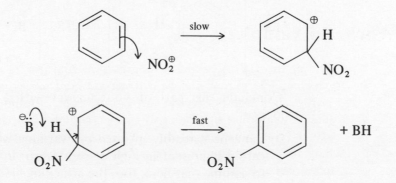

175

The sigma complex intermediate is formed at the expense of loss of most of the resonance energy of the substrate benzene, and this is therefore the slow step. However, activation energy is lower than it would otherwise be because of resonance stabilisation of the intermediate. You should draw limiting forms to represent this, and then construct a labelled energy profile diagram for the whole process.

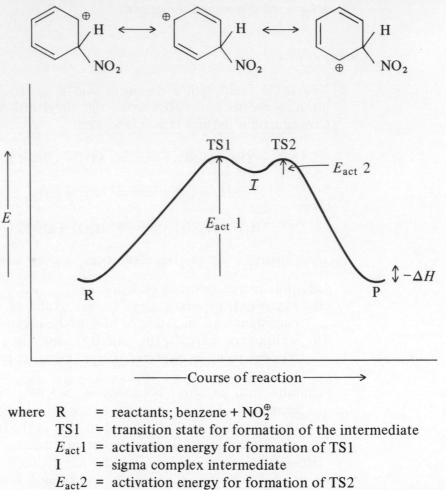

where R = reactants; benzene + $NO_2^{\oplus}$

TS1 = transition state for formation of the intermediate

$E_{act}1$ = activation energy for formation of TS1

I = sigma complex intermediate

$E_{act}2$ = activation energy for formation of TS2

TS2 = transition state for decomposition of the intermediate

P = products; nitrobenzene and $HSO_4^{\ominus}$

$-\Delta H$ = enthalpy of (exothermic) reaction.

The question does say 'discuss in detail', so if your lecture notes show the structures of the two transition states, you should draw them here (although not all first year courses will go into this much detail)

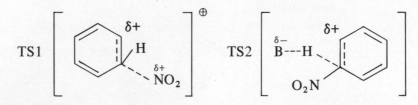

Conclude this part of your answer with a summary of the supporting evidence:

(i) Benzene is readily nitrated on warming with $NO_2^{\oplus}BF_4^{\ominus}$. Since the nitryl cation is present in high concentration in 'mixed acid', it is reasonable to assume that here, too, the attacking electrophile is $NO_2^{\oplus}$.

(ii) As would be expected for the proposed slow step, kinetic studies show that the rate of nitration is first order in concentration of both benzene and nitric acid:

$$\text{Rate} = k\,[\text{benzene}]\,[\text{HNO}_3] = k\,[\text{benzene}]\,[\text{NO}_2^{\oplus}]$$

(iii) No *kinetic isotope effect* is observed with hexadeuterobenzene, which nitrates at the same rate as benzene. It follows that cleavage of the carbon–hydrogen bond is not involved in the rate determining step, as the larger activation energy required to break a carbon–deuterium bond would slow down the reaction of hexadeuterobenzene. This observation also rules out the alternative of a one-step mechanism via a transition state in which cleavage of the carbon–hydrogen bond accompanies formation of the new carbon–electrophile bond.

2. In answering the second part of the question you will be dealing with further supporting evidence for the accepted mechanism, namely that substituents have the expected effect both on the *rate* and *position* of further substitution. Since the rate determining step is formation of a cationic intermediate, electron-releasing substituents (by stabilising this intermediate) should increase the rate of reaction, and electron-withdrawing substituents (having the opposite effect) should *decrease* the rate of reaction — and this is exactly what is observed.

Furthermore, since delocalisation in the intermediate distributes the positive charge over the carbon atoms *ortho* and *para* to the position attacked, substituents should have maximum effect when located in these positions. Thus electron-releasing substituents are more effective in stabilising *ortho* and *para* sigma complexes and are therefore *ortho* and *para* directing. Conversely, electron-withdrawing substituents are most deactivating when attached to carbon atoms bearing positive charge, and so such groups are *meta* directing. Illustrate this by drawing sigma complexes for the two examples in the question.

**Toluene**: the methyl group is electron releasing by the +I effect

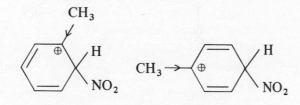

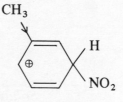

Inductive stabilisation most effective for *ortho* and *para* complexes

Inductive stabilisation less effective for *meta* (but some *meta* substitution does take place)

**Nitrobenzene**: the polar nitro group is destabilising because of juxtaposition of positive charges

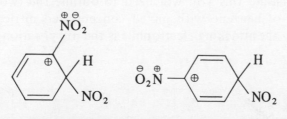

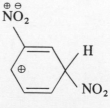

Strongest when *ortho* or *para* to the position attacked

Less strong when *meta* to position attacked

177

Thus, *meta*-dinitrobenzene is formed preferentially, but at a slower rate than for nitration of benzene.

The assumption, of course, is that a more stable intermediate will be formed via a more stable transition state, since it is actually the activation energy for the formation of the transition state which is the controlling factor. You will recall the **Hammond Postulate**, which says that when a transition state and an intermediate are close together on an energy profile, i.e. of similar energies, then they are of closely similar structure. Therefore a substituent will have the same sort of effect (stabilising or destabilising) on both the intermediate and the transition state from which it is formed.

**Example 10.2**

(a) Discuss the factors which govern the directive effects of substituents in the electrophilic substitution of benzene derivatives, illustrating your answer with reference to the following materials:

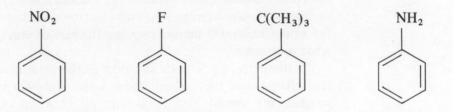

(b) Suggest synthetic routes to THREE of the following compounds starting from benzene:

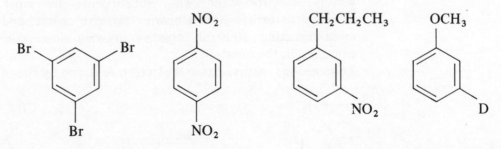

(University of York, 1983)

*Solution 10.2*

(a) In aromatic electrophilic substitution it is found that the majority of substituents are either activating and *ortho/para* directing, or deactivating and *meta* directing. The common exception is halogen which is weakly deactivating but *ortho/para* directing. These reactivity and directive effects result from the ability of a substituent to stabilise or destabilise the transition states for attack by the electrophile on the *ortho*, *meta* and *para* positions.

To illustrate this you will need to outline the two-step mechanism for nitration of benzene with mixed concentrated nitric and sulphuric acids, for which the attacking electrophile is the nitryl cation, $NO_2^{\oplus}$

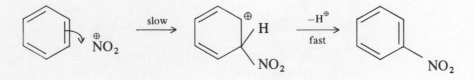

The intermediate cation (sigma complex intermediate) has a resonance stabilised structure in which delocalisation of the remaining $\pi$ electrons distributes the positive charge over ring carbon atoms *ortho* and *para* to the position attacked:

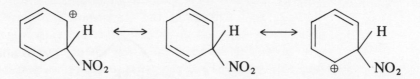

It follows that both stabilising and destabilising substituents will be most effective when located on these ring carbon atoms. Electron-releasing substituents are activating and *ortho/para* directing because they are able to stabilise *ortho* and *para* sigma complexes more than they can the *meta* sigma complex. Conversely, electron-withdrawing substituents are *meta* directing because they have a greater destabilising effect on the *ortho* and *para* sigma complexes than on the *meta* sigma complex. Having made these points as succinctly as possible, you should now consider each of the given examples in turn.

The nitro group has a mesomeric polar structure with the nitrogen atom carrying a unit positive charge:

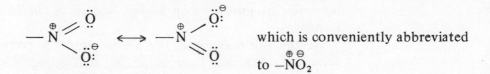

which is conveniently abbreviated to $-NO_2$

Destabilisation is greatest when the nitro group is *ortho* or *para* to the position attacked, as can be seen from the following limiting forms:

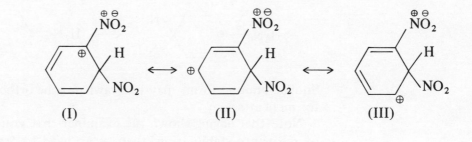

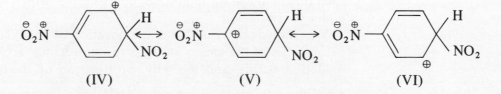

Limiting forms I and V are particularly unstable because of juxtaposition of +ve charges. No equivalent form can be written for *meta* substitution

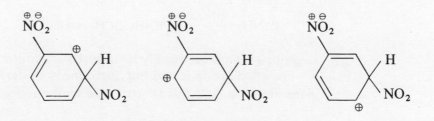

You could stress this point by drawing a comparative energy profile diagram for nitration of benzene and nitrobenzene:

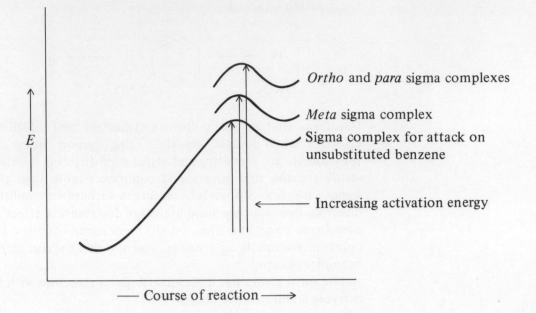

Nitration of nitrobenzene is slow and gives almost exclusively the *meta* isomer.

At the other extreme, the amino group is powerfully activating and *ortho/para* directing. Despite the fact that nitrogen is more electronegative than carbon, the amino group is stabilising because nitrogen can share its non-bonded electron pair with a ring carbon atom carrying positive charge. Show this by drawing the relevant limiting forms, e.g. for para attack

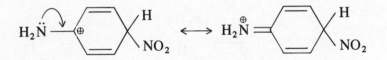

Similar limiting forms may be drawn for the ortho sigma complex, but not for meta attack.

Note that having shown the examiner that you understand the concept of resonance stabilisation, there is no need to draw all the limiting forms each time, but just the important 'contributors' which illustrate the example under consideration.

You should point out that direct nitration of phenylamine is not possible, because firstly the amino group is too powerfully activating and extensive oxidation occurs, and secondly, protonation of the basic amino group in a highly acidic medium converts —$NH_2$ into the ammonium group, —$\overset{\oplus}{N}H_3$ which is powerfully deactivating and *meta* directing. In practice these difficulties are overcome by 'acetylation' of the amino group with ethanoyl chloride:

$$PhNH_2 \xrightarrow{CH_3COCl} PhNHCOCH_3 + HCl$$

a process which can readily be reversed after nitration has been carried out.

The alkyl group in *tert*-butylbenzene is moderately activating and *ortho/para* directing, and this is attributed to inductive stabilisation of the *ortho* and *para* sigma complexes:

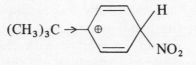

A similar structure can be drawn for the *ortho* sigma complex, but stabilisation is less efficient in the *meta* sigma complex where the inductive effect has to operate over a greater distance:

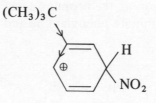

There is another factor here, namely the *size* of the *tert*-butyl group. Because of its bulk, it tends to get in the way of the electrophile approaching the adjacent *ortho* positions, with the result that in nitration of this compound the *para* isomer predominates. Thus the *tert*-butyl group influences the course of reaction by a combination of *electronic* and *steric* effects. Steric hindrance to *ortho* substitution should more correctly be ascribed to *steric compression in the transition state* which increases the activation energy for the formation of the *ortho* sigma complex above that required for the *para* sigma complex. A quick sketch of the energy profiles will again help to clarify your answer:

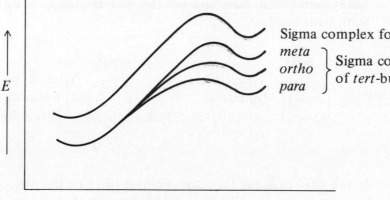

In fluorobenzene there is a closer balance between a destabilising negative inductive effect and a stabilising positive mesomeric effect. As a result of the latter, the fluorine atom is *ortho/para* directing, but because of the strong −I effect, it is also weakly deactivating.

+M effect

 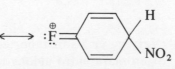

−I effect
(deactivating)

+M effect of the fluorine atom
makes the *ortho* and *para* sigma complexes
less unstable than the *meta*

(Surprisingly, the fluorine atom, which is the most electronegative of the halogen atoms, is also the least deactivating, and fluorobenzene nitrates only marginally less readily than benzene! This is assumed to result from the good overlap possible between the fluorine 2p orbital and the ring carbon 2p orbital. With chlorine, for example, overlap is less effective between a 3p orbital and the carbon 2p orbital.)

Having spent most of·the time on the first part of the question, you can relatively quickly outline the required syntheses in the form of 'annotated equations'.

(b) The symmetrical tribromobenzene cannot be prepared by direct bromination as the bromine substituent is *ortho*/*para* directing. The solution is to brominate phenylamine and then to deaminate via the diazonium salt:

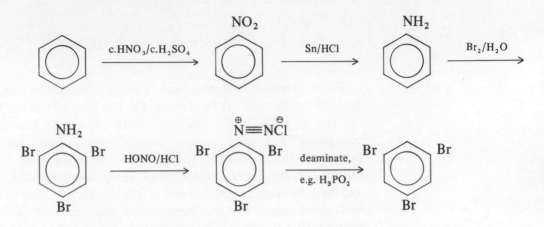

A similar problem arises with the second example in which two *meta* directing groups are *para* to one another. The procedure is to nitrate phenylamine and then convert the amino group to nitro by the Balz–Schiemann reaction:

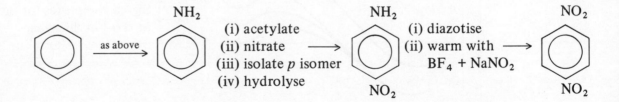

In the next example the alkyl group is *meta* to the nitro group, but nitrobenzene cannot be directly alkylated by the Friedel–Crafts reaction. One possible solution is to acylate benzene, and then after nitration, selectively reduce the keto group to alkyl by the Wolff–Kishner reaction:

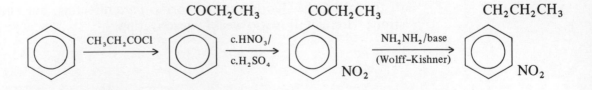

(Varying yields are reported in the literature for the Wolff–Kishner reduction of aromatic nitro aldehydes and ketones. In specific cases it may be necessary to reduce both groups and then convert the amino back to nitro by the Balz–Schiemann reaction.)

182

The last example is the most complicated in terms of the number of operations required, and might therefore be the best one to leave out if you are short of time! However, all the steps involve basic reactions with which you should be quite familiar:

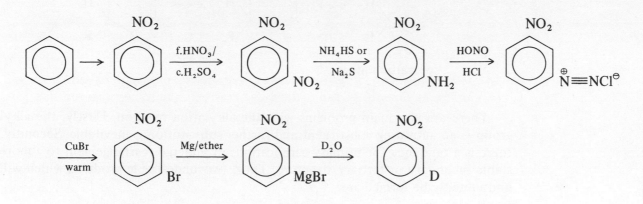

## Example 10.3

Give the mechanism for the alkylation of benzene by an alkyl halide, RX, in the presence of a suitable catalyst.

(Suggest an appropriate catalyst and draw the 3 principal resonance structures for the intermediate containing an sp³ hybridised carbon.)   (*6 marks*)

Why is this method not very satisfactory for producing alkylbenzenes such as n-propylbenzene? Give the structures of two undesirable products you might obtain if you attempted to make n-propylbenzene by this method.   (*6 marks*)

Describe a more satisfactory route for the preparation of n-propylbenzene and explain how this route overcomes the problems associated with the direct alkylation of benzene with an n-propyl halide.   (*8 marks*)

(University of Keele, 1985)

*Solution 10.3*

Alkylation of benzene with a haloalkane (alkylhalide) requires the use of a Lewis acid catalyst such as anhydrous $AlCl_3$ (Friedel–Crafts reaction). The role of the catalyst is to polarise, or even bring about complete ionisation of the haloalkane:

$$RX + AlCl_3 \rightleftharpoons \left[ \overset{\delta+}{R} \text{---} \overset{\delta-}{X} \text{---} AlCl_3 \right] \rightleftharpoons \left[ \overset{\oplus}{R} \ \overset{\ominus}{AlCl_4} \right]$$

Obviously the more stable the carbocation, $R^{\oplus}$, the more likely it is that ionisation will occur. But even then the two ions probably remain closely associated as an ion-pair. The mechanism can be formulated as

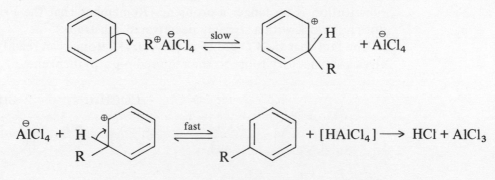

183

Resonance stabilisation of the sigma complex intermediate is represented by the limiting forms

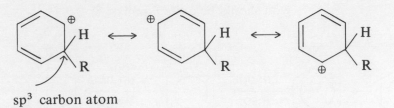

sp³ carbon atom

There are two main problems with the alkylation reaction. Firstly, the alkyl group is an activating substituent and further substitution is inevitable. Secondly, there is a tendency for primary carbocations to undergo rearrangement to a more stable secondary or tertiary structure. Thus, two undesirable products which will undoubtedly be formed are

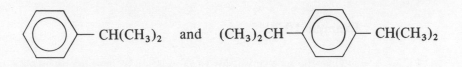

isopropylbenzene               1,4-diisopropylbenzene

(Note that because of the size of the 2-methylethyl (isopropyl) group, disubstitution occurs mainly in the *para* position. Also, because reaction is reversible, substituents may migrate round the ring and trisubstitution gives 1,3,5-triisopropylbenzene.)

These difficulties are overcome by employing the Friedel–Crafts acylation reaction, using an acyl halide in place of the alkyl halide.

$$RCOCl + AlCl_3 \rightleftharpoons [\overset{\oplus}{R}CO\ \overset{\ominus}{AlCl_4}]$$

Unlike an alkyl carbocation, the acylinium ion, $R\overset{\oplus}{C}{=}O$, has a resonance structure with much of the positive charge residing on the oxygen atom

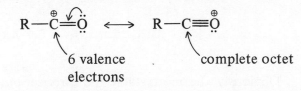

6 valence
electrons                    complete octet

and acylinium ions are not prone to rearrangement. Furthermore, the product is a ketone in which the carbonyl group is a deactivating substituent, and so multiple substitution is no longer a problem. (Remember that the Friedel–Crafts reaction does not work with a strongly deactivated substrate.)

The fact that the CO group in alkyl–aryl ketones can readily be reduced to $CH_2$ gives a satisfactory route to the required n-propylbenzene

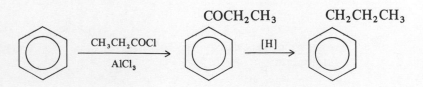

184

Clemmensen reduction, using amalgamated zinc and c.HCl is the preferred method, giving good yields of the alkylbenzene.

## Example 10.4

Give an account of the methods by which aromatic chloro and bromo compounds may be prepared (a) by direct and (b) by indirect substitution. Explain how the chloro-substituent in chlorobenzene affects the reactions of the ring as compared with a hydroxyl substituent.

(University of Bradford, 1983)

*Solution 10.4*

*(a) Direct substitution*

Activated substrates such as phenol and phenylamine (aniline) are readily chlorinated and brominated with the free halogen, or with a halogen carrier such as HO—Cl or HO—Br in aqueous solution. Molecular chlorination of alkylbenzenes is also possible, but reaction is very slow (and precautions may be necessary to avoid radical substitution in the alkyl group).

For most aromatic substrates a Lewis acid catalyst such as $FeCl_3$ or $FeBr_3$ is required to polarise (or perhaps ionise) the halogen:

$$X_2 + FeX_3 \rightleftharpoons \left[ \overset{\delta+}{X} \text{---} X \text{---} \overset{\delta-}{FeX_3} \right] \rightleftharpoons \left[ \overset{\oplus}{X} \ \overset{\ominus}{FeX_4} \right]$$

where X = Cl or Br

Typically, a solution (e.g. in ethanoic acid) of the aromatic compound is heated under reflux with iron powder or $FeCl_3$ ($FeBr_3$) while a solution of the halogen is slowly added from a side arm.

*(b) Indirect substitution*

The principal method here is by reaction of the appropriate diazonium salt. For example, benzenediazonium chloride is readily converted into chlorobenzene or bromobenzene by the **Sandmeyer reaction**. On mixing a solution of the diazonium salt with copper(I) (cuprous) chloride or bromide, nitrogen is slowly evolved, and after several hours the haloarene can be extracted from the reaction mixture, e.g.

A variation which sometimes gives better yields is to warm the diazonium salt solution with freshly precipitated copper powder (the **Gattermann reaction**).

The main limitation to direct halogenation is that, particularly when an activating substituent is present, it may be impossible to avoid the formation of mixtures from which it is difficult to isolate the pure components. (For example, there is only 3°C difference in the boiling points of *ortho* and *para* bromotoluenes.) However, *ortho* and *para* nitrotoluenes are easily separated by fractional crystallisation, and the pure isomers can then be converted to the halo compounds via the diazonium salts.

Another advantage of the indirect method is that by careful phasing of the steps involved, halogen may be introduced into positions not accessible by direct substitution.

An alternative approach of limited application involves *mercuration* of an aromatic nucleus by heating the compound with 'mercuric acetate', $Hg(OCCH_3)_2$, followed by treatment with halogen ($Cl_2$ or $Br_2$) which removes the mercury atom

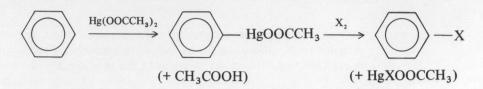

In answering the second part of the question you will need to briefly explain that the rate determining step for aromatic electrophilic substitution involves the formation of a resonance stabilised sigma complex intermediate. Give a mechanistic equation and draw limiting forms to represent the intermediate (as in Example 10.1 above).

Activation and *ortho/para* direction results from stabilisation of the *ortho* and *para* sigma complexes by an electron releasing substituent. Deactivation with *meta* direction is the result of an electron attracting substituent destabilising the *ortho* and *para* sigma complexes more than the *meta* sigma complex.

The OH group in phenol is a powerfully activating and *ortho/para* directing group. Any deactivation from the negative inductive effect of this group is completely swamped by the ability of the oxygen atom to release electrons by a positive mesomeric effect to a ring carbon atom carrying positive charge:

Similar limiting forms may be drawn
for *ortho*, but not for *meta* substitution.

In molecular orbital terms, this is seen as sideways interaction (overlap) of an oxygen 2p orbital containing the non-bonded electron pair with the 2p orbital of the ring carbon atom. The delocalised $\pi$ molecular orbital is extended to include the oxygen atom, which takes a share of the positive charge.

As a result of this additional stabilisation of the sigma complex intermediate, phenol reacts with electrophilic reagents under much milder conditions than are possible with benzene or chlorobenzene. Mononitration occurs with dilute nitric acid at low temperature; with concentrated nitric acid some 2,4,6-trinitrophenol is formed, but in both cases yields are poor and some oxidation is unavoidable. Monochlorination and monobromination can be achieved with a solution of the halogen in $CCl_4$; in aqueous solution the 2,4,6-trihalogenophenol rapidly precipitates out. Phenol also reacts with electrophiles which are too weak to attack benzene or chlorobenzene. Nitrous acid (the nitrosyl cation, $NO^{\oplus}$) gives *para* nitrosophenol, and in alkaline solution the phenoxide ion rapidly 'couples' with benzene diazonium cation

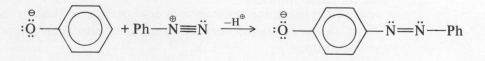

186

Increased activation in the phenoxide ion is the result of the powerful positive mesomeric effect of the negatively charged oxygen atom. Sharing of non-bonded electrons with the *ortho* or *para* ring carbon atoms is unopposed by a negative inductive effect, and without the consequent acquisition of positive charge which occurs with the unionised phenol molecule. Draw the relevant limiting forms to illustrate this:

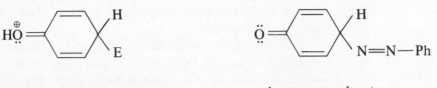

positive charge on oxygen            charges cancel out

Similar electronic effects operate with the chlorosubstituent in chlorobenzene, but with one important difference. Overlap of the larger 3p orbital of the chlorine atom with the ring carbon 2p orbital is much less efficient, with the result that the positive mesomeric effect is weaker than the negative inductive effect. Consequently, although the chloro substituent is *ortho/para* directing, it is also weakly deactivating. Chlorobenzene undergoes nitration, sulphonation, halogenation and the Friedel–Crafts reaction under conditions little different to those used for benzene. In contrast with phenol, it is not halogenated in aqueous solution, nor does it react with nitrous acid or aqueous solutions of diazonium salts.

## Example 10.5

(a) What is meant by the terms kinetic control and thermodynamic control of product formation? (*4 marks*)

(b) Give a brief account of the mechanism of aromatic electrophilic substitution, including an explanation of why the nitro group is *meta*-directing and deactivating, and why the amino-group is *ortho/para*-directing and activating. (*12 marks*)

(c) If toluene (methylbenzene) is treated with isopropyl chloride (($CH_3)_2CHCl$) and aluminium chloride at 25°C, the isopropyl group is introduced largely into the *ortho*- and *para*-positions of the toluene. At higher temperatures, however, the reaction leads entirely to *meta*-isopropyl toluene (*m*-cymene). Why does the toluene methyl group obey normal substituent-effect rules at 25°C, but apparently ignore them at higher temperatures? (*4 marks*)

(University of Sussex, 1983)

*Solution 10.5*

(a) If a reactant or reactants can give rise to two or more products, and the one which predominates is that which is formed the fastest, then the process is said to be under kinetic control. When the major product is the most stable, then the reaction is under thermodynamic control.

A necessary condition for competition between the two forms of control is that the products should be interconvertible, either directly, or via reversion to reactants, i.e. the *reactions must be reversible*. Thus, in a reaction under kinetic control, the product ratio is determined by the difference in energies of activation for the formation of each product. In a reaction under thermodynamic control, the ratio of products is determined by the relative thermodynamic stabilities of the products under the conditions of the reaction. For a more detailed account of this topic see Example 9.3.

(b) The second part of this question is covered in Example 10.2. You should outline the mechanism, e.g. by using the symbol $E^{\oplus}$ to represent an electrophile, and then go on to discuss the effect that the nitro- and amino-substituents have on the stability of the sigma complexes for *ortho/para* and *meta* substitution.

At low temperatures the reaction is under kinetic control. The *ortho* and *para* products predominate because activation energies for formation of the *ortho* and *para* sigma complexes are less than that for formation of the *meta* sigma complex.

Alkylation is a reversible process, and at higher temperatures equilibrium conditions prevail. Differences in activation energies are less significant, and all three isomers could readily be formed. The fact that the *meta* isomer is the sole product must mean that this compound is thermodynamically more stable than the other two. (Theoretical calculations confirm this, and in other experiments there is good agreement between observed product ratios and those predicted from calculated thermodynamic stabilities of the three isomers.)

(c) Finally, although there are only four marks for this section, you could round off your answer with an energy profile diagram showing the relationships between the various isomers.

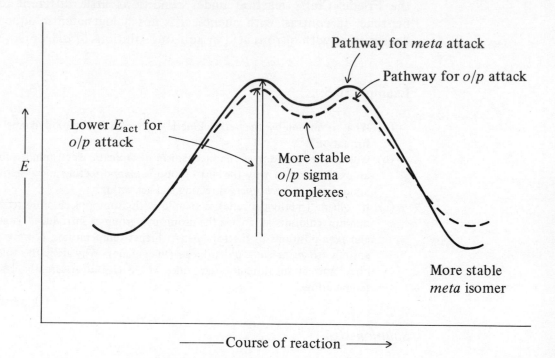

## 10.10 Self-Test Questions

### Question 10.1

How and under what conditions is benzene converted into nitrobenzene in the laboratory?
*(2 marks)*

Outline briefly the mechanism of this reaction. *(5 marks)*

Discuss, in terms of modern theory, the factors which influence the orientation of the product(s) formed in mononitration of EITHER toluene OR nitrobenzene. *(5 marks)*

Starting from benzene or toluene, devise methods for the synthesis of

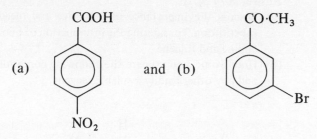

(a) and (b)

(2 × 4 marks)
(Kingston Polytechnic, 1985)

## Question 10.2

Account for the following observations.

(a) Low temperature nitration of an equimolar mixture of toluene and nitrobenzene gives mainly 2- and 4-nitrotoluenes.

(b)

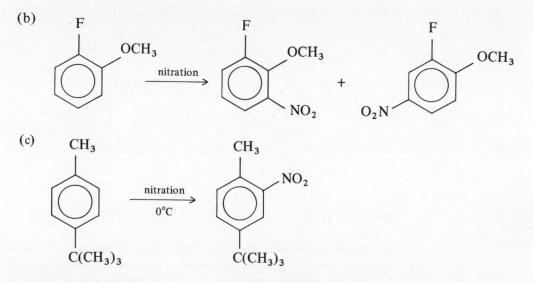

(c)

(d) In mononitration of halobenzenes the *ortho:para* ratio of the products increases from approximately 1:7 in the fluoro compound to approximately 2:3 in the iodo compound.

(Teesside Polytechnic, 1985)

## Question 10.3

Answer *both* parts.

(i) Predict the most favourable position (or positions) for monobromination of $C_6H_5X$, where $X = NO_2$, $NHCOCH_3$, $NH_3^{\oplus}$ and F. In each case explain your reasoning and indicate whether you would expect the rate of substitution to be greater or less than that for benzene.

(ii) Suggest a synthesis of (I) from benzene, giving only essential experimental details

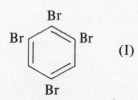

(I)

(University of Durham, 1984)

**Question 10.4**

Answer *both* parts.
(a) Discuss the importance of inductive and mesomeric effects in electrophilic aromatic substitution, considering the bromination of benzoic acid, chlorobenzene, N,N-dimethyl-aniline and toluene.
(b) How would you prepare the following compounds from the indicated starting materials and any other readily available reagent?

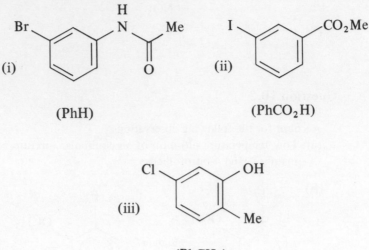

(i)

(PhH)

(ii)

(PhCO$_2$H)

(iii)

(PhCH$_3$)

(University of Southampton, 1985)

**Question 10.5**

Outline a synthesis for *each* of the following compounds by giving a reaction scheme in each case. Use benzene or toluene as your *principal* starting material. Mechanistic details are not required, but reagents and an indication of the reaction conditions should be included in your answer. The minimum number of steps for each synthesis is shown in brackets.

(a)                                   (b)                          (c)            (d)

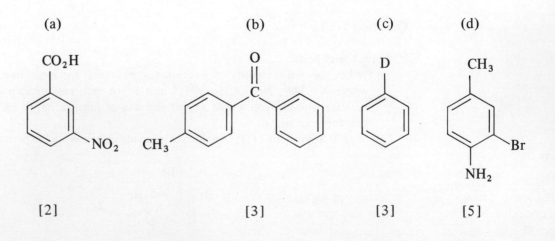

[2]                                   [3]                          [3]            [5]

(University of Exeter, 1985)

190

## Question 10.6

How would you synthesise FOUR of the following compounds, starting from benzene or toluene as seems appropriate? Give reagents and conditions: mechanisms are not required.

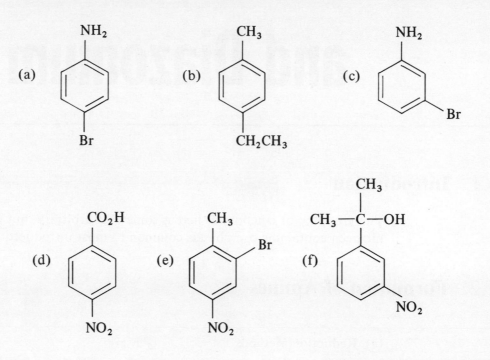

(University of Leeds, 1984)

# 11 Amines, Amino Acids and Diazonium Salts

## 11.1  Introduction

The grouping of compounds here is somewhat arbitrary, but it brings together the nitrogen containing compounds common to most introductory courses.

## 11.2  Formation of Amines

### (a)  Reduction Methods

(i)  $ArNO_2 \xrightarrow[\text{(2) NaOH}]{\text{(1) Fe/HCl}} ArNH_2$

(ii)  $RCN \xrightarrow{[H]} RCH_2NH_2$

e.g. $LiAlH_4$ or Fe/HCl

(iii)  $RCHO \xrightarrow[-H_2O]{NH_3} RCH{=}NH \xrightarrow[\text{T, P}]{H_2/Ni} RCH_2NH_2$

(an imine)

R is alkyl or aryl

### (b)  Replacement of Halogen in Haloalkanes

(i)  $RX + \text{excess } NH_3 \longrightarrow R\overset{\oplus}{N}H_3\overset{\ominus}{X} \xrightarrow{HO^{\ominus}} RNH_2$

R = 1° or 2° (3° haloalkanes eliminate HX)

Limited by formation of $R_2NH$, $R_3N$ and $R_4\overset{\oplus}{N}\overset{\ominus}{X}$

(b)

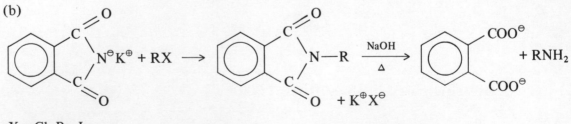

X = Cl, Br, I

**Gabriel synthesis**: gives 1° amines free of 2° and 3° amines.

192

(iii) Degradation of amides

$$RCONH_2 \xrightarrow{Br_2/HO^\ominus} RNH_2$$

**Hofmann rearrangement**

## 11.3 Formation of Amino Acids

**(a) Substitution Methods**

(i) $RCH_2COOH \xrightarrow{PX_3 \mid X_2} RCH(X)COOH \xrightarrow{NH_3} RCH(NH_2)COOH$

X = Cl or Br, **Hell–Volhard–Zelinsky** reaction

(ii) Modification of the Gabriel synthesis

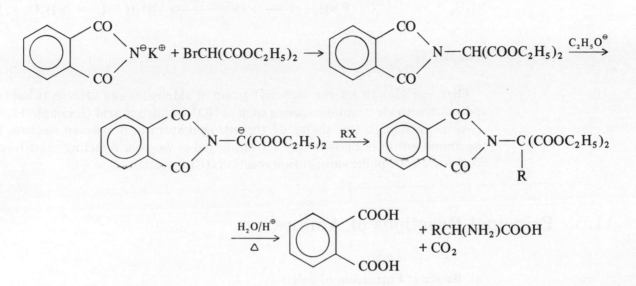

**(b) Reductive Amination**

$$RCOCOOH \xrightarrow{NH_3} \underset{\underset{NH}{\overset{\|}{}}}{RCCOOH} \xrightarrow{[H]} RCH(NH_2)COOH$$

**(c) Strecker Synthesis**

$$RCHO + NH_3 \longrightarrow RCH\!\!\begin{array}{c} OH \\ NH_2 \end{array} \xrightarrow{-H_2O} RCH=\!\!NH \xrightarrow{HCN} \underset{\underset{CN}{\overset{|}{}}}{RC-NH_2} \xrightarrow[\Delta]{H_2O/H^\oplus} RCH(NH_2)COOH$$

## 11.4 Formation of Diazonium Salts

Aryldiazonium salts are prepared by the action of 'nitrous acid' on a primary aromatic amine:

$$Ar-NH_2 + HONO + HCl \xrightarrow{5-10^\circ C} Ar-\overset{\oplus}{N}\equiv N\ \overset{\ominus}{Cl} + 2H_2O$$

193

## 11.5 Structure and Reactivity of Amines

Basic properties of amines are covered in Chapter 4.

The amino group (even when protonated) is a poor leaving group, and there are very few reactions in which it is directly displaced by a nucleophile. One example is thermal decomposition of tetramethylammonium bromide, where $(CH_3)_3N$ is a better leaving group:

$$:\overset{\ominus}{\ddot{B}r}: \quad CH_3 \overset{\oplus}{\frown} \overset{\oplus}{N}(CH_3)_3 \longrightarrow :\ddot{B}rCH_3 + \ddot{N}(CH_3)_3$$

Amines are more nucleophilic than alcohols, and they readily displace halogen from haloalkanes and acyl halides:

$$R\overset{\frown}{\ddot{N}H_2} \quad \overset{|}{\underset{|}{-C}} \overset{\frown}{-\ddot{X}}: \quad R\overset{\oplus}{\ddot{N}}H_2 \overset{|}{\underset{|}{-C-}} + :\ddot{X}:^\ominus \quad \xrightarrow{H\ddot{O}:^\ominus} \quad R\ddot{N}H \overset{|}{\underset{|}{-C-}} \quad (+H_2\ddot{O} + :\ddot{X}:^\ominus)$$

(Chapter 5)

They can also attack the carbonyl group of aldehydes and ketones (Chapter 7) and react with electrophilic species such as $NO^\oplus$ in nitrous acid (Example 11.3).

Again related to the ability of the nitrogen atom to donate an electron pair, the amino group is a powerfully activating and *ortho-para* directing substituent in aromatic electrophilic substitution reactions (Chapter 10).

## 11.6 Principal Reactions of Amines

**(a) Basicity: Formation of Salts**

$$RNH_2 + HX \longrightarrow \overset{\oplus}{R}NH_3 \overset{\ominus}{X}$$

(Chapter 4)

**(b) Reaction as Nucleophiles**

(i) *With Haloalkanes (N-alkylation)*

$$RNH_2 + R'X \longrightarrow \overset{\oplus}{R}NH_2R'\overset{\ominus}{X} \xrightarrow[-HX]{HO^\ominus} RNHR'$$

X = Cl, Br, I

$1° \rightarrow 2° \rightarrow 3° \rightarrow$ quaternary ammonium salts

Also aryl and mixed alkyl–aryl amines

(ii) *With Acyl Halides (N-acylation – formation of acid amides)*

1. $RNH_2 + RCOCl \longrightarrow RCONHR + HCl$
2. $R_2NH + RCOCl \longrightarrow RCONR_2 + HCl$ (initially as salts)

3° amines do not react.

(iii) With Carbonyl Compounds

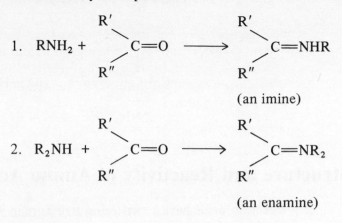

$$1. \quad RNH_2 + \underset{R''}{\overset{R'}{>}}C=O \longrightarrow \underset{R''}{\overset{R'}{>}}C=NHR$$

(an imine)

$$2. \quad R_2NH + \underset{R''}{\overset{R'}{>}}C=O \longrightarrow \underset{R''}{\overset{R'}{>}}C=NR_2$$

(an enamine)

Aliphatic imines are unstable and tend to polymerise. Aromatic imines (Schiff bases) are more stable.

## (c) Reaction with Nitrous Acid

(i) $RNH_2 \xrightarrow{\text{HONO}} [R\overset{\oplus}{N}\equiv N] \longrightarrow N_2 + R^{\oplus} \longrightarrow$ various products

(ii) $ArNH_2 \xrightarrow{\text{HONO}} Ar\overset{\oplus}{N}\equiv N$ (diazonium salts)

(iii) $R_2NH \xrightarrow{\text{HONO}} R_2N-NO$ (N-nitrosamine)

also aromatic 2° amines

(iv)

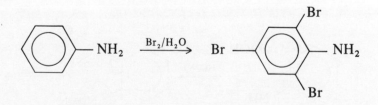

(a C-nitrosamine)

## (d) Electrophilic Substitution in Aromatic Amines

The —NH₂ (—NHR, —NR₂) group is powerfully activating and *ortho/para* directing, e.g.

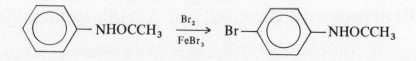

The acetamino group, CH₃CONH—, is more moderately activating, e.g.

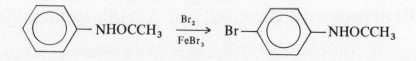

(Chapter 10.)

### (e) Elimination from Quaternary Ammonium Salts (Hofmann Elimination)

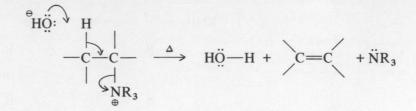

## 11.7 Structure and Reactivity of Amino Acids

α-Amino acids have a **zwitterion** structure in which the amino group is protonated by the carboxyl group:

$$\overset{\oplus}{N}H_3 — CHR — CO\overset{\ominus}{O}$$

They can thus function as both an acid and a base, i.e. are **amphoteric**. The $NH_3^{\oplus}$ group is slightly more acidic than the $COO^{\ominus}$ group is basic. Equilibrium in aqueous solution therefore results in a preponderance of anions:

$$\overset{\oplus}{N}H_3 — CHR — CO\overset{\ominus}{O} + H_2O \rightleftharpoons NH_2 — CHR — CO\overset{\ominus}{O} + H_3\overset{\oplus}{O}$$

Addition of acid displaces the equilibrium to the left, and the pH at which an amino acid carries no net charge is called the **isoelectric point**.

## 11.8 Principal Reactions of Amino Acids

α-Amino acids undergo many of the reactions characteristic of aliphatic carboxylic acids and amines. For example, the carboxyl group can be esterified and on treatment with nitrous acid an α-amino acid forms the corresponding α-hydroxyacid. The amino group can also be acylated, and the amide link between the amino group of one α-amino acid and the carboxyl group of another (similar or different) amino acid is called a **peptide bond**. The chemistry of peptides and proteins is beyond the scope of this text.

Aromatic amino acids exhibit most of the normal reactions of the amino and carboxyl groups. Diazotisation of *ortho*-aminobenzoic acid gives a zwitterion which on heating eliminates nitrogen and carbon dioxide to form **benzyne**, one of a series of **aryne intermediates**:

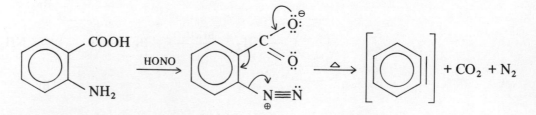

## 11.9 Structure and Reactivity of Aryldiazonium Salts

The instability of diazonium salts is related to the good leaving ability of the nitrogen molecule. There are two types of reaction, those involving loss (replacement) of nitrogen, and those in which the nitrogen is retained.

### (a) Reactions Involving Loss of Nitrogen

Many reactions (such as the formation of phenols) proceed via an intermediate aryl carbocation in what is essentially an $S_N1$ reaction:

$$Ar \overset{\curvearrowright}{N} \overset{\oplus}{\equiv} \ddot{N} \xrightarrow{\text{slow}} \ddot{N} \equiv \ddot{N} + Ar^{\oplus} \xrightarrow[\text{fast}]{\overset{\ddot{Y}^{\ominus}}{}} Ar-Y$$

where $\ddot{Y}^{\ominus}$ is a nucleophile, e.g. $H\ddot{\underset{..}{O}}:$ ($H_2\ddot{O}$), $\overset{\ominus}{C}\overset{..}{N}$, etc.

However, in other cases reaction may be more complicated, involving a *free radical* mechanism.

### (b) Reactions with Retention of Nitrogen

The diazonium cation $Ar-\overset{\oplus}{N}\equiv N$ is a weak electrophile, and as such will attack activated aromatic substrates – '**coupling reactions**'. Thus there is no reaction with benzene itself, but coupling is virtually instantaneous when benzenediazonium chloride is added to an alkaline solution of phenol:

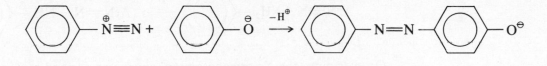

## 11.10 Principal Reactions of Aryldiazonium Salts

### (a) Replacement Reactions

These are summarised in Table 11.1.

**Table 11.1** Replacement reactions of aryldiazonium salts, $Ar\overset{\oplus}{N_2} \overset{\ominus}{X}$

| Displacing group | Reaction conditions | Product |
|---|---|---|
| H | Warm with $H_3PO_2$ | Ar—H* |
| OH | Heat acidified solution | Ar—OH |
| Cl or Br | Heat with CuCl or CuBr (Sandmeyer reaction) | Ar—Cl |
| I | Heat with aq. KI solution | Ar—I |
| F | Heat with $HBF_4$ (Balz–Schiemann reaction) | Ar—F |
| $NO_2$ | Heat with $HBF_4$ + $NaNO_2$ | Ar—$NO_2$ |
| CN | Heat with CuCN/KCN | Ar—CN |

*'Deamination' – useful where an $NH_2$ group initially required as an activating *ortho-para* directing substituent.

### (b) Coupling Reactions

(i) *C-coupling*

$$\overset{\oplus}{ArN}\!\!\equiv\!\!N \ + \ \langle\bigcirc\rangle\!-\!S \ \longrightarrow \ ArN\!\!=\!\!N\!-\!\langle\bigcirc\rangle\!-\!S$$

where S is an activating substituent such as $-NH_2$, $-NHR$, $-OH$ ($\overset{\ominus}{O}$), etc.
Reaction is useful for the synthesis of
1. azo dyes
2. arylamines by reduction (Sn/HCl) of the azo compound

$$ArN\!\!=\!\!N\!-\!\langle\bigcirc\rangle\!-\!S \ \xrightarrow{[H]} \ ArNH_2 \ + \ NH_2\!-\!\langle\bigcirc\rangle\!-\!S$$

(ii) *N-coupling*

$$\overset{\oplus}{ArN}\!\!\equiv\!\!N \ + \ NH_2\!-\!\langle\bigcirc\rangle \ \xrightarrow{-H^{\oplus}} \ ArN\!\!=\!\!N\!-\!NH\!-\!\langle\bigcirc\rangle$$

On warming, the N-(arylazo)aminobenzene so formed readily rearranges to the more stable C-coupled azo compound:

$$ArN\!\!=\!\!N\!-\!NH\!-\!\langle\bigcirc\rangle \ \longrightarrow \ ArN\!\!=\!\!N\!-\!\langle\bigcirc\rangle\!-\!NH_2$$

## 11.11 Worked Examples

**Example 11.1**

The reaction of an alkyl halide (RHal) with ammonia gives rise to a mixture of the compounds $RNH_2$, $R_2NH$, $R_3N$ and $\overset{\oplus}{R_4N}$ $Hal^{\ominus}$. Reaction of RHal with an alkylamine $RNH_2$ similarly yields a mixture of $RNHR$, $R_2NR$ and $\overset{\oplus}{R_3NR}Hal^{\ominus}$.
(a) Explain why and how these mixtures of products are formed.
(b) Suggest a procedure for the conversion of bromoethane to ethylamine without the concurrent formation of secondary and tertiary amines or substituted ammonium salts.
(c) Show how the following transformation could be effected (note: more than one step is necessary)

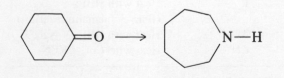

(University of Bradford, 1984)

*Solution 11.1*

(a) When heated with concentrated aqueous ammonia under pressure, primary and secondary haloalkanes undergo bimolecular nucleophilic substitution:

$$\overset{..}{N}H_3 \quad R\overset{}{\longrightarrow}\overset{..}{X}: \quad\longrightarrow\quad \overset{\oplus}{N}H_3-R\overset{\ominus}{\underset{..}{X}}:$$

(With tertiary haloalkanes only elimination products are obtained.)

The first-formed product, an amine salt, can undergo proton exchange with another ammonia molecule:

$$R\overset{\oplus}{N}H_3 \; :\overset{..}{\underset{..}{X}}:^{\ominus} + \overset{..}{N}H_3 \quad\longrightarrow\quad \overset{\oplus}{N}H_4 \; :\overset{\ominus}{\underset{..}{X}}: + R\overset{..}{N}H_2$$

and as the 'free' amine is itself nucleophilic, further reaction with another haloalkane molecule is possible:

$$R\overset{..}{N}H_2 \quad R\overset{}{\longrightarrow}\overset{..}{X}: \quad\longrightarrow\quad R\overset{\oplus}{N}H_2R \; :\overset{..}{\underset{..}{X}}:^{\ominus}$$

Repetition of these steps with the secondary amine salt leads to the formation of $R_3\overset{\oplus}{N}H \; :\overset{\ominus}{\underset{..}{X}}:$ and $R_4\overset{\oplus}{N} \; :\overset{\ominus}{\underset{..}{X}}:$.

If ammonia is replaced by $R\overset{..}{N}H_2$, the same sequence of reactions will result in a similar mixture of products.

You could mention here that to obtain the mixture of free amines, the salts would have to be neutralised, e.g. by addition of NaOH solution. (The use of a large excess of ammonia will give the primary amine as predominant product, but otherwise the reaction is of little value.)

(b) The required procedure is the Gabriel synthesis. Although the question does not specifically ask for the mechanism, you would not lose marks for showing that you understand the principles involved, as well as remembering the practical steps!

The imino group in phthalimide is weakly acidic because loss of a proton gives a resonance stabilised anion:

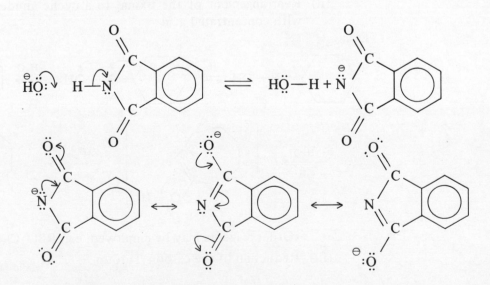

Although the negative charge is shared by the carbonyl oxygen atoms, the anion behaves as though the charge was localised on the nitrogen atom, and is therefore a powerful nucleophile. Reaction of the potassium salt of phthalimide with bromoethane gives N-ethylphthalimide:

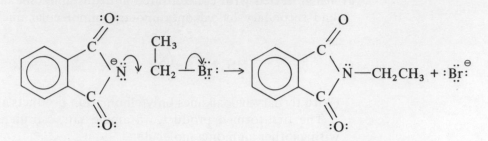

On warming the reaction mixture with aqueous NaOH, hydrolysis occurs, giving good yields of ethylamine uncontaminated by the various products which result from the reaction of bromoethane with ammonia

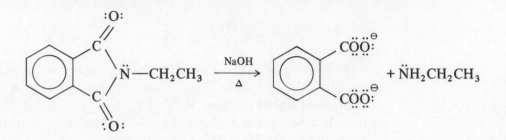

(c) The transformation in part (c) makes use of the Beckmann rearrangement of oximes. Not all first year courses cover this topic, so don't be put off if it is not familiar to you! The necessary steps are as follows:

(i) Formation of cyclohexanone oxime

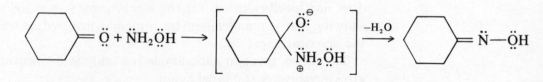

(ii) Rearrangement of the oxime to a cyclic amide (lactam) by warming with concentrated acid

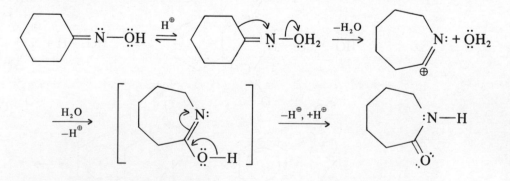

(Other reagents may be employed, e.g. $RSO_2Cl$, $PCl_5$ etc.)

(iii) Reduction of the carbonyl group

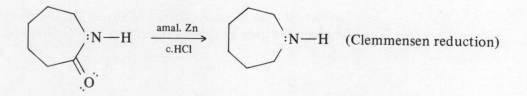

**Example 11.2**

Describe methods for distinguishing between primary, secondary and tertiary amines. Give details of how you would prepare compounds (ii), (iii) and (iv) from aniline (i)

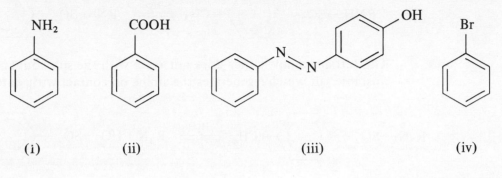

| (i) | (ii) | (iii) | (iv) |

(University of Leeds, 1984)

*Solution 11.2*

The simplest method of distinguishing between primary, secondary and tertiary amines is by their IR spectra. Primary amines exhibit two absorption peaks (near 3400 cm⁻¹ and 3500 cm⁻¹) due to N—H stretching, secondary amines have one peak (near 3350 cm⁻¹). Tertiary amines, having no N—H bonds do not absorb in this region. Furthermore, aromatic compounds are readily distinguished from aliphatic by characteristic peaks (near 1500 cm⁻¹ and 1600 cm⁻¹) due to C═C stretching in the aromatic nucleus.

Of course, an IR spectrometer may not be available in the laboratory, but there are a number of chemical tests which may give useful information.

(i) Acylation

Primary and secondary, but not tertiary, amines are readily acylated, e.g. with $CH_3COCl$ in pyridine:

$$RNH_2 + CH_3COCl \longrightarrow RNHCOCH_3 + HCl$$

Reaction is exothermic and therefore easily detected by a rise in temperature.

(ii) Sulphonylation

Sulphonyl halides such as *para*-toluene sulphonyl chloride react with primary and secondary amines to form (after neutralisation) the corresponding sulphonamides:

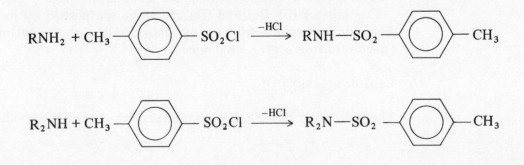

The two sulphonamides may be distinguished by shaking with cold aqueous NaOH solution. No reaction occurs with the secondary derivative, but the primary compound forms a soluble salt

$$RNH-SO_2-\underset{}{\bigcirc}-CH_3 \xrightleftharpoons{HO^{\ominus}} R\overset{\ominus}{N}-SO_2-\underset{}{\bigcirc}-CH_3 + H_2O$$

The tertiary amine either does not react with the sulphonyl chloride, or forms an unstable salt which regenerates the amine on contact with water

$$\underset{\underset{Cl^{\ominus}}{}}{R_3\overset{\oplus}{N}}-SO_2-\underset{}{\bigcirc}-CH_3 \xrightleftharpoons{H_2O} R_3N + HO-SO_2-\underset{}{\bigcirc}-CH_3 + H^{\oplus}Cl^{\ominus}$$

(iii) Reaction with Nitrous Acid ($NaNO_2 + HCl$)

(a) Most aliphatic primary amines are water soluble, and addition to cold nitrous acid solution is accompanied by vigorous evolution of nitrogen

$$RNH_2 \xrightarrow{HONO} [R\overset{\oplus}{N}\equiv N] \longrightarrow N_2 + [R^{\oplus}] \longrightarrow \text{various products}$$

Aromatic primary amines are not very soluble in water, but react with cold nitrous acid to give a soluble diazonium salt which is stable below 5°C.

$$ArNH_2 \xrightarrow{HONO} Ar\overset{\oplus}{N}\equiv N\ Cl^{\ominus}$$

(b) Aliphatic and aromatic secondary amines both form insoluble N-nitroso compounds with nitrous acid

$$R_2NH \xrightarrow{HONO} R_2N-N=O$$

Since most aliphatic secondary amines are fairly soluble in water, the separation of yellow oily drops of nitrosoamine is usually easy to detect. However, the test is less convincing with insoluble aromatic secondary amines.

(c) Tertiary amines are usually described as showing no reaction with aqueous nitrous acid. They may in fact form unstable nitroso salts, $R_3\overset{\oplus}{N}-N=O\ \overset{\ominus}{Cl}$, but the reaction is of little diagnostic value.

The next part of the question requires you to give details, i.e. describe what you would do, as well as outlining the reactions involved.

Compounds (ii), (iii) and (iv) can all be synthesised by reactions of benzene-diazonium chloride obtained by diazotisation of phenylamine (aniline) with aqueous nitrous acid (see Example 11.4).

Benzoic acid (ii)

The diazonium salt solution is warmed with copper(I) cyanide (cuprous cyanide)

$$\underset{}{\bigcirc}-\overset{\oplus}{N}\equiv N \xrightarrow{CuCN} \underset{}{\bigcirc}-CN + N_2$$

Benzonitrile can be solvent extracted, and after removal of the solvent, it is hydrolysed by boiling with acid or alkali. (In the latter case, the solution would be neutralised to liberate the free acid.)

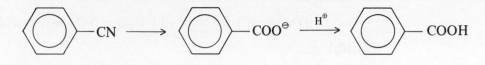

### Phenylazophenol (iii)

The benzenediazonium cation is only weakly electrophilic, but the negatively charged oxygen atom in the phenoxide ion is a powerfully activating substituent. Therefore on adding (slowly, with stirring) a solution of benzenediazonium chloride to an alkaline solution of phenol, the resulting 'coupling reaction' is virtually instantaneous:

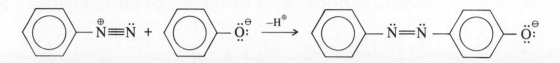

The bright red precipitate of the azo compound should be treated with acid before filtering off, since —OH and not —$\overset{\ominus}{O}\overset{\oplus}{Na}$ is required.

### Bromobenzene (iv)

A benzenediazonium chloride solution is warmed with copper(I) bromide (cuprous bromide), following which it is made alkaline and steam distilled, when bromobenzene will form a separate layer in the distillate.

$+ N_2$  (Sandmeyer reaction)

### Example 11.3

Give a mechanistic account of the reaction of nitric(III) acid (nitrous acid) with aliphatic and aromatic amines. Briefly indicate the importance of the diazotisation reaction and outline an industrial synthesis of N-(4-ethoxyphenyl)ethanamide (phenacetin) (Fig. 1), from 4-ethoxyphenylamine and phenol.

$$CH_3CONH \text{—} \bigcirc \text{—} NHCOCH_3$$

**Figure 1**

(Coventry Polytechnic, 1986)

*Solution 11.3*

Begin by pointing out that, depending on the pH, various electrophilic species may be present in an aqueous solution of nitrous acid ($NaNO_2$ + HCl):

(1) low acidity

$$2 \text{ HONO} \rightleftharpoons H_2O + N_2O_3 \text{ (dinitrogen trioxide)}$$

(2) moderate acidity

$$HONO + HCl \rightleftharpoons H_2O + Cl-NO \text{ (nitrosyl chloride)}$$

(3) higher acidity

$$HONO + HCl \rightleftharpoons Cl + H_2\overset{\oplus}{O}-NO \text{ (protonated acid)}$$

(4) high acidity

$$H_2\overset{\oplus}{O}-NO \rightleftharpoons H_2O + NO^{\oplus} \textbf{ (nitrosyl cation)}$$

Electrophilic strength increases in the series:

$$N_2O_3 < NO-Cl < NO-\overset{\oplus}{O}H_2 < NO^{\oplus}$$

For simplicity, you could refer to the reaction of amines with the nitrosyl cation, $NO^{\oplus}$.

The chemistry of amines is characterised by the ability of the nitrogen atom to donate an electron pair. Thus with $NO^{\oplus}$ the following reaction is possible:

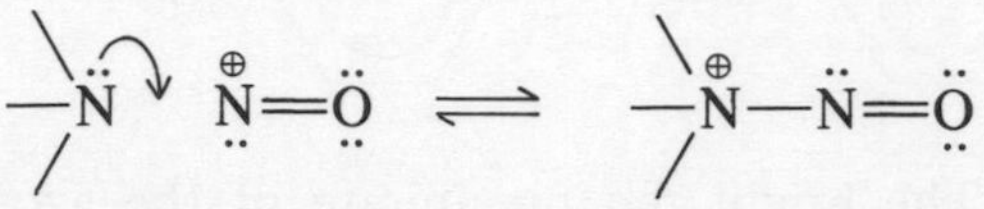

Further course of the reaction is determined by the nature of the amine.

(i) Tertiary aliphatic amines have no means of 'getting rid of' the positive charge, and so either uptake of $NO^{\oplus}$ is reversed ('no reaction'), or the nitroso salt decomposes giving a complex mixture of products.

Tertiary aromatic amines are sterically hindered to attack on the nitrogen atom, which would also involve loss of resonance energy associated with conjugation of the amino group with the aromatic nucleus. Instead, electrophilic substitution occurs at the *para* position to give a C-nitroso compound. Although $NO^{\oplus}$ is a relatively weak electrophile, the dimethylamino group is powerfully activating towards electrophilic substitution:

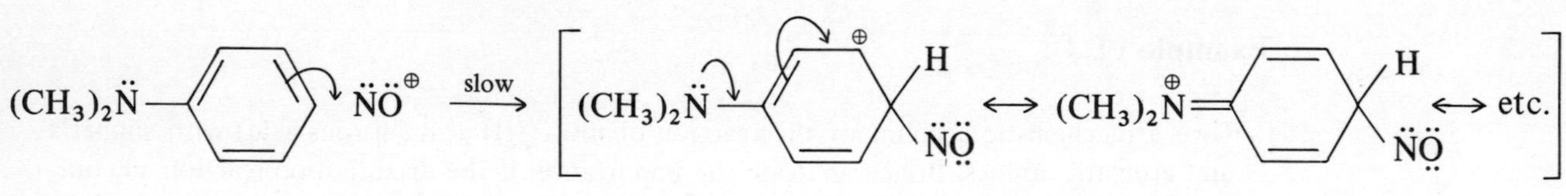

(resonance stabilised sigma complex)

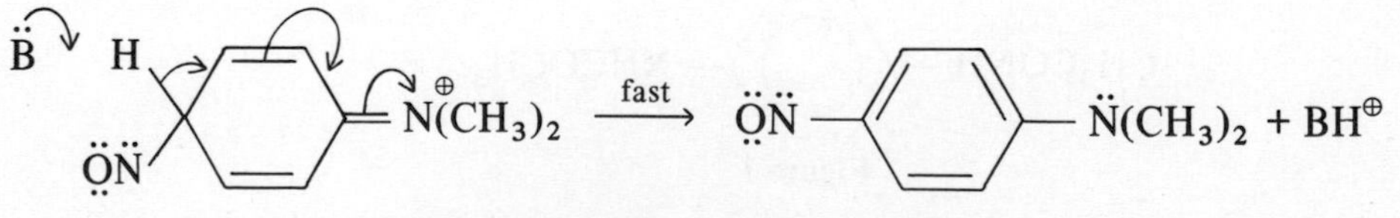

(ii) With secondary amines (aliphatic and aromatic) reaction is completed by loss of a proton to form an N-nitroso compound, e.g.

(iii) Primary amines undergo rearrangement with loss of $H_2O$ to form diazonium salts:

$$R-\overset{..}{N}H_2 \quad \overset{\oplus}{\overset{..}{N}}=\overset{..}{\overset{..}{O}} \longrightarrow R-\overset{\oplus}{\overset{..}{N}}H_2-\overset{..}{N}=\overset{..}{\overset{..}{O}} \xrightarrow{-H^{\oplus}} R-\overset{..}{N}H-\overset{..}{N}=\overset{..}{\overset{..}{O}}$$

$$R-\overset{\overset{\displaystyle H}{|}}{\underset{..}{N}}-\overset{..}{N}-\overset{..}{\overset{..}{O}}: \xrightarrow{-H^{\oplus},\,+H^{\oplus}} R-\overset{..}{N}=\overset{..}{N}-\overset{..}{\overset{..}{O}}H \xrightarrow{+H^{\oplus}} R-\overset{..}{N}=\overset{..}{N}-\overset{\oplus}{\overset{..}{O}}H_2$$

$$R-\overset{..}{N}=\overset{..}{N}-\overset{\oplus}{\overset{..}{O}}H_2 \xrightarrow{-H^{\oplus},\,-H_2O} R-\overset{\oplus}{N}\equiv\overset{..}{N}$$

Where R = alkyl, the diazonium cation immediately loses nitrogen and various products are formed by reaction of the carbocation $R^{\oplus}$

$$R-\overset{\oplus}{N}\equiv\overset{..}{N} \longrightarrow N_2 + R^{\oplus} \longrightarrow \text{various products, e.g. ROH, RCl.}$$

Aromatic diazonium cations (R = aryl) are more stable, at least at temperatures below 5–10°C, and their formation is referred to as '**diazotisation**'.

Aryldiazonium salts undergo two types of reaction, which are important both industrially and in the laboratory for the synthesis of various products.

## (1) Coupling Reactions

Electrophilic substitution with activated substrates — aromatic amines and phenols. Important for the manufacture of azo dyes and indicators

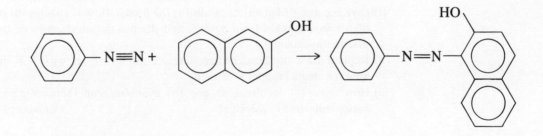

## (2) Displacement Reactions

A whole range of substitution products are formed with the loss of nitrogen

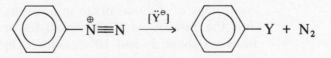

Y = Cl, Br, CN (Sandmeyer, heat with CuX), I (heat with concentrated aqueous KI), F (Balz–Schiemann, heat with $HBF_4$), $NO_2$ (heat with $HBF_4$ + $NaNO_2$), etc.

The industrial synthesis of phenacetin utilises the coupling reaction and cleavage of the azo group which occurs on reduction with metal/acid. It would be sufficient to outline the synthesis by means of annotated equations:

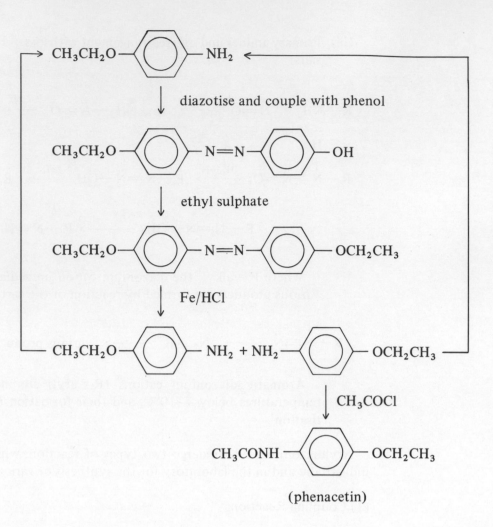

(phenacetin)

**Example 11.4**

(i) Give the essential practical details for the preparation of an aqueous solution of benzene-diazonium chloride from aniline, and discuss the mechanism of the formation of the diazonium cation.

(ii) Explain how the benzenediazonium cation will react with N,N-dimethylaniline, and why the single product is formed.

(iii) How may (a) iodobenzene and (b) cyanobenzene (benzonitrile) be obtained from benzenediazonium chloride? (University of Sheffield, 1985)

*Solution 11.4*

(i) To an ice-cold 'solution' of one equivalent of phenylamine (aniline) in excess 50% hydrochloric acid is added, slowly and with stirring, one equivalent of sodium nitrite in aqueous solution. Reaction is exothermic and the rate of addition should be such as to maintain a temperature between about 5° and 10°C. Only a slight excess of nitrous acid is desirable, so towards the end the mixture should be tested with starch iodide paper (which turns blue with nitrous acid).

Although, depending on the pH, various electrophilic species (such as $N_2O_3$, NOCl, – see Example 11.3) may be present, for simplicity you could illustrate reaction with the nitrosyl cation, $NO^{\oplus}$.

$$2 \; HNO_2 \;\; \rightleftharpoons \;\; NO^{\oplus} + H_2O + NO_2^{\ominus}$$

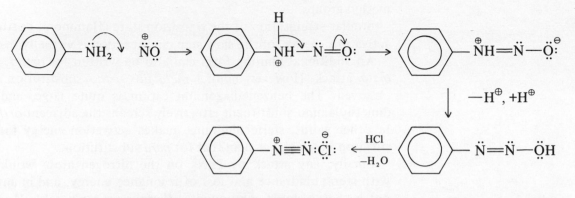

(ii) The benzene diazonium cation is only weakly electrophilic but N,N-dimethylphenylamine (N,N-dimethylaniline) is a powerfully activated substrate. Attack occurs at the *para* position, and reaction follows the normal path for electrophilic substitution:

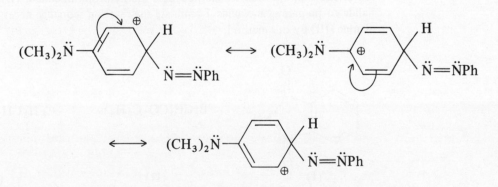

where B could be $H_2\ddot{O}$ or $:\!\ddot{C}l:^{\ominus}$.

The sigma complex intermediate is resonance stabilised, as represented by the limiting forms

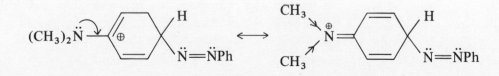

Most importantly, an additional limiting form can be written, involving the dimethylamino group:

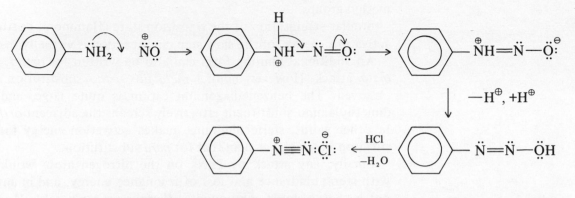

207

This is particularly stable as all atoms (including $N^{\oplus}$) have complete valence octets, and there is additional inductive stabilisation by the two methyl groups.

Similar stabilisation of the transition state (Hammond postulate) lowers activation energy for the slow step, and reaction is virtually instantaneous.

An additional limiting form can also be written for *ortho*, but not for *meta* attack. However, neither *meta* nor *ortho* substitution is normally observed. The benzenediazonium cation is quite large, and the bulky dimethylamino substituent effectively screens the adjacent *ortho* position. In other words, steric hindrance makes activation energy for *ortho* substitution much higher than that for *para* substitution.

Finally, any attack by $PhN_2^{\oplus}$ on the nitrogen atom would also meet with steric hindrance and loss of resonance energy, and in any case could not lead to a stable compound as there is no replaceable H atom present (and $CH_3^{\oplus}$ is not a leaving group).

(iii) The required substitution products can be obtained by warming a solution of benzenediazonium chloride with
(a) concentrated aqueous KI solution and
(b) with copper(I) cyanide (cuprous cyanide – the Sandmeyer reaction).

In both cases it may be necessary to use steam distillation and/or solvent extraction to obtain pure products.

# 11.12 Self-Test Questions

### Question 11.1

How does an alkyl halide react with ammonia;
Why is this not a suitable route for amine preparation?                                 (*10 marks*)
Describe how a primary amine may be separated from a secondary amine with the aid of *para*-toluenesulphonyl chloride.                                 (*8 marks*)
Outline the preparation of pure ethylamine using phthalimide (I) and bromoethane via the Gabriel synthesis.                                 (*7 marks*)
A variation of the Gabriel method uses diethylbromomalonate (II) in place of the alkyl halide to prepare amino acids. Formulate the reaction sequence necessary to prepare phenylalanine (III) by this method.

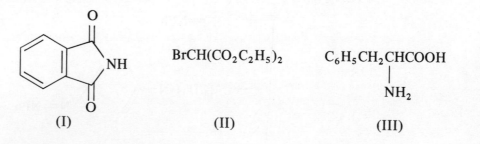

(I)                    (II)                    (III)

(*8 marks*)
(Thames Polytechnic, 1984)

## Question 11.2

(a) Describe the preparation of an aqueous solution of an aryldiazonium salt. *(5 marks)*

(b) Outline the important features of diazonium coupling reactions. *(5 marks)*

(c) Outline the synthesis of the following compounds starting from benzene.

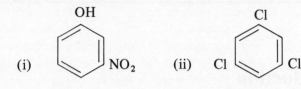

(i)    (ii)

*(2 × 5 marks)*
(The Polytechnic, Huddersfield, 1985)

## Question 11.3

(a) Describe the preparation of an aqueous solution of an aryldiazonium salt. *(5 marks)*

(b) Using any necessary reagents, show how two of the following compounds can be prepared from benzene via a diazonium salt intermediate:

(i) *m*-nitrophenol      (ii) *m*-bromobenzoic acid

(iii) *m*-dibromobenzene     (iv) *p*-methylbenzoic acid     *(6 marks each)*

(c) Show by means of equations how methyl orange (below) can be prepared from aniline (benzenamine) and sulphanilic acid (p-aminobenzene sulphonic acid)

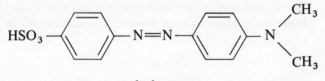

methyl orange

*(4 marks)*
(University of Keele, 1985)

## Question 11.4

Discuss the reactivity of aliphatic and aromatic amines under the following headings:

(a) Basicity. *(6 marks)*

(b) Acidity *(4 marks)*

(c) Nucleophilic character *(6 marks)*

(d) Substrates for nucleophilic substitution and elimination *(4 marks)*

(Coventry Polytechnic, 1987)

# 12 Free Radical Reactions

## 12.1 Introduction

A free radical may be defined as a *high energy species with an unpaired ('odd')
electron*. This chapter covers methods of formation, characteristic reactions, and
illustrative substitution, addition and polymerisation processes.

## 12.2 General Methods of Formation

Radicals may be generated:

(1) **thermally** (application of heat) or
(2) **photolytically** (irradiation with ultraviolet light or bright sunlight).

At a wavelength of 270 nm the quantum size is equivalent to 425 kJ mol$^{-1}$ –
sufficient energy to disrupt most sigma bonds.

The temperature required for thermolysis varies with structure. Chlorine
molecules begin to dissociate above 250°C (523 K), whereas the carbon–carbon
bond in alkanes is only broken homolytically at temperatures above about 475°C
(748 K).

### (a) Use of Initiators

Certain compounds containing relatively weak sigma bonds may undergo homolysis at much lower temperatures. Such compounds may be used to 'trigger off'
radical reactions, when they are known as **initiators**. The most commonly used
compounds are organic peroxides or aliphatic azo compounds. For example,

*Dialkyl peroxides*

$$R-\ddot{O}-\ddot{O}-R \xrightarrow{\Delta} 2\,R-\dot{\ddot{O}}: \qquad (\Delta H \sim 150\,\text{kJ mol}^{-1})$$

e.g. R = (CH$_3$)$_3$C, 'di-tert-butyl peroxide'

Note the use of 'half-headed' arrows. As a reminder, both lone-pairs and unpaired
electrons are represented, but it is customary (and quite acceptable in examinations) to show only the unpaired electron. Thus the alkoxy radical is normally
drawn as RȮ.

*Diacyl peroxides*

$$R-\underset{\underset{\displaystyle :\!O:}{\|}}{C}-\ddot{O}-\ddot{O}-\underset{\underset{\displaystyle :\!O:}{\|}}{C}-R \xrightarrow{\Delta} 2\,R-\underset{\underset{\displaystyle :\!O:}{\|}}{C}-\dot{\ddot{O}}:$$

When R = phenyl, further decomposition occurs fairly rapidly to give carbon dioxide and phenyl radicals:

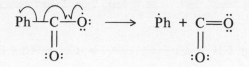

(Displacement of carbon dioxide is energetically favourable because it is a very stable molecule.)

*Aliphatic azo compounds*

$$R\text{—}\ddot{N}\text{=}\ddot{N}\text{—}R \longrightarrow \dot{R} + \ddot{N}\text{≡}\ddot{N} + \dot{R}$$

The 'driving force' is again the expulsion of a very stable species, the nitrogen molecule.

The compound where $R = NC\text{—}C(CH_3)_2$ is known as 'AIBM', azo-bis-iso-butyronitrile (2,2-azo-bis-methylpropanonitrile), a much used initiator in the polymer industry. Homolysis occurs well below 100°C (373 K) and the rate of dissociation, i.e. concentration of initiating radicals, can be controlled to some extent by varying the temperature.

## 12.3    Structure and Reactivity

Most radicals have a planar or nearly planar structure. The carbon atom in alkyl radicals is $sp^2$ hybridised and the unshared electron is contained in the 'spare' 2p atomic orbital oriented at right angles to the plane of the radical:

Although free radicals are neutral, they are *electron deficient species* which react by seeking to 'pair off' the odd electron. This may be achieved in a number of ways.

## 12.4    General Reactions of Free Radicals

### (a)  Radical Pairing

$$\dot{R} \quad \dot{R}' \longrightarrow R\text{—}R'$$

Where R = R, the process is called **dimerisation**, e.g.

$$\dot{C}H_3 + \dot{C}H_3 \longrightarrow CH_3\text{—}CH_3$$

## (b) Radical Transfer

$$\dot{R} \quad X \quad \dot{R'} \longrightarrow R—X + \dot{R'}$$

e.g. $\dot{C}H_3 + Cl—Cl \longrightarrow CH_3—Cl + \dot{C}l$

The methyl group removes a chlorine atom (radical) and in so doing effectively transfers the odd electron to a new radical (the chlorine atom).

An important special case of this is where the removed radical is a hydrogen atom, known as **hydrogen abstraction**,

e.g. $\dot{C}l \quad H \quad CH_3 \longrightarrow Cl—H + \dot{C}H_3$

## (c) Radical Addition to an Alkene

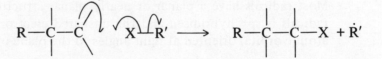

The new radical may further react in one of two ways
    (i) Radical abstraction

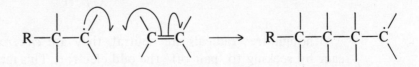

    The net result is radical addition of RX to the alkene double bond.
  (ii) Addition to another molecule of alkene

Repetition of this step will build up a longer and longer chain until the sequence is terminated in some way. This is the basis of radical polymerisation.

## (d) Disproportionation

This is a special case of hydrogen abstraction, e.g.

$$CH_3—\dot{C}H_2 \quad H—CH_2—\dot{C}H_2 \longrightarrow CH_3—CH_3 + CH_2{=}CH_2$$

212

## 12.5 Evidence for Radical Intermediates

### (a) Spectroscopic Evidence

Normal spectroscopic methods (UV, IR and PMR) are of little value, because (with the exception of certain long-lived radicals) the concentration of radicals seldom exceeds $10^{-8}$ M.

The technique of electron spin resonance (ESR) spectroscopy is widely used for the investigation of radicals, and in favourable cases will detect radicals in concentrations as low as $10^{-9}$ M.

Techniques have been developed for examining very short-lived radicals which are otherwise difficult to detect. These include continuous generation flow techniques, 'freezing' radicals in a matrix at low temperatures, and a technique known as 'spin trapping'. This involves the addition of a compound which will react with radicals to generate more stable (i.e. longer-lived) radicals. One compound used in this way is 2-methyl-2-nitrosopropane,

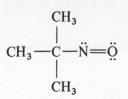

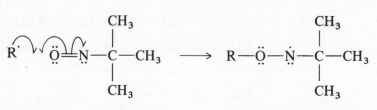

short-lived    'spin trap'    longer-lived
radical    radical

### (b) Chemical Evidence

Reactions involving radical intermediates often exhibit mixed order kinetics and are relatively insensitive to changes in the polarity or pH of the medium. Addition of known radical initiators will usually increase the rate of a radical reaction. Conversely, **radical inhibitors** will decrease the rate, or even stop a radical process altogether. Inhibitors are of two types:
  (i) Those reacting so as to give a neutral molecule, e.g. nitrogen oxide (nitric oxide)

$$\dot{R} \quad \ddot{N}=\ddot{O}: \longrightarrow \quad R-\underset{\cdot}{N}=\ddot{O}:$$

(The addition of a compound (radical) which will react with a radical intermediate to form a recognisable by-product is known as radical trapping.)
  (ii) Compounds forming more stable radicals unable to continue a radical chain. These include alkenes, phenols, aromatic amines, aromatic disulphides, iodine and oxygen.
  For example hydrogen abstraction from a phenol gives a relatively unreactive resonance-stabilised radical:

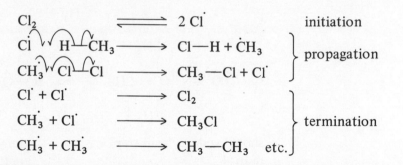

(see also anti-oxidants)

The following examples illustrate some of the more common radical processes.

## 12.6 Radical Substitution Reactions

### (a) Halogenation of Alkanes

#### (i) *Chlorination of Methane*

$$
\begin{array}{lll}
Cl_2 & \rightleftharpoons & 2\,Cl\cdot & \text{initiation} \\
Cl\cdot \; H\text{---}CH_3 & \longrightarrow & Cl\text{---}H + \cdot CH_3 \\
CH_3\cdot \; Cl\text{---}Cl & \longrightarrow & CH_3\text{---}Cl + Cl\cdot
\end{array} \Big\} \text{propagation}
$$

$$
\begin{array}{lll}
Cl\cdot + Cl\cdot & \longrightarrow & Cl_2 \\
CH_3\cdot + Cl\cdot & \longrightarrow & CH_3Cl \\
CH_3\cdot + CH_3\cdot & \longrightarrow & CH_3\text{---}CH_3 \quad \text{etc.}
\end{array} \Big\} \text{termination}
$$

Evidence for the accepted mechanism:
(1) Independent experiments show that chlorine does dissociate into chlorine atoms when heated to 250°C or irradiated with UV light.
(2) Chlorination occurs at lower temperatures with the use of free radical initiators.
(3) In the photochemical process, many thousands of chloromethane molecules are produced for each photon of light absorbed, as would be expected for a chain propagated process.
(4) The addition of a small amount of oxygen (or other radical scavenger) slows down or may even inhibit reaction. Oxygen has a diradical structure and is able to react with a methyl radical to form the much less reactive peroxide radical:

$$
CH_3\cdot \; \dot{O}=\dot{O} \longrightarrow CH_3\text{---}\dot{O}=\dot{O}
$$

This effectively terminates the chain and so reaction slows down until all the oxygen has been used up in this way.
(5) Analysis of the product mixture shows the presence of small amounts of ethane, formed by combination of methyl radicals but no hydrogen, thus ruling out the alternative, energetically unfavourable, step

214

$$\overset{\cdot}{Cl} \;\overgroup{\;}\; \overset{\frown}{CH_3} \overgroup{\;} H \;\xrightarrow{\;\;\times\;\;}\; Cl{-}CH_3 + \overset{\cdot}{H}$$

### (ii) *Reaction of the Other Halogens*

Fluorination is highly exothermic and difficult to control. Bromine atoms are less energetic and bromination is slow. This reaction is really only useful for replacement of more reactive 3° hydrogen. Iodination is too slow to be of any practical value.

### (iii) *Chlorination of Propane*

Monochlorination of propane gives two products, 1-chloropropane and 2-chloropropane, in the ratio of 45:55. Two factors influence the ratio of isomers. These are:

(1) the **probability factor**: there are three times as many 1° H atoms (two $CH_3$ groups as 2° H atoms (one $CH_2$ group), and

(2) the **reactivity factor**: 2° H atoms are more easily removed than 1° H atoms, because activation energy for the formation of the secondary propyl radical is less than that for formation of the primary propyl radical. The simplest explanation for this is that free radicals, like carbocations, are electron deficient species, and stabilisation results from the positive inductive effect of alkyl groups attached to the electron deficient carbon atom:

$$\overset{\cdot}{CH_3} \;<\; CH_3{\rightarrow}\overset{\cdot}{CH_2} \;<\; \underset{CH_3}{\overset{CH_3}{\diagdown}}\overset{\diagup}{\underset{}{\overset{\cdot}{CH}}} \;<\; \underset{CH_3}{\overset{CH_3}{\diagdown}}\overset{\diagup}{\underset{}{\overset{\cdot}{C}}}{\leftarrow}CH_3$$

### (iv) *Bromination of Propane: the Concept of Selectivity*

At 125°C (398 K) monobromination of propane gives 1-bromopropane and 2-bromopropane in the ratio of 2:98. Bromine atoms are formed more readily than chlorine atoms, but are therefore correspondingly less reactive (less energetic). Consequently, the collision of a bromine atom with a methyl group is less likely to generate the relatively high activation energy necessary for the abstraction of a 1°H atom. A much higher proportion of collisions with the central methylene group (where less activation energy is necessary for the abstraction of a 2°H atom) are successful, and the bromine atom is said to be *more selective* than the chlorine atom.

This is an important general principle which applies to both ionic and radical processes, namely that the less reactive an attacking reagent, the more selective it is in reacting with a substrate.

### (v) *Chlorination of Methylbenzene (Toluene)*

Substitution in the methyl group occurs under relatively mild conditions – chlorine gas is passed into boiling methylbenzene in bright sunlight. Replacement of successive hydrogen atoms is smooth and progressive, and reaction can be monitored simply by observing the increase in mass.

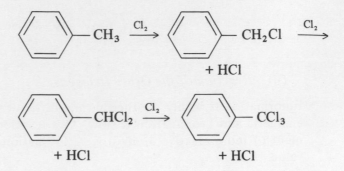

$+ HCl$

$+ HCl$         $+ HCl$

Activation energy for the abstraction of hydrogen is quite low as the resulting 'benzyl' free radical is resonance stabilised:

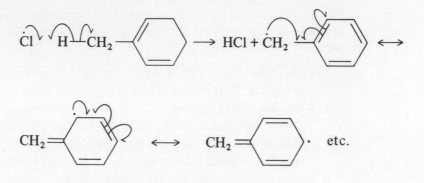

### (vi) *Vapour Phase Chlorination of Propene*

When a mixture of chlorine and propene are heated to about 450°C (723 K), substitution occurs in the methyl group. The mechanism is similar to that for substitution in an alkane, but hydrogen abstraction is facilitated by the formation of a resonance-stabilised allylic radical intermediate:

$$\dot{C}l \quad H—CH_2—CH=CH_2 \longrightarrow HCl + \dot{C}H_2—CH=CH_2 \longrightarrow CH_2=CH—\dot{C}H_2$$

Some free radical addition to the double bond may also take place, but the first step is readily reversible and at high temperature equilibrium lies well to the left. (Trace amounts of iodine or bromine catalyse *cis–trans* isomerisation of alkenes even at room temperature.)

### (vii) *N-Bromosuccinimide (N-Bromobutandioic Acid Imide)*

'NBS' is used specifically for allylic and benzylic bromination. The reagent contains a small amount of residual bromine which, with the aid of bright sunlight or a radical initiator, dissociates into bromine atoms

$$Br_2 \rightleftharpoons 2\,\dot{Br}$$

Substitution proceeds in the usual way

$$\dot{Br} \quad H—CH—CH=CH— \longrightarrow Br—H + \dot{C}H—CH=CH—$$

$$—CH=CH—\dot{C}H \quad Br—Br \longrightarrow —CH=CH—CH—Br + \dot{Br}$$

and the role of 'NBS' is to convert the by-product HBr into $Br_2$ molecules, thus maintaining a constant low concentration of bromine.

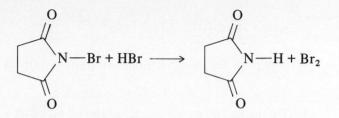

### (b) Aromatic Homolytic Substitution

#### (i) *Phenylation of Benzene*

The accepted mechanism is a special case of radical transfer, and the steps are analogous to those in electrophilic substitution of benzene, e.g.

benzoyl peroxide $\xrightarrow{\Delta}$ Ph˙

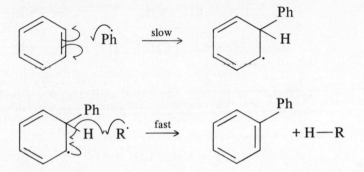

where R is a radical acceptor, analogous to the proton acceptor in aromatic electrophilic substitution.

Like the sigma complex intermediate in electrophilic substitution, the free radical intermediate is resonance stabilised:

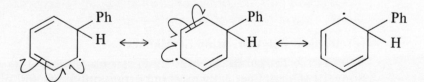

## 12.7   Radical Addition Reactions

### (a) Addition of Chlorine to Propene

$Cl_2 \rightleftharpoons$ 2 Cl˙ (thermolytic or photolytic)

$$CH_3-CH{=}CH_2 \quad Cl˙ \longrightarrow CH_3-\dot{C}H-CH_2-Cl$$

(as in ionic reactions of propene, addition occurs so as to give the more stable of the two possible intermediates)

$$Cl-CH_2-\overset{\cdot}{C}H-CH_3 \quad Cl\frown Cl \longrightarrow Cl-CH_2-CHCl-CH_3 + \overset{\cdot}{C}l$$

followed by the usual termination steps. Addition is reversible at high temperature.


### (b)  Addition of Hydrogen Bromide to Propene

In a non-polar solvent and in the presence of a small amount of a peroxide, HBr adds to propene to give 1-bromopropane, the anti-Markownikov product (Kharasch peroxide effect).

$$R-O \frown O-R \longrightarrow 2\,R-\overset{\cdot}{O}$$

$$R-\overset{\cdot}{O} \quad H\frown Br \longrightarrow R-O-H + \overset{\cdot}{B}r$$

$$\overset{\cdot}{B}r \quad CH_2{=}CH-CH_3 \longrightarrow Br-CH_2-\overset{\cdot}{C}H-CH_3$$

$$Br-CH_2-\overset{\cdot}{C}H-CH_3 \quad H\frown Br \longrightarrow Br-CH_2-CH_2-CH_3 + \overset{\cdot}{B}r$$

Termination follows the usual pattern, and the mechanism is supported by the isolation of some 1,2-dibromopropane and 2,3-bisbromomethylbutane:

$$Br-CH_2-\overset{\cdot}{C}H-CH_3 + \overset{\cdot}{B}r \longrightarrow Br-CH_2-CH(Br)-CH_3$$

$$2\,CH_3-\overset{\cdot}{C}H-CH_2-Br \longrightarrow \begin{array}{c} CH_3-CH-CH_2-Br \\ | \\ CH_3-CH-CH_2-Br \end{array}$$


### (c)  Radical Polymerisation of Alkenes

Some of the commercially most important polymers and plastics are manufactured from alkene monomers by a process of addition polymerisation. This may involve an ionic or a radical chain mechanism, depending on the reaction conditions employed.

Well known examples of polymers produced by radical polymerisation are 'PVC', 'PVA', 'Polyacrylonitrile', 'Polystyrene', 'Perspex', and 'Teflon'. The IUPAC names and structures of these compounds are given in Table 12.1.

Reaction conditions vary, but these different 'vinyl' (ethenyl) monomers all polymerise by essentially the same free radical mechanism, involving initiation, propagation and termination steps.

For simplicity in discussing mechanisms, it is convenient to represent a growing polymer chain with a 'wiggly' line in front of the structural unit responsible for propagating the chain:

**Table 12.1** Some vinyl polymers

| Name of polymer | Monomer structure | Repeat unit in polymer |
|---|---|---|
| Polyvinyl chloride, PVC, Poly(chloroethene) | $CH_2\!=\!CH\!-\!Cl$ | $-CH_2-\underset{\underset{Cl}{\mid}}{CH}-$ |
| Polyvinylacetate, PVA Poly(ethenylethanoate) | $CH_2\!=\!CH\!-\!OOCCH_3$ | $-CH_2-\underset{\underset{OOCCH_3}{\mid}}{CH}-$ |
| Polyacrylonitrile, Poly(propenonitrile) | $CH_2\!=\!CH\!-\!CN$ | $-CH_2-\underset{\underset{CN}{\mid}}{CH}-$ |
| Polystyrene, Poly(phenylethene) | $CH_2\!=\!CH\!-\!Ph$ | $-CH_2-\underset{\underset{Ph}{\mid}}{CH}-$ |
| Polymethylmethacrylate, Perspex, Poly(methyl 2-methylpropenoate) | $CH_2\!=\!\underset{\underset{CH_3}{\mid}}{C}\!-\!OOCCH_3$ | $-CH_2-\overset{\overset{COOCH_3}{\mid}}{\underset{\underset{CH_3}{\mid}}{C}}-$ |
| Polytetrafluoroethylene, Teflon, etc., Poly(tetrafluoroethene) | $CF_2\!=\!CF_2$ | $-\overset{\overset{F}{\mid}}{\underset{\underset{F}{\mid}}{C}}-\overset{\overset{F}{\mid}}{\underset{\underset{F}{\mid}}{C}}-$ |

$$\text{wwwww } CH_2-\overset{\bullet}{C}H$$

polymer chain of unspecified length → growing end of chain

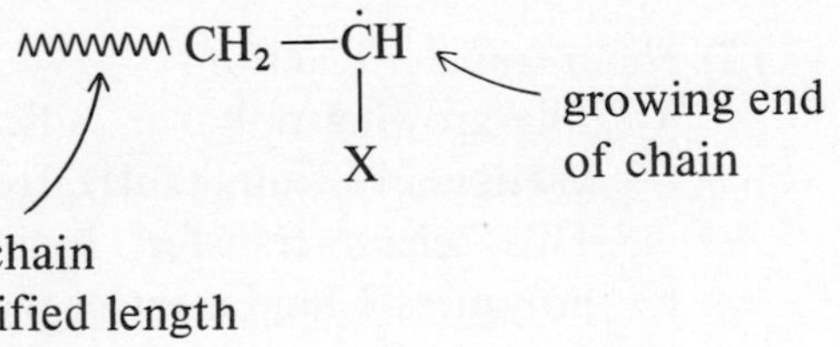

where X = Cl, CN, etc.

*Initiation*

$$\text{Peroxides or 'AIBN'} \xrightarrow{\Delta} \text{radical R}^{\bullet}$$

*Propagation*

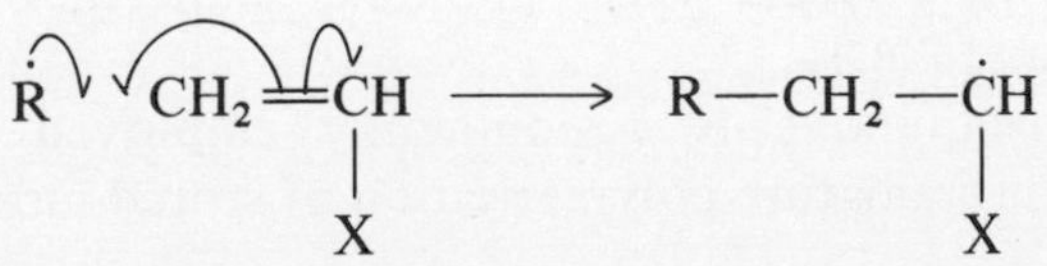

$$\overset{\bullet}{R} + CH_2\!=\!\underset{\underset{X}{\mid}}{CH} \longrightarrow R-CH_2-\underset{\underset{X}{\mid}}{\overset{\bullet}{C}H}$$

$$R-CH_2-\overset{\cdot}{C}H + CH_2{=}CH \rightarrow R-CH_2CH-CH_2-\overset{\cdot}{C}H \xrightarrow{repeat} \text{\Large\char"2DC\char"2DC}CH_2-\overset{\cdot}{C}H$$

with X substituents below each CH group.

Two factors are involved in the 'head to tail' mode of addition:

(i) **Steric**: it is easier for the polymer radical to add to the vinyl carbon atom adjacent to the one carrying the substituent.

(ii) **Electronic**: the growing radical is stabilised by the substituent X. When X = Ph, the radical is resonance stabilised:

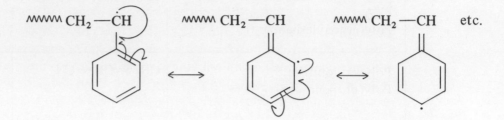

Termination

(1) dimerisation

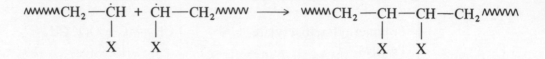

(2) disproportionation

$$\text{wwwCH}_2-\overset{\cdot}{C}H \quad X \quad + \quad H-CH \text{ www} \longrightarrow \text{wwwCH}_2-CH_2-X + CH \text{ www}$$

(3) chain transfer reactions

(i) The growing polymer radical abstracts hydrogen (benzylic hydrogen is particularly vulnerable) from its own or an adjacent polymer chain. This '**chain transfer**' limits chain length and gives branched chain polymers. Chain transfer may also occur to a monomer molecule.

(ii) Addition of a '**chain transfer reagent**' (such as dodecylthiol, R—SH) is a method of controlling polymer size:

$$\text{wwwCH}_2-\overset{\cdot}{C}H \quad + \quad H-SR \longrightarrow \text{wwwCH}_2-CH_2 + \overset{\cdot}{S}R$$

with X below the CH groups.

The radical $\overset{\cdot}{S}R$ is insufficiently reactive to initiate a new chain by attacking another vinyl monomer molecule. Other reagents include $CCl_4$ and $PhCH_3$.

Benzoquinone is a commonly employed inhibitor also used to prevent premature polymerisation of stored monomer.

(i) *Stereochemistry of Vinyl Polymers*

The position of the 'side chain' group X may be
  (1) stereorandom = **atactic**, or
  (2) stereoregular = **isotactic** (all the same side), or
              **syndiotactic** (regular alternation of sides).
Isotactic polymers are highly crystalline, hence tend to be hard and brittle.


(ii) *Copolymerisation*

Properties of polymers may be appreciably modified by copolymerisation of two different monomers. For example, 'toughened' Perspex is produced by copolymerisation of but-1,3-diene with phenylethene in the ratio of 1:3. The resulting polymer is more flexible and less brittle than polystyrene. The copolymer is not simple a mixture of poly(buta-1,3-diene) and poly(phenylethene), but consists of polymer chains with a random arrangement of the two monomer units.

Natural rubber is essentially polyisoprene (poly(2-methylbuta-1,3-diene)), in a stereoregular 1,4-addition with a *cis* arrangement of the residual double bonds:

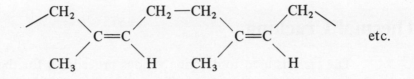

The synthetic rubber 'SBN' is a copolymer of phenylethene and buta-1,3-diene in the ratio 1:3. Addition is mainly 1,4- with some 1,2-, but the residual double bonds have a *trans* arrangement:

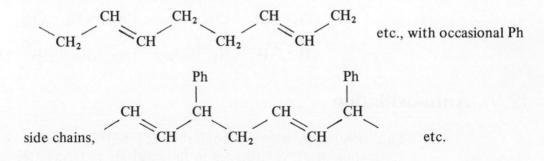

(iii) *Plasticisers*

Polymers such as poly(chloroethene) (PVC) which in 'pure' form may be too rigid or brittle, can be made more flexible by the incorporation of a plasticiser. These compounds form a kind of 'internal' lubricant, enabling polymer molecules to more easily slide over one another. An example is dibutyl phthalate (dibutyl benzene-1,2-dicarboxylate). Depending on the use to which the material is to be put, anything from 10% to 45% of this compound may be added to the raw polymer.


**(d) Addition of Chlorine to Benzene**

When irradiated with bright sunlight, or in the presence of peroxides, chlorine reacts additively with benzene to give the 'hexachloride'. The white crystalline

221

solid which eventually separates is in fact a mixture of geometrical isomers of 1,2,3,4,5,6-hexachlorocyclohexane. The essential steps are

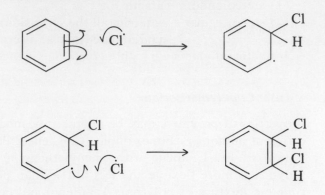

repetition ⟶ 'hexachloride'

Benzene reacts similarly with bromine, but not with iodine. Direct fluorination is too violent to be of practical value.

## 12.8  Thermal Cracking

The term is used to describe various procedures for the breakdown of petroleum fractions into mixtures of alkanes and alkenes of lower relative molecular masses. Various catalysts may be used and reactions may then proceed by ionic mechanisms. Simple thermal cracking is initiated by homolysis of a carbon–carbon bond. A simple example is the formation of methane and ethene when propene is heated to 500–600°C (773–873 K):

$$CH_3\!-\!CH_2\!-\!CH_3 \xrightarrow{\Delta} CH_3\!-\!\dot{C}H_2 + \dot{C}H_3$$

$$\dot{C}H_3 \quad H\!-\!CH_2\!-\!\dot{C}H_2 \longrightarrow CH_4 + CH_2\!=\!CH_2$$

## 12.9  Auto-oxidation

This slow, spontaneous air-oxidation of certain organic compounds involves abstraction of active hydrogen in the substrate to form a free radical, which then reacts with oxygen. Representing the substrate by SUB — H, the mechanism can be summarised as

Initiation    Initiator ⟶ Radical, Ṙ

$$SUB\!-\!H \quad R \longrightarrow R\!-\!H + SU\dot{B}$$

Propagation

$$SU\dot{B} \quad \dot{O}\!-\!\dot{O} \longrightarrow SUB\!-\!O\!-\!\dot{O}$$

$$SUB\!-\!O\!-\!\dot{O} \quad H\!-\!SUB \longrightarrow SUB\!-\!O\!-\!O\!-\!H + SU\dot{B}$$

Termination

$$SU\dot{B} + SU\dot{B} \longrightarrow SUB\!-\!SUB$$

$$SU\dot{B} + SUB\!-\!O\!-\!\dot{O} \longrightarrow SUB\!-\!O\!-\!O\!-\!SUB, \text{ etc.}$$

Examples include the perishing of rubber tubing, degradation of other high polymers exposed to bright sunlight for prolonged periods, the 'drying' of paints and varnishes, etc., air-oxidation of benzaldehyde to benzoic acid, and the formation of peroxides in old samples of alkoxy-alkanes (ethers).

Reaction may be initiated by heat or light, and in the case of drying of paints, etc., by the addition of various metallic catalysts (salts of cobalt, manganese, iron, etc.) or free radical initiators.

Substances particularly prone to auto-oxidation (such as unsaturated fats and oils, and alkene monomers and polymers, all of which contain allylic hydrogen) may be protected to some extent by the incorporation of **anti-oxidant**. These radical inhibitors are compounds such as phenols and aromatic amines which can readily react with radical intermediates to form less reactive radicals unable to continue a chain reaction. Two 'permitted' additives for polyunsaturated butter substitutes are esters of gallic acid, and 'BHT'('butylated hydroxy toluene'):

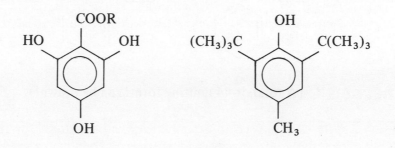

an ester of gallic acid     'BHT'
(2,4,6-trihydroxybenzoic acid)     (2,6-bis(dimethylethyl)-4-methylphenol)

Vegetable oils do not auto-oxidise as readily as animal oils and fats, as they contain a natural anti-oxidant called α-tocopherol (Vitamin E):

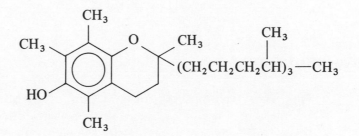

α-tocopherol

# 12.10   Long-lived Radicals

The relative molecular mass of hexaphenylethane in solution is lower than the theoretical value, a phenomenon attributed to dissociation of the compound into triphenyl free radicals:

$$Ph_3C \text{---} CPh_3 \rightleftharpoons 2\ Ph_3\dot{C}$$

The extent of dissociation varies with the nature of the solvent, concentration and temperature. A 5% solution in benzene is about 3% dissociated.

Enthalpy of dissociation is very low, less than 50 kJ mol$^{-1}$, compared with 340 kJ mol$^{-1}$ for the carbon–carbon bond in ethane. Two factors are involved:

(i) The central carbon–carbon bond length is a little longer than usual (158 pm) indicating some steric compression in the molecule. This is relieved on dissociation, and has to be overcome on recombination of the triphenylmethyl radicals.

(ii) Triphenylmethyl radicals have a resonance structure which makes them particularly stable.

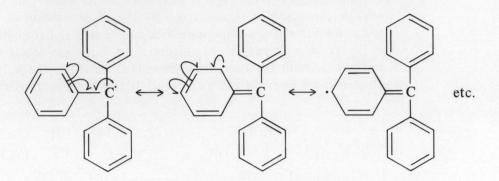

(A total of 44 limiting forms can be drawn!)

# 12.11  Worked Examples

### Example 12.1

Outline a mechanism for monochlorination of methane, and show that this is consistent with experimental observation. *(6 marks)*

Indicate the steps that are taken to exercise control over the halogenation process in industrial applications. *(4 marks)*

Account mechanistically for the following observations:

(i) monochlorination of propane gives approximately equal quantities of 1-chloropropane and 2-chloropropane; *(5 marks)*

(ii) side chain monochlorination of ethylbenzene gives 1-chloroethylbenzene (90%) and 2-chloroethylbenzene (10%), whereas monobromination of ethylbenzene gives only 1-bromoethylbenzene. *(5 marks)*

(Coventry Polytechnic, 1987)

*Solution 12.1*

Monochlorination of methane takes place in three stages:

Initiation $\qquad\qquad Cl_2 \; \rightleftharpoons \; 2 \, \overset{\centerdot}{C}l$

Propagation $\quad \overset{\centerdot}{C}l + CH_4 \longrightarrow HCl + \overset{\centerdot}{C}H_3$

$\qquad\qquad\quad \overset{\centerdot}{C}H_3 + Cl_2 \longrightarrow CH_3Cl + \overset{\centerdot}{C}l$

Termination $\quad \overset{\centerdot}{C}l + \overset{\centerdot}{C}l \longrightarrow Cl_2$

$\qquad\qquad\quad \overset{\centerdot}{C}H_3 + \overset{\centerdot}{C}l \longrightarrow CH_3Cl$

$\qquad\qquad\quad \overset{\centerdot}{C}H_3 + \overset{\centerdot}{C}H_3 \longrightarrow CH_3{-}CH_3 \; \text{etc.}$

The proposed mechanism is consistent with the following observations.

(1) No reaction occurs in the dark at room temperature, but reaction is rapid when a mixture of methane and chlorine is heated above about 250°C (423 K), or is exposed to bright sunlight (UV radiation) at room temperature.

Only on absorption of UV radiation or at high temperature can chlorine molecules dissociate into chlorine atoms (radicals).

(2) If a small amount of oxygen is added to the alkane/halogen mixture, reaction slows down or stops, but after a delay, picks up again to the previous rate.

Oxygen has a diradical structure, and readily pairs off an electron with the odd electron of a methyl radical:

$$\overset{\centerdot}{O}{-}\overset{\centerdot}{O} + \overset{\centerdot}{C}H_3 \longrightarrow CH_3{-}O{-}\overset{\centerdot}{O}$$

The resulting peroxide radical is much less reactive and is unable to participate in the chain propagating steps. When all of the oxygen molecules have been consumed in this way, propagation can continue.

(3) GLC analysis of the product mixture reveals the presence of small amounts of ethane, but no hydrogen. This is consient with the formation of methyl radicals which occasionally combine to form ethane, and rules out the possible alternative propagation step in which a chlorine atom abstracts methyl rather than hydrogen:

$$\overset{\centerdot}{C}l + CH_4 \longrightarrow\!\!\!\!\times\!\!\!\!\longrightarrow CH_3Cl + \overset{\centerdot}{H}$$

(4) Finally, in light initiated chlorination, many thousands of product molecules

are formed for each photon of light absorbed — exactly what would be expected for a chain propagated process.

Direct halogenation of alkanes is not a suitable laboratory synthesis, but vast quantities of haloalkanes are required in industry both as solvents and as starting materials for other compounds. Mixed products are unavoidable, but the proportion of monosubstitution can be increased by having a large alkane:halogen ratio.

Chlorination, and more particularly fluorination, are highly exothermic processes, and the major concern is the rate of halogenation which can readily become explosively fast. Control is exerted by:

(1) diluting the reactants with an inert gas such as nitrogen;
(2) utilising a reactor containing a large surface area of conducting metal (e.g. copper);
(3) having very short reaction times, e.g. the mixed gases are passed through a reaction vessel into a cooling chamber at such a rate that the reactants are only at a high temperature for 1/10th s or less.

The ease of removal of hydrogen from an alkane is related to the stability of the resulting alkyl free radical. Thus it is easier to abstract hydrogen from the 2-position in propane (to form the secondary propyl radical) than from the 1-position (to form the less stable primary radical):

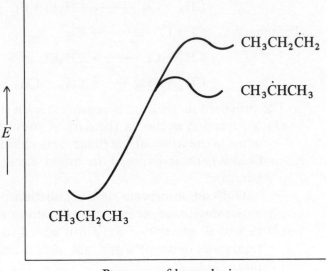

— Progress of homolysis ⟶

Thus despite the fact that there are 6 primary hydrogen atoms to only 2 secondary hydrogen atoms (the **probability factor**), a higher proportion of collisions between chlorine atoms and hydrogen atoms in position 2 are successful (the **reactivity factor**). Hence the observation that 1-chloropropane and 2-chloropropane are formed in nearly equal amounts.

Similar considerations apply to side chain halogenation of ethylbenzene. Hydrogen is abstracted preferentially from the 1-position because this leads to a resonance stabilised (benzylic) radical:

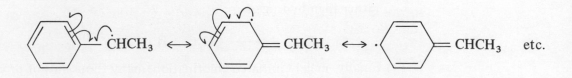

226

Bromine molecules dissociate more readily than chlorine molecules, but the resulting bromine atoms are less energetic (less reactive) than chlorine atoms. Consequently, in bromination of ethylbenzene, only those collisions with the more reactive secondary position (C1) are successful in achieving the necessary activation energy for hydrogen abstraction, and no 2-bromoethylbenzene is formed. Bromine atoms are said to be more **selective**.

**Example 12.2**

What name is given to the enthalpy change associated with homolytic bond cleavage?

By reference to the following reactions discuss the factors which determine whether a chemical process proceeds by a homolytic or heterolytic mechanism.

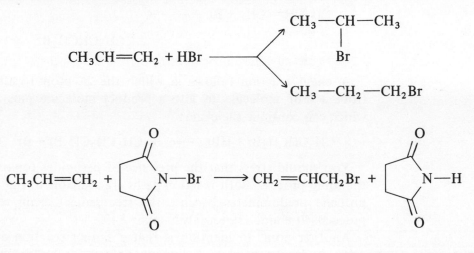

(University of York, 1984)

*Solution 12.2*

Bond dissociation energy.

Formation of 2-bromopropane is the result of *ionic* addition of HBr, which involves the formation of the more stable of the two possible carbocation intermediates:

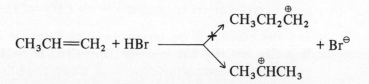

Protonation of the alkene is slow: fast uptake of $Br^\ominus$ completes the reaction:

$$CH_3\overset{\oplus}{C}HCH_3 + Br^\ominus \longrightarrow CH_3CH(Br)CH_3$$

Hydrogen bromide is bubbled into a solution of the alkene, preferably in a moderately polar solvent such as ethanoic acid. Its use will minimise the activation energy for protonation of the alkene by stabilisation (more effective solvation) of the intermediate carbocation. (More strictly, by stabilisation of the transition state from which the carbocation is formed.)

Concentrated aqueous HBr can be used, but mixed products result from reaction of the intermediate carbocation with water molecules.

227

In the presence of peroxides (or even of dissolved oxygen), homolytic addition leads to 1-bromopropane as the main product:

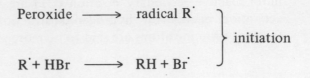

Then $\overset{\cdot}{Br}$ adds preferentially to the terminal carbon atom, as less activation energy is required for the formation of the more stable secondary propyl radical:

$$CH_3CH{=}CH_2 + \overset{\cdot}{Br} \nearrow \begin{array}{l} CH_3CH(Br)\overset{\cdot}{C}H_2 \qquad (1°) \\[1em] CH_3\overset{\cdot}{C}HCH_2Br \qquad (2°) \end{array}$$

A chain reaction follows in which the 2° propyl radical abstracts hydrogen from a HBr molecule to give a product molecule plus another bromine atom which can continue the chain:

$$CH_3\overset{\cdot}{C}HCH_2Br + HBr \longrightarrow CH_3CH_2CH_2Br + \overset{\cdot}{Br}$$

You should stress that the presence of peroxides (or oxygen) does not inhibit the ionic process. Both reaction paths are available to the reactants, but 1-bromopropane predominates because the free radical chain reaction, once initiated, proceeds at a much faster rate.

Another point to mention is that a similar reaction does not occur with the other hydrogen halides. Activation energy for hydrogen abstraction from HCl and HF is too high for this to occur, and iodine atoms (readily formed from HI) are too unreactive to attack the double bond.

The reagent N-bromosuccinimide contains a small amount of residual bromine, and in the presence of peroxides or bright sunlight, substitution occurs in the methyl group by the normal free radical chain mechanism

$$Br_2 \overset{h\nu}{\rightleftharpoons} 2 \overset{\cdot}{Br}$$

$$CH_2{=}CH{-}CH_3 + \overset{\cdot}{Br} \longrightarrow CH_3{=}CH{-}\overset{\cdot}{C}H_2 + HBr$$

$$CH_2{=}CH{-}\overset{\cdot}{C}H_2 + HBr \longrightarrow CH_2{=}CH{-}CH_2Br + \overset{\cdot}{Br}$$

and the chain will be terminated, for example by radical combination such as $\overset{\cdot}{Br} + \overset{\cdot}{C}H_2{-}CH{=}CH \longrightarrow Br{-}CH_2CH{=}CH_2$.

There is a competing process in which the $\overset{\cdot}{Br}$ adds to the alkene double bond:

$$CH_3{-}CH{=}CH_2 + \overset{\cdot}{Br} \longrightarrow CH_3{-}\overset{\cdot}{C}H{-}CH_2Br$$

However, the bromopropyl radical is relatively unstable, and with only very low concentrations of bromine, it dissociates again into propene and $\overset{\cdot}{Br}$ before collision with a molecule of $Br_2$ can complete the addition process.

Conversely, the propenyl radical is resonance stabilised

$$\overset{\cdot}{C}H_2CH{=}CH_2 \longleftrightarrow CH_2{=}CH{-}\overset{\cdot}{C}H_2$$

and has a much longer lifetime in which to await the arrival of a $Br_2$ molecule to complete the substitution reaction.

The role of the 'NBS' is to supply and maintain a constant low concentration of bromine atoms by conversion of the HBr by-product into bromine molecules:

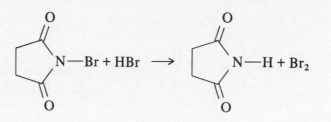

Conclude your answer with a summary of the conditions which favour one process or the other.

Heterolytic addition of HBr is normal in a moderately polar solvent, but precautions may be necessary to avoid competition from the homolytic process. Reaction should be carried out away from bright sunlight, oxygen or peroxides should be excluded, or their effect nullified by the inclusion of a radical inhibitor such as hydroquinone.

Homolytic substitution (bromination) is usually insensitive to the polarity of the solvent if one is used, and is initiated simply by the absorption of bright sunlight (UV radiation). Homolytic addition predominates in the presence of peroxides, and may lead to appreciable amounts of anti-Markownikov product if oxygen is not excluded.

**Example 12.3**

List the products resulting from monochlorination of n-butane.

Give the products from a further chlorination step applied to 2-chlorobutane in its various stereochemical forms. Where possible give the appropriate R/S name, and comment upon any stereochemical relationship between products. (Brunel University, 1983)

*Solution 12.3*

At first glance this might appear to be a rather short question. However, under examination conditions it would take you quite a while to work out the stereochemical relationships and check that you have made the correct R and S assignments.

The products are 1-chlorobutane (achiral) and racemic 2-chlorobutane (major product).

A racemic mixture is obtained because attack by a chlorine molecule can occur either side of the planar secondary butyl radical intermediate:

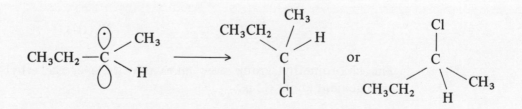

The introduction of a second chlorine atom will give a mixture of 1,2-. 2,2-. 2,3- and 1,3-dichlorobutanes.

Stereochemical relationships between these various products are probably most easily demonstrated by means of Fischer projection formulae.

Begin by checking that you have the correct structures for the R and S isomers, e.g.

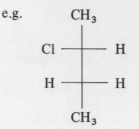

The order of decreasing priority about C2
is $Cl > CH_2CH_3 > CH_3 > H$

Thus the arrangement about C2 is

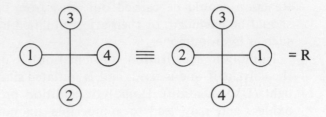

and the two isomers are

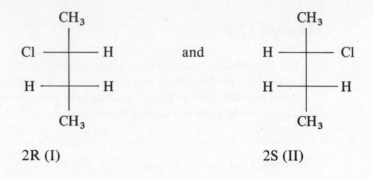

Formation of 1,2-dichlorobutane

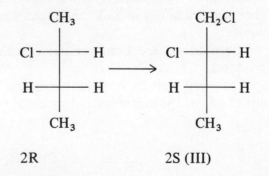

The chloromethyl group now takes priority over the ethyl group, and so the arrangement about C2 is

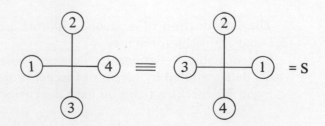

230

and consequently,

2S 2-chlorobutane → 2R, 1,2-dichlorobutane

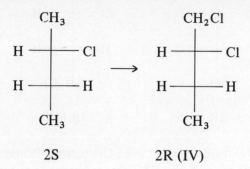

2S           2R (IV)

Formation of 2,2-dichlorobutane

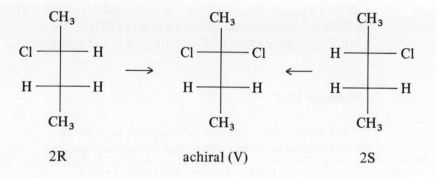

2R       achiral (V)       2S

Formation of 2,3-dichlorobutane

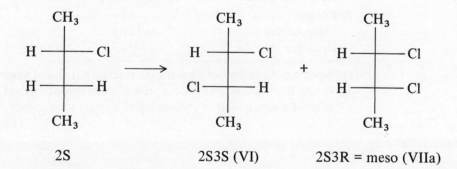

2S       2S3S (VI)       2S3R = meso (VIIa)

Similarly,

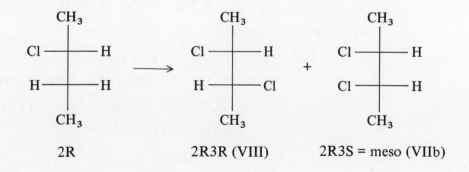

2R       2R3R (VIII)       2R3S = meso (VIIb)

Formation of 1,3-dichlorobutane

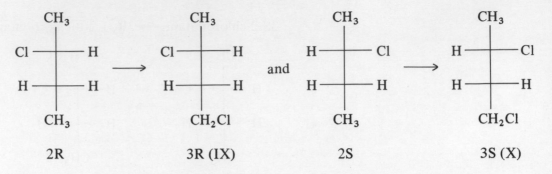

2R        3R (IX)        2S        3S (X)

Relationships between the various isomers may be summarised as follows:
2R (I) and 2S (II) 2-chlorobutanes are enantiomeric.
2S (III) and 2R (IV) 1,2-dichlorobutanes are enantiomeric.
2,2-dichlorobutane (V) is achiral.
2S3S (VI) and 2R3R (VIII) 2,3-dichlorobutanes are enantiomeric.
2S3R (VIIa) = 2R3S (VIIb) 2,3-dichlorobutane is an optically inactive meso form which is a diastereoisomer of (VI) and (VIII).
3R (IX) and 3S (X) 1,3-dichlorobutanes are enantiomeric.

**Example 12.4**

(a) Explain what is meant by 'homolytic bond fission'.
(b) Calculate the enthalpy change for the reaction

$$CH_3CH_3 + Cl_2 \longrightarrow CH_3CH_2Cl + HCl$$

using the following data

| Substance | $\Delta H_f^{\ominus}$/kJ mol$^{-1}$ at 298 K |
|---|---|
| Ethane | −84.6 |
| Chloroethane | −111.7 |
| Hydrogen chloride | −92.3 |

(c) Write out the individual steps for the reaction (b) above, naming the steps appropriately.
(d) Would the reaction in (b) above provide a convenient method for the laboratory preparation of a pure sample of chloroethane? Explain your answer.

(University of London, 1984)

*Solution 12.4*

(a) Cleavage of an electron pair bond such that each of the bonded atoms receives one electron:

$$A \overset{\frown \frown}{—} B \longrightarrow A^{\cdot} + B^{\cdot}$$

(b) Using the relationship

$$\Delta H = \Delta H_f^{\ominus} (\text{product}) - \Delta H_f^{\ominus} (\text{reactants})$$
$$= \Delta H_f^{\ominus} CH_3CH_2Cl + \Delta H_f^{\ominus} HCl - \Delta H_f^{\ominus} CH_3CH_3$$
$$= (-111.7) + (-92.3) - (-84.6) \text{ kJ mol}^{-1}$$
$$= -204.0 + 84.6 \text{ kJ mol}^{-1}$$
$$= -119.4 \text{ kJ mol}^{-1}$$

(c) Initiation

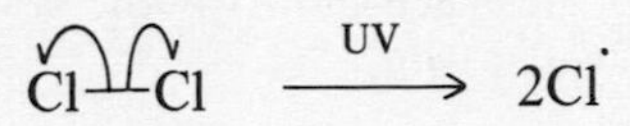

$$Cl \overset{\frown}{\phantom{.}} \overset{\frown}{\phantom{.}} Cl \xrightarrow{\text{UV}} 2Cl^{\cdot}$$

Chain propagation

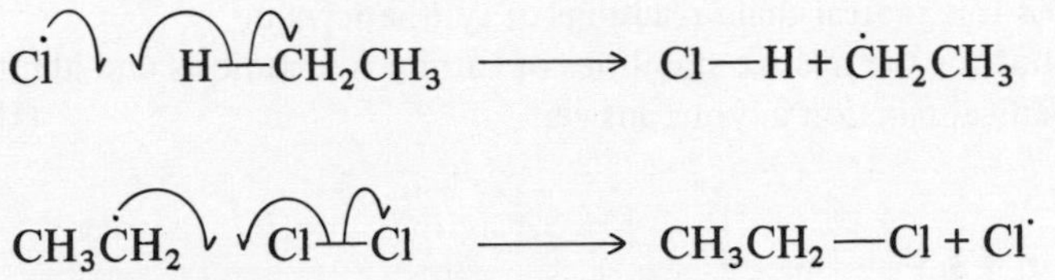

$$Cl^{\cdot} \quad H \overset{\frown}{\phantom{.}} CH_2CH_3 \longrightarrow Cl{-}H + \dot{C}H_2CH_3 \quad \text{(hydrogen abstraction)}$$

$$CH_3\dot{C}H_2 \quad Cl \overset{\frown}{\phantom{.}} Cl \longrightarrow CH_3CH_2{-}Cl + Cl^{\cdot}$$

Termination

$$\text{e.g.} \quad Cl^{\cdot} \quad Cl^{\cdot} \longrightarrow Cl_2$$

$$CH_3\dot{C}H_2 \quad Cl^{\cdot} \longrightarrow CH_3CH_2{-}Cl \quad \text{etc.}$$

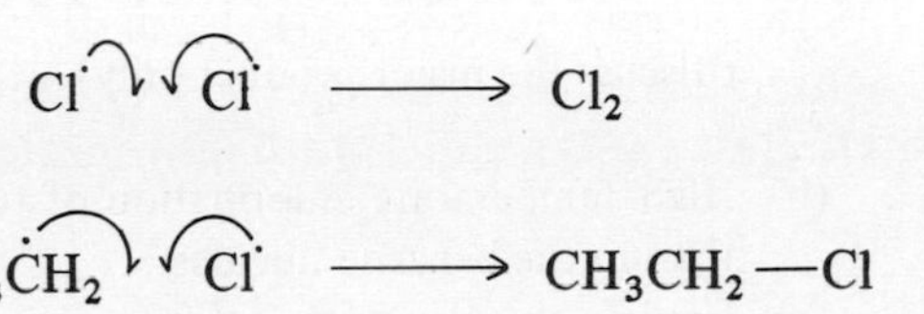

(d) No, because it is not possible to stop at monosubstitution. The use of a large excess of ethane would give a higher proportion of monochloroethane, but even then appreciable amounts of other products, from dichloroethane through to hexachloroethane are unavoidable. Isolation of chloroethane from such a mixture is not feasible in the ordinary laboratory!

# 12.12  Self-Test Questions

**Question 12.1**

(a)  Give the systematic names of the following alkanes:

(i)
$$\underset{\phantom{x}}{\overset{\displaystyle CH_2CH_3}{\underset{|}{CH_3{-}CH{-}CH_2{-}CH_2{-}CH_3}}}$$

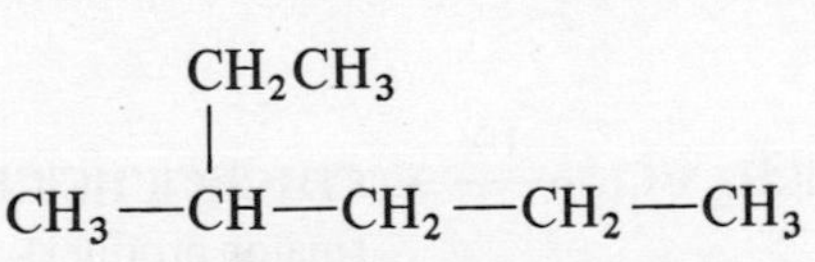

(ii)
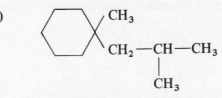

(iii)
$$CH_3{-}\underset{\underset{CH_3}{|}}{\overset{\overset{CH_3}{|}}{C}}{-}CH_2{-}\underset{\underset{CH_2{-}CH_3}{|}}{CH}{-}\overset{\overset{CH_2{-}CH_2{-}CH_3}{|}}{CH}{-}CH_2{-}CH_3$$

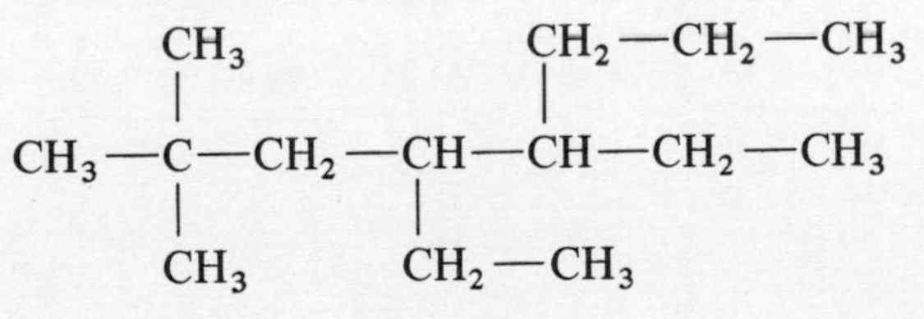

(b) What would be the major product from the bromination of alkane (i) with bromine and why?

(c) Give the mechanism for the reaction of methane with bromine which is initiated by ultraviolet light. (*9 marks*)

(University of Keele, 1984)

## Question 12.2

Discuss free radical chain reactions of synthetic value.
Account for the relative stabilities of carbon free radicals and identify the four basic types of free radical reaction in your answer. (University of Exeter, 1985)

## Question 12.3

(a) Complete the following and describe the mechanism of the reaction involved.

$$CH_3.CH_2.CH(CH_3).CH_2.CH_3 \xrightarrow{\text{Br}_2/\text{UV light}} (H)$$

(discuss the major product only)

(City University, 1984)

(b) High temperature chlorination of toluene leads to substitution in the side-chain rather than in the benzene nucleus. (Thames Polytechnic, 1984)

(c) Explain the observation that vapour phase chlorination of propane gives 1-chloropropane and 2-chloropropane in the ratio of 45:55, whereas bromination under the similar conditions gives almost exclusively 2-bromopropane. (Coventry Polytechnic, 1983)

(part questions)

## Question 12.4

Give a mechanistic explanation for the following observations.

(a) $CH_3CH_2CH{=}CH \xrightarrow[\text{sunlight}]{\text{NBS}} CH_3CH(Br)CH{=}CH_2 + CH_3CH{=}CHCH_2Br$
(major product) (minor product)

(NBS = N-bromosuccinimide)

(b) [cyclopentadiene structure] $+ CBrCl_3 \xrightarrow[\text{low temp}]{\text{UV}}$ Br [cyclopentene structure] $CCl_3$

(c) $CH_3CH{=}CHCH_3 + Cl_2 \xrightarrow{\text{UV}} CH_3CHClCHClCH_3 + CH_3CHClCH{=}CH_2$
(major product) (minor product)

(Coventry Polytechnic (part questions compiled from various progress tests))

# 13 Guided Route and Short Answer Questions

## 13.1 Introduction

'Guided route' questions appear quite frequently on first year examination papers. They enable the examiner to test your knowledge of a variety of reactions from different classes of compounds. And they give you, the student, an opportunity not only to demonstrate such knowledge, but perhaps more importantly, to show that you can think logically and make correct deductions from experimental observations!

Questions of this type from some Chemistry Departments can be quite sophisticated, often requiring detailed interpretation of spectroscopic data. Examples are not included here, as they could prove rather difficult for students taking less specialist courses.

There is no rigid approach to answering guided route questions, but it is usually helpful to begin by summarising the information in tabular form as a flow diagram, alongside which any immediate or obvious deductions can be written. The problem may then unfold itself in a forward direction, but quite often it will be necessary to work backwards from known compounds or final products of easily deduced structure. What is certain, is that the more practice you have at this kind of question, the better you get at answering them!

Many first year examination papers now include a compulsory section consisting of a number of 'Short Answer' questions. Usually the instruction is to 'use the spaces provided' for your answers — which gives you a good indication of the length of answer expected. As with guided route questions, the short answer format enables the examiner to cover a wide range of topics, but frequently the purpose is to test the candidates' knowledge of really basic ideas and concepts, with which all would-be chemists should be familiar before embarking on later years of the course! If short answer questions are a feature of your own first year examinations, you will obviously wish to work through a number of old papers and discuss the answers with your tutor. In the meantime, the examples included in this chapter should help to familiarise you with this type of question.

## 13.2 Worked Examples (Guided Route)

### Example 13.1

A neutral organic compound (A) is found to contain 62.07% C and 10.34% H. Hydrolysis of (A) in dilute aqueous acid leads to the isolation of (B) $C_4H_{10}O$. (B) can be oxidised firstly to an aldehyde and further to an acid (C) $C_4H_8O_2$. When (B) is heated in a mixture of NaBr and $H_2SO_4$ (D) $C_4H_9Br$ is formed, which on treatment with alcoholic KOH yields a gaseous compound (E) $C_4H_8$ which is collected as a liquid at low temperature. Treatment of (E) with HBr yields (F) $C_4H_9Br$ which is not the same as (D). On dissolving (F) in aqueous acetone a compound (G) $C_4H_{10}O$, which is not the same as (B) is formed. (G) is not oxidised under mild conditions. Compound (E) undergoes ozonolysis to give formaldehyde (methanal) and (H) $C_3H_6O$. (H) gives no reaction with Fehlings reagent or Tollens reagent. The RMM of compound (A) is 116.

Briefly explain each step of the above sequence, and identify compounds (A) to (H).

*(20 marks)*

(Sunderland Polytechnic, 1985)

*Solution 13.1*

Begin as suggested above by summarising the information in flow diagram form:

(A)   RMM 116, 62.07% C, 10.34% H, neutral

⬇ hydrolysis

(B)   $C_4H_{10}O$ $\xrightarrow{[O]}$ Aldehyde $\xrightarrow{[O]}$ Acid (C) $C_4H_8O_2$

⬇ $NaBr/H_2SO_4$

(D)   $C_4H_9Br$

⬇ alc. KOH

(E)   $C_4H_8$, low boiling liquid, $\xrightarrow{O_3}$ HCHO + (H) $C_3H_6O$, non-reducing

⬇ HBr

(F)   $C_4H_9Br$, isomer of (D)

⬇ aq. acetone

(G)   $C_4H_{10}O$, isomer of (B), not easily oxidised

(A) is a neutral compound which undergoes hydrolysis and could therefore be a haloalkane or an ester, etc.

(E) give HCHO and (H) $C_3H_6O$ on ozonolysis. As (H) is a non-reducing carbonyl compound, then from the formula it can only be $CH_3COCH_3$. Knowing that the carbonyl groups in the ozonolysis products come from the originally double bonded carbon atoms of an alkene, (E) must be 2-methylpropene:

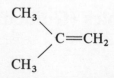

Since (E) was formed by elimination of HBr from (D), the latter compound must have the formula

236

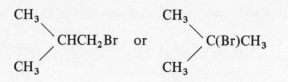

and (B) therefore is

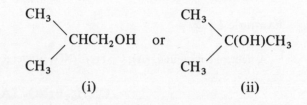

$$\text{(i)} \qquad\qquad\qquad \text{(ii)}$$

However, since (B) can be further oxidised, it cannot have the tertiary structure, (ii).

It is now apparent the (A) is an ester having the structure

$$(CH_3)_2CHCH_2OOCR$$

and the nature of the group R can be deduced from the given RMM (116) and the relative atomic masses of the atoms involved:

$$(CH_3)_2CHCH_2O = 73, \therefore RCO = 116-73 = 43$$

Thus RCO can only be $CH_3CO$, and (A) has the molecular formula $C_6H_{12}O_2$.

Confirmation is obtained by calculation of the percentage of carbon and hydrogen present,

$$\%C = \frac{72}{116} \times 100 = 62.07 \qquad \%H = \frac{12}{116} \times 100 = 10.34$$

when the figures agree with those given in the question.

You can now complete your answer by outlining the whole reaction sequence.

$$(CH_3)_2CHCH_2OOCCH_3 \xrightarrow{\text{hydrolysis}} (CH_3)_2CHCH_2OH$$
$$\text{(A)} \qquad\qquad\qquad\qquad\qquad \text{(B)}$$

On oxidation, (B) gives the aldehyde $(CH_3)_2CHCHO$, and then the acid $(CH_3)_2CHCOOH$ (C).

Reaction of (B) with $NaBr/H_2SO_4$ replaces OH by Br to give (D), which on treatment with alcoholic KOH undergoes elimination to an alkene (E):

$$(CH_3)_2CHCH_2Br \xrightarrow{-HBr} (CH_3)_2C{=}CH_2$$
$$\text{(D)} \qquad\qquad\qquad\qquad \text{(E)}$$

Ozonolysis of (E) gives HCHO and $CH_3COCH_3$ (non-reducing) and addition of HBr to (E) generates the haloalkane (F) via the more stable of the two possible carbocation intermediates

$$(CH_3)_2C{=}CH_2 \longrightarrow [(CH_3)_2\overset{\oplus}{C}CH_3] \longrightarrow (CH_3)_2C(Br)CH_3$$
$$\text{(E)} \qquad\qquad\qquad (3°) \qquad\qquad\qquad \text{(F)}$$
$$\text{isomeric with (D)}$$

(F) is a tertiary haloalkane, and as such is readily hydrolysed in aqueous acetone

$$(CH_3)_2C(Br)CH_3 \xrightarrow{HO^{\ominus}} (CH_3)_2C(OH)CH_3$$

$$\text{(F)} \qquad\qquad \text{(G)}$$

isomeric with (B)
and as a 3° alcohol is
resistant to oxidation

**Example 13.2**

A neutral compound, (A), $C_{22}H_{32}BrNO_2$, undergoes the following reactions:

$$C_{22}H_{32}BrNO_2 \text{ (A)}$$

↓ $O_3$/oxidative work-up

$C_9H_{18}O$      +      $C_5H_8O_3$ (B)      +      $C_8H_6BrNO_4$ (C)

↓ $NaBH_4$          ↓ $NaOH/I_2$        ↓ $KMnO_4$

$C_9H_{20}O$        $HO_2CCH_2CH_2CO_2H + CHI_3$      $C_7H_4BrNO_4$ (D)

↓ conc. $H_2SO_4$                        ↓ 1. $Br_2/Fe$
                                             2. $KMnO_4$

$C_9H_{18}$                                            Me

↓ $O_3$/oxidative work-up

$C_5H_{10}O_2$ (E) + $CH_3COCH_2CH_3$

*(structure: para-substituted benzene ring with Me at top and $NO_2$ at bottom)*

The following information may also be useful:
1. Compounds (B), (C), (D) and (E) are acidic; (E) is optically active.
2. The nuclear magnetic resonance spectrum of (A) exhibits multiplets owing to two olefinic protons; one of the coupling constants observed is large ($J$ = 18 Hz), indicative of a *trans*-disubstituted alkene.

Deduce the structure of A.             (University of Leeds, 1983)

*Solution 13.2*

The examiner here has been kind in tabulating much of the data for you! Taking first the lefthand column, the compound $C_9H_{18}O$ is an ozonolysis product, and therefore contains a carbonyl group. Treatment with $NaBH_4$ will reduce the

$\diagdown$C=O to —CH(OH)—, and subsequent treatment with c.$H_2SO_4$ dehydrates

the alcohol to an alkene, $C_9H_{18}$. Ozonolysis (with oxidative work-up) of this gives butanone plus (E), which you are told is acidic and optically active. Having the formula $C_4H_9COOH$, the structure can only be

$$\begin{array}{c} CH_3 \\ \diagdown \\ \overset{*}{C}HCOOH \quad (\overset{*}{C} \text{ is chiral}) \\ \diagup \\ CH_3CH_2 \end{array}$$

The alkene from which these products were obtained must have the structure

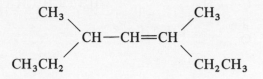

and the alcohol from which this was formed is

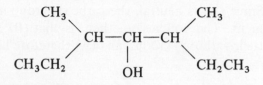

Thus the original ozonolysis product, $C_9H_{18}O$, must have been

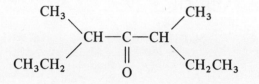

Turning next to data in the righthand column, you are told that compound (D) can be synthesised by bromination of 4-methylnitrobenzene (*para*-nitrotoluene). Since the methyl group is *ortho/para* directing and the nitro group is *meta* directing, monobromination will occur adjacent to the methyl group, giving the structure

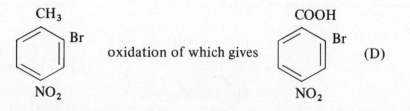

Compound (C) has one more carbon atom, and is an acidic ozonolysis product. Its structure must therefore be

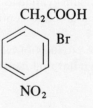

The third ozonolysis product (B) is converted into butandioic acid by the iodoform reaction ($I_2$/NaOH) which cleaves off one carbon atom from a $CH_3CO-$ grouping. Thus (B), $C_5H_8O_3$, and acidic, has the structure

$$CH_3COCH_2CH_2COOH$$

In order to piece together the structure of (A) from the three ozonolysis products

239

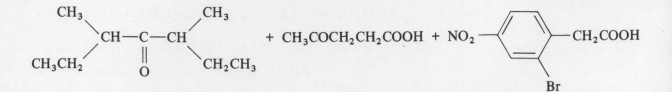

you have to take into account the additional information that (A) has two olefinic protons and a large coupling constant indicative of a *trans*-disubstituted alkene. Since (A) is neutral, the carboxyl group in (B) resulted from the ozonolysis treatment. The conclusion, then, is that (B) and (C) must have been linked through their carboxyl groups, and (A) therefore has the structure

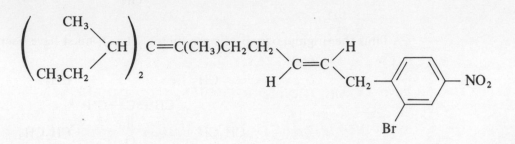

Make a final check that the structure is in agreement with the molecular formula $C_{22}H_{32}BrNO_2$!

**Example 13.3**

Reaction of (−) 1-phenylethanol (*C*) with sodium hydride and then with ethyl iodide gave (−) (*D*), $C_{10}H_{14}O$.

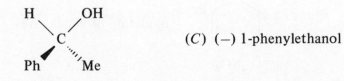

(*C*) (−) 1-phenylethanol

Reaction of the sodium salt of (−) mandelic acid ($PhCH(OH)CO_2Na$) with sodium hydride and then ethyl iodide gave, after acidification, an intermediate acid (*E*), which was reduced with lithium aluminium hydride to give (−) (*F*) $C_{10}H_{14}O_2$. Reaction of (*F*) with toluene-*p*-sulphonyl chloride ($MeC_6H_5SO_2Cl$) gave an intermediate (*G*) which was reduced with lithium aluminium hydride to give (−) (*D*).

Deduce the absolute configuration of (−) mandelic acid and the structures of (*D*) to (*G*).

Predict the products of reaction of (*C*) with (i) neat $SOCl_2$ and (ii) toluene-*p*-sulphonyl chloride in pyridine, followed by sodium acetate and acetic acid.

(University of Southampton, 1984)

*Solution 13.3*

Conversion of (C) into (D) does not involve breaking a bond to the chiral carbon atom,

$$—OH \xrightarrow{NaH} —\overset{\ominus}{O} \xrightarrow{EtI} —OEt$$

and so (−) (D) had the same configuration as (−) (C):

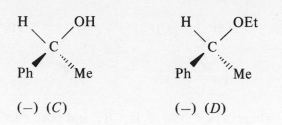

(−) (C)          (−) (D)

Similarly,

PhCH(OH)CO$_2$Na    (−) mandelic acid

    1. NaH
    2. EtI
    3. H$^{\oplus}$

PhCH(OEt)CO$_2$H    (E)

    LiAlH$_4$  (reduces —COOH to —CH$_2$OH)

PhCH(OEt)CH$_2$OH  (−) (F)   (which agrees with the molecular formula C$_{10}$H$_{14}$O$_2$

    toluene-$p$-sulphonyl chloride

PhCH(OEt)CH$_2$OSO$_2$C$_6$H$_4$Me

    LiAlH$_4$   (removes the sulphonate group)

PhCH(OEt)CH$_3$  (−) (D)

From which it follows that (−) mandelic acid has the same configuration as (−) (D),

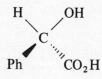

The use of sulphur dichloride oxide (thionyl chloride) converts —OH into —Cl with retention of configuration (S$_N$i mechanism)

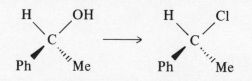

Reaction of (C) with toluene-$p$-sulphonyl chloride gives the 'tosylate' ester with retention of configuration, but further reaction with the ethanoate (acetate) anion is a bimolecular process (S$_N$2) which will involve inversion of configuration

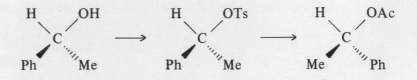

where OTs = $OSO_2C_6H_4CH_3$, and OAc = $OCOCH_3$.

**Example 13.4**

Identify the compounds (*A*)-(*H*) in the reaction sequence outlined below.

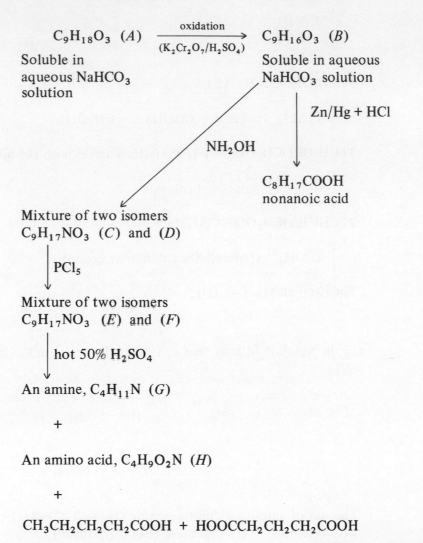

$C_9H_{18}O_3$  (*A*) $\xrightarrow[\text{(K}_2\text{Cr}_2\text{O}_7/\text{H}_2\text{SO}_4)]{\text{oxidation}}$ $C_9H_{16}O_3$  (*B*)

Soluble in
aqueous $NaHCO_3$
solution

Soluble in aqueous
$NaHCO_3$ solution

Zn/Hg + HCl

$NH_2OH$

$C_8H_{17}COOH$
nonanoic acid

Mixture of two isomers
$C_9H_{17}NO_3$  (*C*) and (*D*)

| PCl$_5$

Mixture of two isomers
$C_9H_{17}NO_3$  (*E*) and (*F*)

| hot 50% $H_2SO_4$

An amine, $C_4H_{11}N$  (*G*)

+

An amino acid, $C_4H_9O_2N$  (*H*)

+

$CH_3CH_2CH_2CH_2COOH$  +  $HOOCCH_2CH_2CH_2COOH$

(University of Bradford, 1983)

*Solution 13.4*

Solubility of (*A*) in $NaHCO_3$ indicates that it is a carboxylic acid. Oxidation gives (*B*) which has the formula of (*A*) minus 2H, implying a change from —CH(OH)— to $\diagdown$C=O. This is confirmed by Clemmensen reduction (Zn/Hg + HCl) of $\diagdown$C=O to —CH$_2$—. Thus (*A*) is a hydroxyacid and (*B*) is a ketoacid.

242

Reaction of (B) with $NH_2OH$ converts it into a mixture of *syn* and *anti* oximes, (C) and (D), which on treatment with $PCl_5$ undergo the Beckmann rearrangement to give a mixture of two different amides, (E) and (F). These are subsequently hydrolysed by hot 50% $H_2SO_4$ to a mixture of amines and carboxylic acids.

Bearing in mind that in the Beckmann rearrangement it is the group *anti* to the OH group which migrates, the reaction may be summarised as follows:

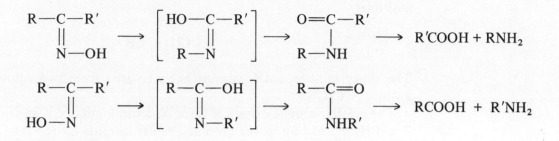

From the earlier observations, either R or R' contains a carboxyl group. Let this be R', then

$$R'COOH = HOOC(CH_2)_3COOH \text{ and } RNH_2 = CH_3(CH_2)_3NH_2$$

$$RCOOH = CH_3(CH_2)_3COOH \text{ and } R'NH_2 = HOOC(CH_2)_4NH_2$$

Then the oximes (C) and (D) must have the structures

$$CH_3(CH_2)_3\overset{\displaystyle \text{N—OH}}{\overset{\|}{—C—}}(CH_2)_3COOH \quad \text{and} \quad CH_3(CH_2)_3\overset{\displaystyle \text{HO—N}}{\overset{\|}{—C—}}(CH_2)_3COOH,$$

and the corresponding amides (E) and (F) are

$$HOOC(CH_2)_3CONH(CH_2)_3CH_3 \quad \text{and} \quad CH_3(CH_2)_3CONH(CH_2)_3COOH$$

Don't forget to check at each stage that the proposed structure is consistent with the given molecular formula.

Finally, the keto-acid (B) from which the oximes were formed must have the structure $CH_3(CH_2)_3CO(CH_2)_3COOH$, and the hydroxyacid (A) is $CH_3(CH_2)_3CH(OH)(CH_2)_3COOH$.


**Example 13.5**

An optically active hydrocarbon (B), $C_9H_{16}$, gave an optically inactive product, $C_9H_{18}$, on hydrogenation. Ozonolysis of (B), followed by reduction with Zn in $CH_3COOH$, gave ethanal and compound (C), $C_7H_{12}O$. Compound (C) was optically inactive and showed an infra-red absorption at 1745 $cm^{-1}$, and 4 lines in its proton-decoupled $^{13}C$ NMR spectrum. Its $^1H$ NMR spectrum showed absorptions at $\delta 1.0$ (6H, doublet), $\delta 1.5$ (4H, multiplet) and $\delta 2.5$ (2H, multiplet). On shaking with alkaline $D_2O$, two protons in (C) were exchanged for deuterium.

Suggest structures for (B) and (C) which are in accord with all the information above.

*(18 marks)*

The hydrogenation product of (B) contained two stereoisomers in 9:1 ratio — suggest a structure for the major isomer. *(4 marks)*

How many stereoisomers of (B) are possible (you need not draw them)? *(3 marks)*

(University of Bristol, 1985)

*Solution 13.5*

Begin with the spectroscopic data for (C). The infra-red absorption at 1745 cm$^{-1}$ points to a carbonyl compound, and the molecular formula $C_6H_{12}CO$, indicates that it is unsaturated. Thus, compound (C) has a double bond or is cyclic.

The upfield doublet of 6H in the $^1$H NMR spectrum suggests two equivalent $CH_3$ groups, each adjacent to a CH group which in turn is responsible for the 2H multiplet:

$$CH_3—CH<$$

The other multiplet, 4H, is most likely to arise from two equivalent methylene groups, $—CH_2—$.

The relative atomic masses of these structural units, $>C{=}O$, 2 $CH_3CH—$ and 2 $—CH_2—$, add up to the given molecular formula for (C). The compound must therefore have a cyclic structure, which can only be

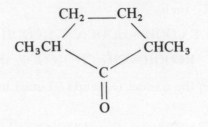

Note that this is consistent with the $^{13}$C NMR spectrum of 4 lines, corresponding to four 'different' carbon atoms (i.e. carbon atoms in four different environments, $CH_3$, $CH_2$, CH and CO).

The two hydrogen atoms adjacent to the carbonyl group will be weakly acidic, accounting for the deuterium exchange which occurs with $D_2O$.

You are told that compound (C) is optically inactive, and so the molecule must have a plane of symmetry, indicating that it is the *cis*-isomer

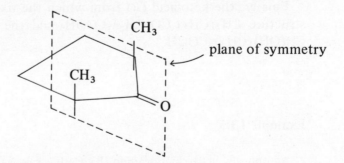

It follows that since ethanal and (C) are the ozonolysis products of (B), the latter compound has the structure

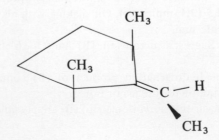

With no plane of symmetry in the molecule, (B) is optically active, but its reduction product is not:

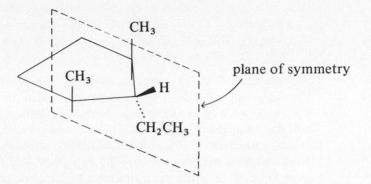

The two stereoisomers referred to in the question are those where the ethyl group is either *cis* or *trans* (*syn* or *anti*) to the two methyl groups. The major product will be the thermodynamically more stable (sterically less crowded) *trans* isomer, drawn above.

Finally, there are 4 stereoisomers of (B): 2 geometrical isomers, each of which can exist as an enantiomeric pair. Although not required by the question, make sure that you are clear about this point by drawing the four structures. These are given at the end of the solutions to the self-test questions, page 335.

# 13.3 Self-Test Questions (Guided Route)

### Question 13.1

An alcohol, A, $C_4H_{10}O$ was oxidised by $CrO_3$ to yield a carbonyl compound, B, $C_4H_8O$. On treatment of compound B with $I_2$/aq. NaOH, there was obtained a precipitate of iodoform, together with an aqueous solution of sodium propanoate.

Deduce structures for compounds A and B, and give a mechanism for the reaction of B with $I_2$/aq. NaOH. (Polytechnic of North London, 1984)

### Question 13.2

From the information given below, deduce the structures of A–D.

$$A \xrightarrow{LiAlH_4} B \xrightarrow{H_2/Pd} C \xrightarrow{c.H_2SO_4} D$$

| Compound | A | B | C | D |
|---|---|---|---|---|
| Formula | $C_{10}H_{10}O$ | $C_{10}H_{12}O$ | $C_{10}H_{14}O$ | $C_{10}H_{12}$ |
| Reaction with Brady's reagent | +ve | −ve | −ve | −ve |
| Result of iodoform test | +ve | +ve | +ve | −ve |
| Effect on bromine water | Decolorised | Decolorised | No reaction | Decolorised |

Ozonolysis of **D** gives phenylethanal and ethanal. Compound **A** can be synthesised by the reaction of propanone with benzaldehyde in the presence of base. (*15 marks*)
Give a mechanism for the above reaction used to synthesise **A**. (*5 marks*)

(Coventry Polytechnic, 1986)

### Question 13.3

A compound, **A**, $C_3H_6O$, gives an orange precipitate with a solution of 2,4-dinitrophenyl-hydrazine and reacts with methylmagnesium iodide to give compound **B**, $C_4H_{10}O$, after hydrolytic work-up of the reaction mixture. Oxidation of compound **B** with chromic acid afforded compound **C**, $C_4H_8O$, which on treatment with aqueous $KOH/I_2$ afforded yellow crystals. Reaction of compound **B** with $PBr_3$ afforded compound **D**, $C_4H_9Br$, which on treatment with AgCN gave compound **E**, $C_5H_9N$. Reduction of **E** with $LiAlH_4$ gave compound **F**, $C_5H_{13}N$, which on treatment with excess iodomethane gave compound **G**, $C_7H_{18}NI$. Reaction of compound with silver oxide gave **H**, $C_7H_{19}NO$, which on dry distillation afforded compound **I**, $C_4H_8$, and trimethylamine as the organic products.

Identify compounds **A–I** and explain the reactions occurring in each of the transformations described. (Loughborough University of Technology, 1985)
(Examiner's note, kindly supplied to the author: examinees should be aware that the use of silver cyanide leads to the *iso*nitrile as the main product.)

### Question 13.4

A hydrocarbon, *S*, was synthesised by the following route

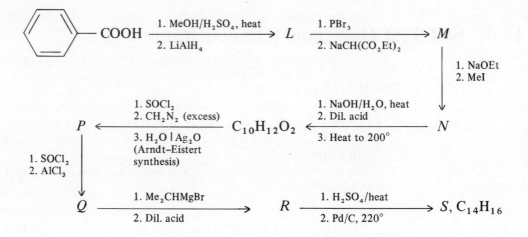

Elucidate the structure of *S* and give structures of the intermediates *L* to *R*.

(University of Leeds, 1984)

### Question 13.5

Compound *A*, $C_5H_8O$, reacts with an excess of hydrogen in the presence of a metal catalyst to give $C_5H_{10}O$, and gives a yellow precipitate with 2,4-dinitrophenylhydrazine reagent. With phenylmagnesium bromide, followed by aqueous acid, *A* gives *B*, $C_{11}H_{14}O$.

Hot phosphoric acid converts *B* to *C*, $C_{11}H_{12}$. Esterification of *D* with methanol and sulphuric acid produced *E*, $C_{12}H_{14}O_3$, which gave *F*, $C_{29}H_{28}O_2$, after treatment with phenylmagnesium bromide followed by aqueous acid.

Hot phosphoric acid converted *F* into *G*, $C_{29}H_{24}$, ozonolysis of which gave two moles of benzophenone, $(C_6H_5)_2C{=}O$, and a compound which, on oxidation ($K_2Cr_2O_4$ in hot acid) yielded acetic acid (one mole) and carbon dioxide (one mole).

On hydrogenation, compound *C* absorbed one mole of hydrogen and compound *G* absorbed two.

Deduce structures for compounds *A–G*. Give your reasoning.

(University of Sussex, 1983)

### Question 13.6

An achiral compound (*A*), $C_8H_{14}O$, yields (*B*), $C_5H_8O_2$, and acetone when treated with ozone followed by zinc and water. Compound (*A*) also reacts with iodine in sodium hydroxide solution to give a yellow precipitate and a water soluble product which can be obtained from the alkaline solution by treatment with hydrochloric acid. When compound (*B*) is warmed with sodium hydroxide solution, it undergoes reaction to yield (*C*), $C_5H_6O$.

Reaction of (*A*) with hot aqueous potassium permanganate yields (*D*), $C_5H_8O_3$, which is converted to (*E*), $C_7H_{12}O_3$, on reaction with ethanol and dry hydrogen chloride.

Deduce the structures of the compounds (*A*)–(*E*) and predict the product which would be obtained on reaction of (*B*) with hydrazine ($H_2NNH_2$).

(University of Southampton, 1983)

### Question 13.7

Compounds (H) and (I) have the same molecular formula, $C_8H_8O$. These compounds gave major absorption bands at the following infra-red frequencies ($cm^{-1}$):

(H) 3030 (m), 2900 (m), 1720 (m), 1600 (m).
(I) 2820 (m), 2730 (m), 1695 (s), 1605 (s).

The $^1H$ NMR spectra showed the following resonances with multiplicities and integrals given:

(H) at 3.66δ (doublet, 2H), 7.36δ (multiplet, 5H), 9.71δ (triplet, 1H),

and

(I) at 2.41δ (singlet, 3H), 7.19δ (doublet, 2H), 7.66δ (doublet, 2H) and 9.81δ (singlet, 1H).

When treated with dilute aqueous alkali, compound (H) gave an aldol condensation product, (J), $C_{16}H_{16}O_2$, whereas reaction of compound (I) with concentrated aqueous alkali (10M) resulted in a Cannizzaro reaction which yielded (K), $C_8H_{10}O$, and (L), $C_8H_8O_2$.

Compounds (H) and (I) both reacted with phenylhydrazine in glacial acetic acid to give phenylhydrazones (M) and (N) respectively. Comment on the infra-red and NMR data and suggest possible structures for (H) and (I); identify compounds (J) to (N).   (*10 marks*)

Using as examples the compounds which you have identified, suggest possible mechanisms for TWO of the reactions mentioned in the question.   (*10 marks*)

(The City University, 1985)

## 13.4   Worked Examples (Short Answer Questions)

### Example 13.6

(i) Which, if any, of the following exhibit enantiomerism?

$$CH_3-\underset{\underset{Cl}{|}}{CH}-CH{=}CH_2, \quad CH_3-\underset{\underset{CH_3}{|}}{\overset{\overset{OH}{|}}{C}}-CH_2-CH_3, \quad CH_2{=}CH-\underset{\underset{Cl}{|}}{CH}-CH_2-CH_2Cl$$

(*3 marks*)

(ii) Predict the product formed in each of the two reactions given. Briefly explain your answer.

$$CH_3—CH=CH_2 \xrightarrow{H_2O/H_2SO_4} ?$$

$$CH_3—CH=CH_2 \xrightarrow[\text{(ii) } H_2O_2/NaOH]{\text{(i) } (BH_3)_2} ?$$

*(3 marks)*

(iii) Give the IUPAC name of each of the following

$$CH_3—CH_2—CH_2—CH—CH_2—CH_2—CH_3, \quad CH_2=CH—CH_2—CH_2OH, \quad ClCH_2—CH_2—CHO$$
$$\underset{\displaystyle CH_3—CH—CH_3}{|}$$

*(3 marks)*

(iv) What reagent(s) would be used to accomplish the following?

$$trans\ CH_3—CH_2—CH=CH—CH_3 \xrightarrow{?} CH_3—CH_2—CH=O + CH_3—CH=O$$

What product(s) would you expect to find from the *cis* isomer using the same reagent(s)? *(3 marks)*

(v) What product would you expect to obtain when (R)-2-bromo-2-phenyloctane is allowed to react in a mixture of water and acetone? *(3 marks)*

(vi) List three different reagents for the conversion of propan-2-ol to 2-chloropropane. Indicate with a brief explanation which method you think is the best.

(Sunderland Polytechnic, 1985)
(Part-question, total marks 35, time allowed, 45 min)

*Solution 13.6*

(i)

$$CH_3—\overset{*}{C}H—CH=CH_2 \quad and \quad CH_2=CH—\overset{*}{C}H—CH_2—CH_2Cl \quad (* \text{ chiral carbon atom})$$
$$\underset{\displaystyle Cl}{|} \qquad\qquad\qquad\qquad \underset{\displaystyle Cl}{|}$$

The other compound has no chiral carbon atom.

(ii) $CH_3CH(OH)CH_3$ via the more stable of the two possible carbocation intermediates, $CH_3—\overset{\oplus}{C}H—CH_3$.

$CH_3CH_2CH_2OH$. Reaction is not ionic, but proceeds via a cyclic transition state in which the electron-deficient boron atom receives electrons from the pi bond:

$$CH_3—CH=CH_2 \quad\longrightarrow\quad \left[ CH_3—CH\cdots CH_2 \atop \quad\ \ H\cdots BH_2 \right] \quad\longrightarrow\quad CH_3—CH_2—CH_2—BH_2$$
$$\underset{\displaystyle H—BH_2}{\ }$$

Repetition gives $(CH_3—CH_2—CH_2)_3B$, which on treatment with $H_2O_2/$ NaOH gives $3 \times CH_3—CH_2—CH_2OH$.

(iii) 4-(methylethyl)heptane, but-3-en-1-ol, 3-chloropropanal.

(iv) Ozone followed by decomposition of the ozonide with zinc dust and water (or hydrogen plus a catalyst). *Cis* and *trans* isomers give the same ozonolysis products.

(v) As a tertiary haloalkane able to form a resonance stabilised carbocation

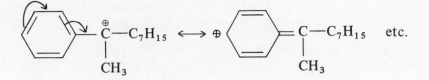

reaction in the highly ionising solvent mixture will be *unimolecular*. Attack by a water molecule can occur with equal ease on either side of the planar carbocation intermediate, leading to an equimolar mixture of (R) and (S) 2-phenyloctan-2-ol (racemisation).

(vi) 1. Conc. aq. $HCl/ZnCl_2$
2. $NaCl/H_2SO_4$
3. $PCl_3$

Protonation of the alcohol in acidic media may lead to the formation of some secondary propyl carbocation ($CH_3—\overset{\oplus}{C}H—CH_3$) which in turn could expel a proton to form propene. Such competition from elimination does not occur with the phosphorus halide, which is therefore the preferred reagent.

**Example 13.7**

1. Define the stereochemical term configuration. With the help of a clear three-dimensional representation of a suitable structural formula, explain the changes in stereochemistry that are involved when a change of configuration occurs.
2. Complete the following equation: $CH_3Br + NaOH =$ ..............
   What type of reaction is this?
3. Write structural formulae for ALL compounds represented by (a) $C_3H_8O$, (b) $C_3H_6$.
4. Give an example of a condensation reaction. Show through writing out a mechanistic scheme, that the example you have chosen involves a sequence of addition and elimination processes.
5. Illustrate the terms 'aromatic compound' and 'aromatic substitution' by giving one example of each.
6. Distinguish carefully between the terms 'hydrogenation' and 'reduction'. Give ONE example for each process involving the formation of a saturated compound, and include experimental details and reagents that you would use to carry out the examples you have chosen.

(Oxford Polytechnic, 1985)

(One compulsory section of a 2-h paper)

*Solution 13.7*

1. 'Configuration' relates to the gross spatial arrangement of the atoms in a molecule which can only be changed by breaking and reforming bonds, e.g.

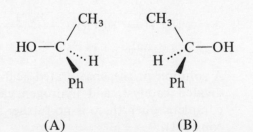

      (A)              (B)

Stereoisomer (A) can only be converted into (B) by interchanging positions of two of the ligands attached to the central (chiral) carbon atom. The process is said to involve *inversion* of configuration, and the two isomers (enantiomers) are non-superposable.

2.  $CH_3Br + NaOH \longrightarrow CH_3OH + NaBr$

Nucleophilic substitution (hydrolysis).

3. (a)  $CH_3CH_2CH_2OH, CH_3CH(OH)CH_3, CH_3OCH_2CH_3$

  (b)  $CH_3CH{=}CH_2, \quad CH_2{-}CH_2$
  
  $\qquad\qquad\qquad\qquad\qquad CH_2$

4. You have a wide choice here! A suitable example would be reaction of ethanal with hydroxylamine to form ethanal oxime:

$$CH_3CHO + NH_2OH \longrightarrow CH_3CH{=}NOH + H_2O$$

  (i) nucleophilic addition to the carbonyl group

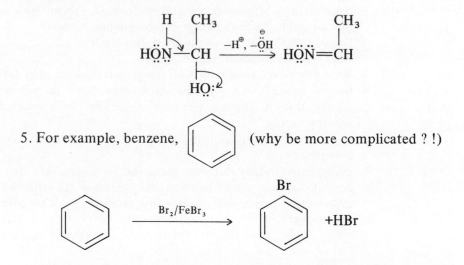

  (ii) elimination of a molecule of water

5. For example, benzene, ⬡ (why be more complicated ? !)

+HBr

Aromatic electrophilic substitution, H replaced by Br. The wording of the question does not imply that you are expected to write out the mechanism.

6. 'Hydrogenation' involves the addition of elemental hydrogen to an unsaturated compound — a process normally requiring the aid of a catalyst, e.g.

$$CH_3CH{=}CHCOOH \xrightarrow{H_2/Pd} CH_3CH_2CH_2COOH$$

but-2-enoic acid          butanoic acid

A solution of the unsaturated acid is stirred with a suspension of the finely divided catalyst, and hydrogen gas passed in under pressure. Reaction is complete when there is no further absorption of hydrogen (pressure remains constant).

Reduction by other methods usually involves addition of hydrogen atoms to an unsaturated group (e.g. reduction of $\diagup\!\!\!\diagdown C\!\!=\!\!O$ with $LiAlH_4$ or $NaBH_4$ to $-CH(OH)-$) or replacement of an atom by one or more hydrogen atoms (e.g. reduction of $\diagup\!\!\!\diagdown C\!\!=\!\!O$ with amalgamated zinc and concentrated hydrochloric acid to $-CH_2-$). To show that you appreciate the wider application of the concept, you could take as an example the reduction of a halogen compound via the formation of a Grignard reagent, e.g.

$$PhCH_2Br + Mg \longrightarrow PhCH_2MgBr \xrightarrow{\ H_2O\ } PhCH_3 + Mg(OH)Br$$

A solution of the compound in dry ether, contained in a flask fitted with a reflux condenser, is stirred with a slight excess of magnesium turnings. On completion of reaction, the solution is decanted from any residual magnesium and the Grignard reagent decomposed by dropwise addition of water. The organic layer is subsequently dried over anhydrous $MgSO_4$, and the product recovered by removal of the ether in a rotatory evaporator.

**Example 13.8**

Answers to this question require only one structure or one short sentence. Answer ALL parts.
(a) Draw the structure corresponding to (E)-1,1-dichloropent-3-en-2-one.
(b) Give a three-dimensional representation of (R)-2-methylbutanal.
(c) Draw a Newman projection of compound (I).

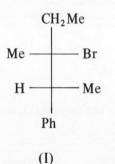

(I)

(d) Paying particular attention to stereochemistry, draw the product resulting from $S_N2$ attack of hydroxide ion on compound (I).
(e) Why is 2,2,2-trifluoroethanol more acidic than ethanol?
(f) Draw the major canonical (resonance) structure of ethanamide, which can be used to explain its low basicity.
(g) What effects would the addition of a small amount of an organic impurity have on the melting point of an otherwise pure organic compound?
(h) Why is ethyne considerably more acidic than ethane?
(i) Why does 2-nitrophenol have a lower boiling point than 4-nitrophenol?
(j) Draw the enol tautomer of 2-phenylethanal ($Ph-CH_2-CHO$).

(University of York, 1984)

*Solution 13.8*

(a) Students often find it easier to construct an organic structure from the systematic name than vice versa. Look first at the suffix for the number of carbon atoms in the parent compound (penta = 5). Draw a row of carbon atoms, numbered appropriately,

$$\overset{5}{C}-\overset{4}{C}-\overset{3}{C}-\overset{2}{C}-\overset{1}{C}$$

and then put in the functional groups and substituents,

$$\overset{5}{CH_3}-\overset{4}{CH}=\overset{3}{CH}-\overset{2}{CO}-\overset{1}{CHCl_2}$$

The (E) isomer (see page 24 if you have forgotten the convention!) has the two top-priority groups on opposite sides of the double bond, so the required structure is

(b) The order of decreasing priority (R—S system, see page 20) of ligands on the chiral carbon atom is

$$CHO > CH_3CH_2 > CH_3 > H$$

For the (R) isomer these must be placed in clockwise descending order when the molecule is viewed with the ligand of lowest priority facing away:

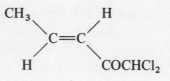

Thus the required structure is

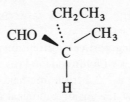

(c) Draw the view along the C1–C2 axis (represented by the intersections of the crossed lines in the Fischer projection). No particular conformation is asked for, but the Fischer projection is in the eclipsed form:

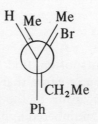

252

(d) Bimolecular reaction will invert the configuration about C2, so the resulting alcohol will have the structure

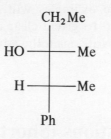

(e) Because the powerful negative inductive effect of the three fluorine atoms stabilises the conjugate alkoxide ion, and lowers the activation energy for its formation. (In a longer answer, you would explain that it is the *differential* effect of the fluorine atoms in stabilising the conjugate anion relative to the undissociated alcohol which determines the position of equilibrium, i.e. the acid strength — see Chapter 4).

(f)

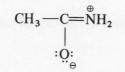

If the lone pair of electrons on the nitrogen atom are involved in resonance, they are less available for donation to a proton. (Alternatively, uptake of a proton can only be accomplished at the expense of loss of resonance energy associated with conjugation of the amino and carbonyl groups. And protonation of the nitrogen atom sets up an unfavourable repulsion between adjacent $N^{\oplus}$ and $C^{\delta+}$ atoms:

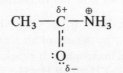

(g) Even small amounts of organic impurities would depress (i.e. lower) the melting point of a pure organic compound. (Hence the use of melting points as a criterion of purity.)

(h) The sp hybridised carbon atom of an alkyne is more electronegative than the $sp^3$ carbon atom in ethane, and so is better able to accommodate a negative charge in the conjugate anion. Another way of putting this would be to say that in the conjugate anion, the electron pair is held more strongly in a compact sp hybrid orbital (with high 's' character) than in a more diffuse $sp^3$ orbital, and is therefore less readily donated to a proton).

(i) Hydrogen bonding between an oxygen atom of the nitro group and the hydrogen atom of the hydroxyl group is *intra*molecular in 2-nitrophenol, but *inter*molecular in 4-nitrophenol. This raises the boiling point of the latter compound. Draw the structure of the 2-nitro isomer:

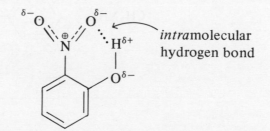

*intra*molecular hydrogen bond

(j) Ph—CH=CH—OH

As these examples will have shown, short-answer questions are often wide ranging and to give a clear concise answer in just one or two sentences is not all that easy. Note again the advice given at the beginning of this chapter!

# 13.5 Self-Test Questions (Short Answer Questions)

### Question 13.8

1. Draw diagrams to show ALL the stereoisomers of the following (if you use any special conventions in representing the isomers, please identify them):

   (i) $CH_3$—$CH(OH)$—$CHO$        (ii) $CH_3$—$CH=CH$—$C_2H_5$        (*5 marks*)

2. Sketch the most stable and the least stable conformations of the following molecules:

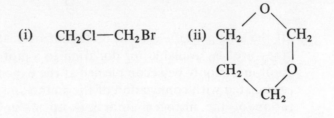

   (*5 marks*)

3. Identify each of the following as electrophile, nucleophile, free radical or none of these:

   (i) $CH_3COO^{\ominus}$   (ii) $\overset{.}{Br}$   (iii) $NH_4^{\oplus}$   (iv) $\overset{\ominus}{Cl}$        (*2 marks*)

4. Indicate, by means of an arrow, the atom in the following molecules which you would expect a nucleophile to attack:
   (i) $CH_3$—$CH=CHCH_2CH_2Cl$   (ii) $CH_3$—$CO$—$O$—$CH=CH_2$        (*2 marks*)

5. Complete the following equations:

   $$CH_3—COOC_2H_5 + NaOH \longrightarrow$$

   $$CH_3—CHO + C_2H_5OH \xrightarrow{\ H^{\oplus}\ }$$

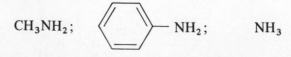

   (*9 marks*)

6. Briefly describe one test to distinguish between an alcohol and a phenol.   (*3 marks*)
7. Place the following bases in order of increasing basic strength:

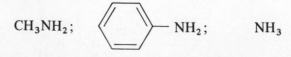

   (*2 marks*)
   (Wolverhampton Polytechnic, 1985)
   (part of a compulsory section)

254

## Question 13.9

1. The typical reaction of carbonyl compounds is:
   A electrophilic attack at carbonyl carbon
   B nucleophilic attack at carbonyl carbon
   C free radical attack at carbonyl carbon
   D electrophilic attack at carbonyl oxygen
2. Which of the following would you predict to be aromatic?

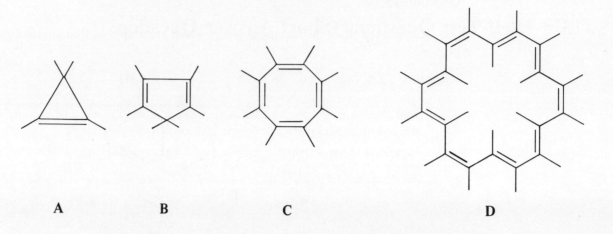

   A        B        C        D

3. The nitronium ion ($NO_2^{\oplus}$) commonly functions in organic chemistry as:
   A a free radical
   B an electrophile
   C a nucleophile
   D a Lewis base
4. The formulae represent

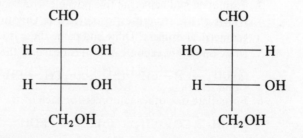

   A diastereoisomers
   B enantiomers
   C conformers
   D geometrical isomers
5. Applying the Sequence Rules, which of the following priority orders is correct in determining the R—S configuration of

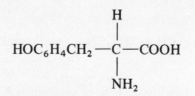

   A $NH_2 > HOC_6H_4CH_2 > COOH > H$
   B $NH_2 > COOH > HOC_6H_4CH_2 > H$
   C $HOC_6H_4CH_2 > COOH > NH_2 > H$
   D $COOH > HOC_6H_4CH_2 > NH_2 > H$

6. Which of the following compounds on hydrolysis produces butan-2-ol and ethanoic acid?

    A $CH_3CH_2CH(CH_3)_2$
    B $CH_3CH_2CO_2CH(CH_3)_2$
    C $CH_3CO_2CH(CH_3)CH_2CH_3$
    D $CH_3CH_2CO_2CH(CH_3)CH_2CH_3$

(Plymouth Polytechnic, 1984)<br>(part of compulsory section)

## Question 13.10

1. Encircle and name the functional groups which are present in the morphine molecule (I)

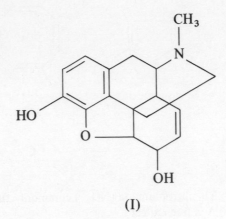

(I)

2. Write structural formulae for the following compounds:
   (a) 3-methylpentan-2,4-dione;
   (b) 2-methylhex-3-yn-1-ol.
3. Formulate and name the five possible structural isomers of molecular formula $C_4H_8$.
4. Which of the structures formulated in Question 2 may be further represented as a pair of geometrical isomers? Draw and name these isomers.
5. Draw one other canonical form for each of the following species:

    (a) $CH_3-\overset{\oplus}{C}H-CH{=}CH_2$    (b) $CH_3-\overset{\ominus}{C}H-CHO$    (c) $CH_2{=}CH-\overset{\cdot}{C}H_2$

6. Formulate one other tautomer for each of the following structures:

    (a) $CH_3-CO-CH_3$    (b) $CH_2{=}CHOH$    (c) $CH_3-CO-NH_2$

7. Draw the structure for the preferred conformation of methylcyclohexane, labelling the axial and equatorial bonds.
8. Draw the Newman projection formula for (a) the most stable, and (b) the least stable conformation of 1-bromopropane.
9. Give one simple chemical test in each case which could be used to distinguish between the compounds in the following pairs:
   (a) $CH_3CH(OH)CH_3$   and   $CH_3CH_2CH_2OH$
   (b) $CH_3CH_2CHO$   and   $CH_3COCH_3$

    (c) $CH_3(CH_2)_4CH_3$       and

10. Identify the lettered compounds A–D obtained in the following reaction sequences:

    (a) $(CH_3)_2CHCH{=}CH_2$ $\xrightarrow[\text{(ii) }H_2O_2,\ HO^{\ominus}]{\text{(i) }B_2H_6}$ A $\xrightarrow{H_2CrO_4}$ B

    (b) $CH_3CH{=}CH_2 + HBr$ $\xrightarrow{\text{peroxides}}$ C $\xrightarrow{CH_3O^{\ominus}}$ D

256

11. Formulate the two alkenes obtained from dehydrobromination of 3-bromo-2-methyl-butane. Indicate which is the major product, and explain why it is formed.

(Thames Polytechnic, 1984)
(part of compulsory section)

**Question 13.11**

1. $CH_3CH_2\overset{\oplus}{\underset{\underset{CH_3}{|}}{CHN}}(CH_3)_3 \ \overset{\ominus}{OH}$    (I)

(a). Write down the structure of the principal product obtained by heating compound (I).
(*1 mark*)

(b) Write down a mechanism for this E2 reaction, using a Newman projection to show the reacting conformation of (I).    (*2 marks*)

2. When 2-bromopropane (isopropyl bromide) is treated with (a) ethanol, or (b) sodium ethoxide in ethanol, two different distributions of products are obtained as follows:

$$CH_3\underset{\underset{Br}{|}}{CH}CH_3 \quad \overset{55°}{\underset{\text{(b) NaOC}_2\text{H}_5/\text{C}_2\text{H}_5\text{OH}}{\overset{\text{(a) C}_2\text{H}_5\text{OH}}{\diagdown}}} \quad CH_3CH{=}CH_2 \ + \ (CH_3)_2CHOC_2H_5$$

| (a) | 3% | 97% |
| (b) | 79% | 21% |

What explanation can you offer for this change in product distribution brought about by changing the reagents?    (*3 marks*)

3. Write down the product(s) obtained by the addition of DBr to *cis*-but-2-ene, showing fully, with standard three-dimensional formulae, the stereochemistry of each one. State the relationship between the products if there is more than one.    (*3 marks*)

4. Write down the structure of the principal products obtained in each of the following addition reactions.

(a)    $(CH_3)_2C{=}CH_2 + HCl \quad \longrightarrow$    (*2 marks*)

(b)    $-CH_3 + H_2O \quad \overset{H^{\oplus}}{\longrightarrow}$    (*2 marks*)

(c)    $(CH_3)_3\overset{\oplus}{N}CH{=}CH_2 \ \overset{\ominus}{I} \ + \ HI \longrightarrow$    (*2 marks*)

5. (a) Give the structures of the Diels–Alder products obtained from the following sets of compounds.

(i)    $+ CH_2{=}CH{-}CH{=}CH_2$    (*2 marks*)

(ii)    $CH_2{=}CH{-}CH{=}CH_2 + C_6H_5CH{=}CHNO_2$    (*2 marks*)

257

(b) Which diene and dienophile combination will produce the following Diels–Alder adducts?

*(2 marks)*

(i)

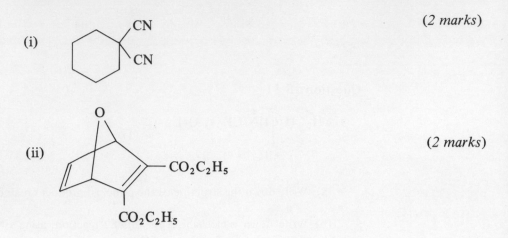

(ii)

*(2 marks)*

6. (a) Write a simple mechanism to account for the following reaction, making use of 'curly arrows', and showing clearly the role of the base catalyst.

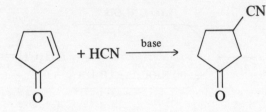

*(3 marks)*

(b) What other product might you have expected from this combination of reagents and catalysts. *(1 mark)*

7. Which of the following molecules or ions would you expect to show aromatic character?

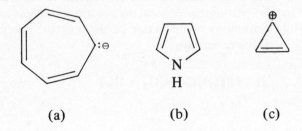

   (a)             (b)        (c)

8. Give the major organic product which you would expect from each of the following sets of reagents.

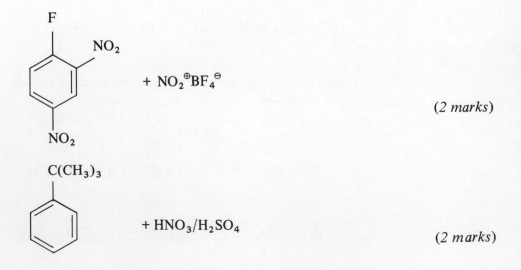

$+ NO_2^{\oplus}BF_4^{\ominus}$

*(2 marks)*

$+ HNO_3/H_2SO_4$

*(2 marks)*

258

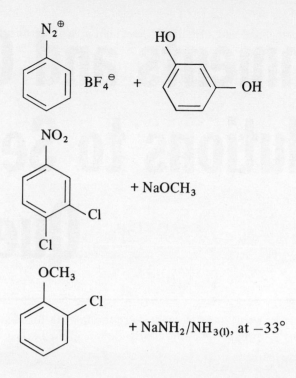

(2 marks)

+ NaOCH$_3$

(2 marks)

+ NaNH$_2$/NH$_{3(l)}$, at $-33°$

(2 marks)

(University of Sussex, 1982)
(compulsory section)

# 14 Comments and Outline Solutions to Self-Test Questions

**Chapter 1**

*Question 1.1*

(a)

(i)

(ii)

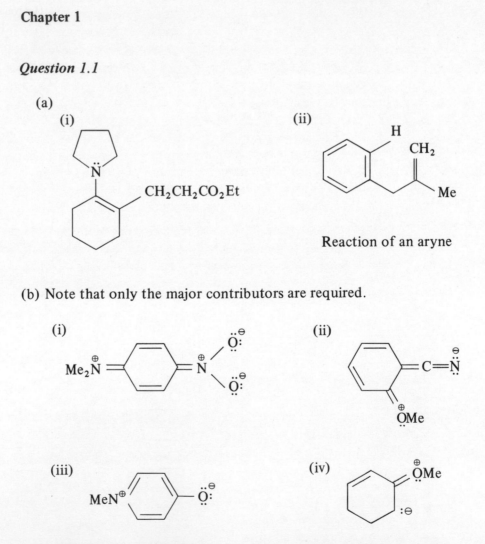

Reaction of an aryne

(b) Note that only the major contributors are required.

(i)                                    (ii)

(iii)                                  (iv)

Unshared electron pairs are shown so that you can check the electron distribution, but are not strictly required as they have been omitted in the question.

(c) Chlorine in the 2-position destabilises the acid,

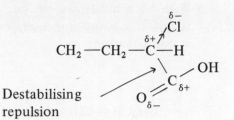

Destabilising
repulsion

but stabilises the conjugate anion,

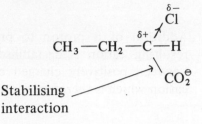

Stabilising
interaction

thus increasing the acidity. The effect diminishes as the chlorine atom is moved further from the functional group. See Section 4.10 and Example 4.1.

(d) The +I effect of the ethyl group attached to the positively charged nitrogen atom is stabilising. Steric compression between three ethyl groups and the three methyl groups on the boron atom outweighs inductive stabilisation and the triethyl compound is the least stable of the three.

(e) This topic may not have been discussed in your course. Briefly, lithium chloride dissolves in polar solvents to give solvated ion pairs in equilibrium with separately solvated ions.

(i) Ethanol is a moderately polar, *protic* solvent, able to solvate anions by hydrogen bonding, and cations by ion–dipole interaction (electron pair donation by the oxygen atom).

(ii) Acetonitrile ($CH_3\overset{\delta+}{C}{\equiv}\overset{..\,\delta-}{N}$) is an *aprotic* polar solvent of higher relative permittivity (dielectric constant) than ethanol. It will solvate lithium ions by ion–dipole interaction, but is less effective in solvating chloride ion, which is therefore much more reactive in this solvent.

### Question 1.2

(a) $H_2C{=}\overset{\oplus}{O}{-}H$ (One extra electron-pair bond, and both C and O have complete octets.)

(b) $H_2C{=}CH{-}\overset{..\,\ominus}{O}{:}$ (Oxygen is better able to accommodate the negative charge.)

(c) $H_2\overset{\oplus}{N}{=}CH_2$ (N and C both have complete octets, and extra electron pair bond.)

(d) $H_2C{=}\overset{\oplus}{N}{=}\overset{\ominus}{\underset{..}{N}}$ (Nitrogen is better able to accommodate the negative charge.)

(e) $H_2\overset{\oplus}{N}{=}CH{-}\overset{..}{O}{-}CH_3$ (Nitrogen is better able to accommodate the positive charge.)

*Question 1.3*

(a) With both substrates, the rate determining step is protonation of the alkene to give a carbocation intermediate. Propene preferentially forms the more stable 2° cation:

$$CH_3CH{=}CH_2 \xrightarrow{\quad H^\oplus \quad}$$

$$CH_3CH_2{\rightarrow}\overset{\oplus}{C}H_2 \qquad (1°)$$

$$CH_3{\rightarrow}\overset{\oplus}{C}H{\leftarrow}CH_3 \qquad (2°)$$

Addition of a proton to propenoic acid is more difficult because the resulting cation is destabilised by the polar carboxyl group. Repulsion between positively charged carbon atoms is minimised in the 1° carbocation which is therefore formed in preference to the 2° ion:

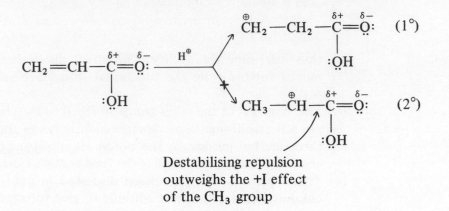

Destabilising repulsion outweighs the +I effect of the $CH_3$ group

(b) Bromomethoxymethane readily ionises in a polar solvent since loss of $Br^\ominus$ gives a resonance stabilised cation:

$$CH_3{-}\ddot{O}{-}CH_2{-}\ddot{B}\ddot{r}{:} \;\rightleftharpoons\; CH_3{-}\overset{\curvearrowleft}{\ddot{O}}{\overset{\oplus}{-}}CH_2 \;\longleftrightarrow\; CH_3{-}\overset{\oplus}{\ddot{O}}{=}CH_2 + {:}\ddot{B}\ddot{r}{:}^\ominus$$

$$\qquad\qquad\qquad\qquad\qquad\qquad\qquad (I) \qquad\qquad\qquad\qquad (II)$$

Limiting form (II) is particularly stable as it has one additional electron pair bond, and the oxygen and carbon atoms all have complete octets.

Bromopropanone shows less tendency to ionise in this way because the resulting carbocation would be destabilised by having positive charge on adjacent atoms:

$$CH_3{-}\overset{\delta+}{C}{-}CH_2{-}\ddot{B}\ddot{r}{:} \;\not\rightleftharpoons\; CH_3{-}\overset{\delta+}{C}{-}\overset{\oplus}{C}H_2 + {:}\ddot{B}\ddot{r}{:}^\ominus$$

$$\overset{\|}{{:}\ddot{O}{:}^{\delta-}} \qquad\qquad\qquad\quad \overset{\|}{{:}\ddot{O}{:}^{\delta-}} \quad \text{destabilising repulsion}$$

On the other hand, the hydrogen atoms on the α-carbon atom are activated (i.e. made weakly acidic) by the presence of the two electron-withdrawing substituents, carbonyl and bromine. Removal of an α-hydrogen atom by a base such as hydroxide ion is favoured by the formation of a resonance stabilised anion:

$$\overset{\ominus}{HO}: \overset{\curvearrowright}{\downarrow} H \overset{\frown}{-CH} \overset{}{-C} = O \;\rightleftharpoons\; \overset{\cdot\cdot}{HO} - H + \overset{\ominus}{:}\overset{\cdot\cdot}{\underset{}{C}} \overset{\frown}{H} \overset{}{-C} \overset{\curvearrowright}{=} \overset{\cdot\cdot}{O}: \;\longleftrightarrow\; CH = C - \overset{\cdot\cdot}{\underset{\cdot\cdot}{O}}{:}^{\ominus}$$

$$\qquad\quad \underset{:\overset{\cdot\cdot}{Br}:}{|}\; \underset{CH_3}{|} \qquad\qquad\qquad\qquad \underset{:\overset{\cdot\cdot}{Br}:}{|}\; \underset{CH_3}{|} \qquad\qquad\qquad \underset{:\overset{\cdot\cdot}{Br}:}{|}\; \underset{CH_3}{|}$$

$$\qquad\qquad\qquad\qquad\qquad\qquad\qquad\qquad\qquad (I) \qquad\qquad\qquad\qquad (II)$$

Limiting form (I) is stabilised by the negative inductive effect of the bromine atom, and (II) makes a large contribution because the oxygen atom is better at accommodating the negative charge than is the less electronegative carbon atom.

(c) Addition of a proton to ethene gives a relatively unstable primary carbocation,

$$CH_2 \overset{\curvearrowright}{=\!=} CH_2 \overset{\curvearrowright}{\downarrow} H \overset{\frown}{-} \overset{\cdot\cdot}{\underset{\cdot\cdot}{I}}: \;\rightleftharpoons\; \overset{\oplus}{C}H_2 - CH_3 + :\overset{\cdot\cdot}{\underset{\cdot\cdot}{I}}:^{\ominus}$$

and reaction with HI is slow and incomplete.

However, chloroethene reacts even less readily because the negative inductive of the chlorine atom is destabilising and therefore increases the activation energy for protonation:

$$:\overset{\cdot\cdot}{\underset{\cdot\cdot}{C}}l - CH_2 \overset{\curvearrowright}{=\!=} CH_2 \overset{\curvearrowright}{\downarrow} H \overset{\frown}{-} \overset{\cdot\cdot}{\underset{\cdot\cdot}{I}}: \;\rightleftharpoons\; :\overset{\cdot\cdot}{\underset{\cdot\cdot}{C}}l \leftarrow \overset{\oplus}{C}H_2 - CH_3 + :\overset{\cdot\cdot}{\underset{\cdot\cdot}{I}}:^{\ominus}$$

$$\qquad\qquad\qquad\qquad\qquad\qquad\qquad\qquad\qquad \overset{}{\underset{\text{destabilising}}{\diagdown -I \text{ effect of chlorine is}}}$$

In principle, reaction could proceed via either of the two carbocation intermediates (I) or (II):

$$CH_2 = CH - \overset{\cdot\cdot}{\underset{\cdot\cdot}{C}}l: \xrightarrow{H^{\oplus}} \begin{cases} \overset{\oplus}{C}H_2 - CH_2 - \overset{\cdot\cdot}{\underset{\cdot\cdot}{C}}l: \quad (I) \\[2ex] CH_3 - \overset{\oplus}{C}H - \overset{\cdot\cdot}{\underset{\cdot\cdot}{C}}l: \quad (II) \end{cases}$$

Since the product is 1-chloro-2-iodoethane, protonation occurs only at C2 to give what must be the more stable of the two intermediates, carbocation (II).

Reaction is very slow because of the strong negative inductive effect of the chlorine atom. Greater activation energy is required for the formation of $CH_3\overset{\oplus}{C}HCl$ from $CH_2\!=\!CHCl$ than $CH_3\overset{\oplus}{C}H_2$ from $CH_2\!=\!CH_2$.

It might initially be expected that carbocation (I), where the positive charge is further removed from the chlorine atom, would be more stable than (II). The explanation is that (II) has a resonance structure not possible for carbocation (I):

$$:\overset{\cdot\cdot}{\underset{}{C}}l \overset{\curvearrowleft}{-} \overset{\oplus}{C}H - CH_3 \;\longleftrightarrow\; \overset{\oplus}{:\overset{\cdot\cdot}{C}l} = CH - CH_3$$

$$\qquad\qquad (III) \qquad\qquad\qquad\qquad (IV)$$

Although the second limiting form (IV) has the positive charge on the chlorine atom, it still makes a significant contribution since it has an addi-

263

tional electron pair bond, and both the chlorine and carbon atoms have complete octets.

### Question 1.4

Aromatic electrophilic substitution involves the formation of a cationic intermediate (sigma complex) in which the positive charge is shared by carbon atoms *ortho* and *para* to the position attacked

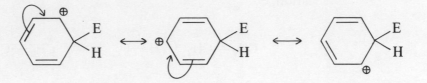

summarised by the composite formula,

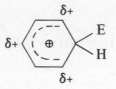

It follows that substituents (whether activating or deactivating) will have their maximum effect when in those positions.

Electron releasing substituents will stabilise the sigma complexes for *ortho* and *para* attack and will therefore be *ortho/para* directing.

Electron attracting substituents will destabilise the sigma complexes for *ortho* and *para* substitution, and will therefore be *meta* directing.

Give some examples of stabilising groups, e.g.

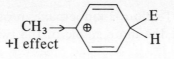

+I effect

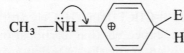

+M effect

N.B. +M ≫ −I

e.g.

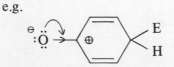

+I effect
+M effect

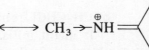

additional limiting form

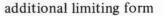

very stable additional limiting form, charges neutralised

264

and of destabilising (*meta* directing) groups, e.g.

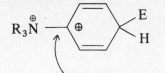

destabilising repulsion

e.g.

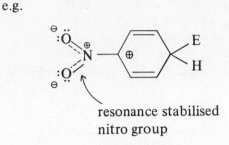

resonance stabilised
nitro group

Finally, you should mention that in order to maximise resonance interaction of a substituent with the aromatic ring, the relevant p orbitals have to be correctly aligned. Quote as an example the observation that the dimethylamino group is *ortho/para* directing and N,N-dimethylphenylamine couples with the diazonium cation ($PhN_2^{\oplus}$) in the *para* position. The presence of methyl groups adjacent to the dimethylamino group restricts overlap of the nitrogen and carbon 2p orbitals, and 2,6-dimethyl-N,N-dimethylphenylamine does not react with diazonium cation.

For a fuller discussion, see Chapter 10, Example 10.2.

## Chapter 2

### *Question 2.1*

Notice that although the question says 'explain concisely', the examiner is expecting more than a simple definition of the two terms. You must show that you understand their meaning and the difference between them. *Configuration* relates to the spatial arrangement of the atoms or groups in a stereoisomer and can only be changed by breaking and reforming bonds. *Conformation* deals with different spatial arrangements of atoms or groups in a molecule which are interconvertible by rotations about sigma bonds. A stereoisomer has only one configuration but many conformations may be possible. Illustrate your answer with specific examples, e.g. represent the two configurations of glyceraldehyde by Fischer projections and show eclipsed and staggered conformations of one of them by means of Newman projections along the C2–C3 axis:

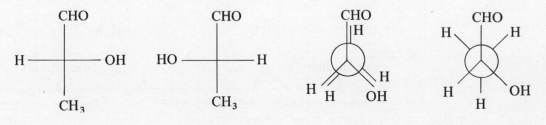

Begin the next section by writing down the formula of each compound, noting the position of chiral carbon atoms and/or double bonds present. Then represent the stereoisomers by appropriate projections, labelling enantiomers, diastereo-isomers, etc.

(a)

$CH_3CH(Br)CH(Br)CH_3$

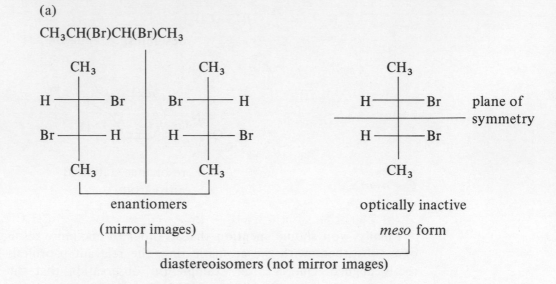

enantiomers

(mirror images)

optically inactive

*meso* form

diastereoisomers (not mirror images)

(b) $CH_3CHClCHClCH_2CH_3$ There are now two different chiral centres, so there will be $2^2 = 4$ isomers = 2 enantiomeric pairs:

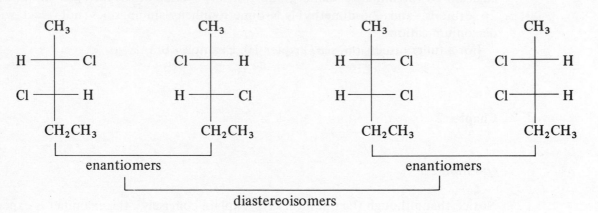

enantiomers

enantiomers

diastereoisomers

(c) $CH_3CH{=}CHCHBrCH_2CH_3$ This compound has a chiral centre and a double bond. There are two geometric isomers, *cis* (Z) and *trans* (E), each of which exists as an enantiomeric pair:

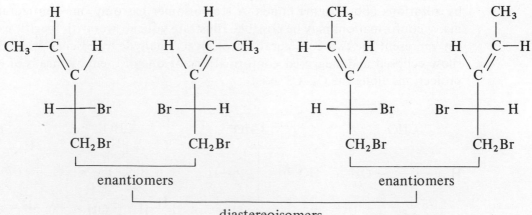

enantiomers

enantiomers

diastereoisomers

(d) $CH_3CH\!=\!CH\!-\!CH\!=\!CHCH_2CH_3$ There are no chiral carbon atoms, but the presence of two double bonds means that four geometric isomers are possible. These are *cis-cis*, *cis-trans*, *trans-cis* and *trans-trans* (Z-Z, Z-E, E-Z and EE), e.g.

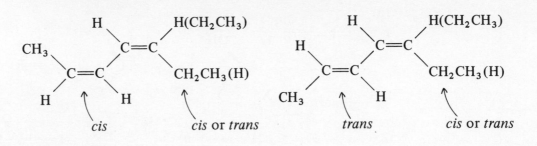

cis        cis or trans        trans        cis or trans

## Question 2.2

(a) The order of priority is $-Br > -CH\!=\!CH_2 > -CH_2CH_3 > -H$

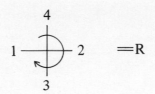

1 ——— 2    = R

(b) Only the top carbon atom is chiral. The order of priority is
$-F > -CH(CH_3)_2 > -CH\!=\!CH_2 > -CH_3$

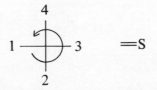

1 ——— 3    = S

Note that in (a), $-CH\!=\!CH_2 \equiv -C\!\!\begin{smallmatrix} C-\\ \\ C- \end{smallmatrix}$ has priority over $-CH_2CH_3$,

but in (b) the 'real' $-C\!\!\begin{smallmatrix} C-\\ \\ C- \end{smallmatrix}$ has priority over $-CH\!=\!CH_2$.

(c) Priorities are $-OH > -CHO\ (\equiv -CH\!\!\begin{smallmatrix} O-\\ \\ O- \end{smallmatrix}) > -CH_2OH > -H.$

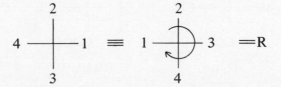

(d) The top left-hand carbon atom of the ring (labelled X) is $-\overset{\displaystyle H}{\underset{\displaystyle H}{C}}\!-\!C\!-$

267

and the top right-hand carbon (labelled Y) is equivalent to —$\overset{\displaystyle H}{\underset{\displaystyle H}{C}}$—$\overset{\displaystyle |}{\underset{\displaystyle |}{C}}$—.

Thus Y has priority over X and the order of decreasing priorities is

—OH > —Y > —X > —H

### Question 2.3

(a)

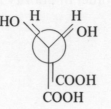

(b) Priority of ligands on C-1 is —OH > —COOH > — CH(OH)COOH > —H, therefore the arrangement is

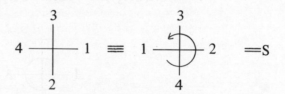

Priorities on C-2 are the same, but the ligands are arranged the other way round, giving C-2 the R-configuration.

(c) For easier comparison, it is a good idea to convert all the formulae into Fischer projections (remembering that they represent eclipsed conformations)

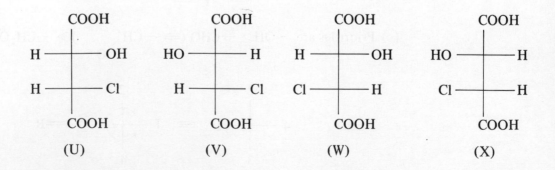

Thus, U and X, and V and W are enantiomeric pairs, and the two pairs are diastereoisomers.

(d) Use one enantiomer of the optically active alcohol, say the (+) form, (+)-2-methylbutan-1-ol. Reaction with the racemic acid will give a diastereoisomeric pair of esters:

2 (±)-acid + 2 (+)-alcohol $\longrightarrow$ (+) (+)-ester + (−) (+)-ester

Diastereoisomers have different physical properties and may therefore be separated by fractional distillation under reduced pressure, or by preparative GLC.

### Question 2.4

Begin by explaining the concept of optical activity and the non-superposable mirror image relationship of enantiomeric isomers. Give a simple diagram of a polarimeter to illustrate the idea of rotation of the plane of plane polarised light. Quote examples of the importance of optical isomers for example in the physiological activity of enzymes, etc.

   (a) This compound has two different chiral centres, giving rise to two enantiomeric pairs of isomers which are diastereoisomers of one another:

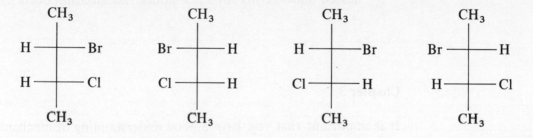

   (b) 1,2-dichlorocyclopropane has two geometric isomers, *cis* and *trans*. The former has a plane of symmetry, but the *trans* isomer has enantiomeric forms:

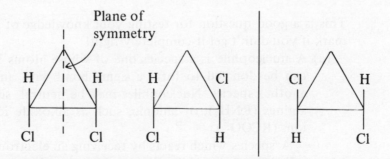

   (c) There is no chiral carbon atom present, but since the planes of the two pi bonds are at right angles, there is no overall plane of symmetry and the molecule can exist in enantiomeric forms:

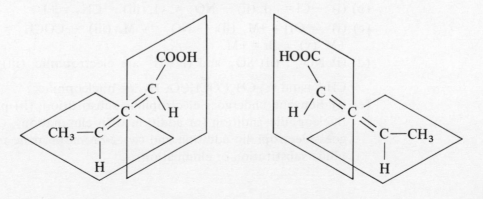

(d) The bulky *ortho* substituents have a twofold effect: they ensure that the the two rings cannot be coplanar, and they restrict rotation of one ring relative to the other. Thus the molecule has no overall plane of symmetry and consequently exists in two mirror image spatial arrangements, i.e. an enantiomeric pair:

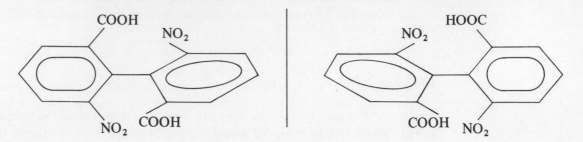

Non-superposable mirror images

(Incidentally, if the $NO_2$ groups are replaced by smaller F atoms, the energy barrier to rotation is such that if a solution of one optical isomer is heated under reflux for a few hours, racemisation occurs.)

## Chapter 3

It is important that you have a good understanding of mechanistic principles and answers here are in greater detail than for other self-test questions.

### Question 3.1

This is a good question for testing your knowledge of really basic material. Black mark if you didn't get it completely right!

(a) A nucleophile is a species one of whose atoms has an electron pair which can be donated to form a sigma bond with an atom (usually carbon) of another species. Nucleophiles may be neutral, such as alcohols (RÖH) and amines (RN̈H$_2$), or anionic, such as alkoxide ions (RÖ$^\ominus$), or carboxylate ions (RCOO$^\ominus$).

A species which reacts by receiving an electron pair from a nucleophile is called an electrophile. Electrophiles may be cationic, such as carbocations (R$^\oplus$), or the nitryl cation (NO$_2{}^\oplus$), or neutral polar or polarisable molecules such as hydrogen bromide $(\overset{\delta+}{H}\!-\!\overset{\delta-}{Br}\!:)$ or carbonyl compounds (e.g. $(\overset{\delta+}{R}CH\!=\!\overset{\delta-}{\ddot{O}})$.

(b) (i) —Cl = −I, (ii) —NO$_2$ = −I, (iii) —CH$_3$ = +I.

(c) (i) —OH = +M, (ii) —NO$_2$ = −M, (iii) —COCH$_3$ = −M, (iv) —SCH$_3$ = +M, (v) —Br = +M.

(d) (i) NO$_2{}^\oplus$, (ii) SO$_3$ and (iv) H$^\oplus$ are electrophilic, (iii) CH$_3$MgI, (reacts as CH$_3{}^\ominus$) and (v) CH$_3$COĊHCOCH$_3$ are nucleophilic.

(e) (i) Benzene undergoes electrophilic substitution, (ii) propanone undergoes nucleophilic addition (or addition with elimination), (iii) but-2-ene undergoes electrophilic addition and (iv) 2-bromopropane may undergo nucleophilic substitution or elimination.

(a) Ethyl hydrogensulphate, $CH_3CH_2OSO_2OH$, which may go on to form diethyl sulphate, $(CH_3CH_2)_2SO_4$.

Electrophilic addition. In the absence of solvent, the only nucleophile present is $HSO_4^{\ominus}$.

$$CH_2{=}CH_2 \quad H{-}\ddot{O}SO_2\ddot{O}H \xrightleftharpoons{\text{slow}} CH_3CH_2^{\oplus} + {:}\ddot{O}S\ddot{O}_2\ddot{O}H$$

$$H\ddot{O}S\ddot{O}_2{-}\ddot{O}{:}^{\ominus} \quad {}^{\oplus}CH_2CH_3 \xrightarrow{\text{fast}} H\ddot{O}S\ddot{O}_2\ddot{O}CH_2CH_3$$

(b) 2-Bromo-2-methylpropane, $(CH_3)_2C(Br)CH_3$.

Free radical substitution.

Initiation $\qquad {:}\ddot{B}r{-}\ddot{B}r{:} \rightleftharpoons 2\ {:}\ddot{B}r{\cdot}$

Propagation ${:}\ddot{B}r{:} \quad H{-}C(CH_3)_3 \longrightarrow {:}\ddot{B}r{-}H + (CH_3)_3\ \dot{C}$

$\qquad (CH_3)_3\ \dot{C} \quad {:}\ddot{B}r{-}\ddot{B}r{:} \longrightarrow (CH_3)_3\ C{-}\ddot{B}r{:} + {:}\ddot{B}r{:}$

Termination $\qquad (CH_3)_3\ \dot{C} + {:}\ddot{B}r{:} \longrightarrow (CH_3)_3\ C{-}\ddot{B}r{:}$ etc.

Bromine atoms are relatively unreactive ('selective') radicals and are only successful in removing the hydrogen atom which leads to the most stable of the possible free radical intermediates. (Order of stability of alkyl radicals is $1° < 2° < 3°$.)

(c) 1-Bromopentane, $CH_3(CH_2)_3CH_2Br$.

Nucleophilic substitution in the protonated alcohol.

$$CH_3(CH_2)_3CH_2OH + H_2SO_4 \rightleftharpoons CH_3(CH_2)_3CH_2\overset{\oplus}{O}H_2 + HSO_4^{\ominus}$$

Protonation converts the poor leaving OH group into the good leaving $\overset{\oplus}{O}H_2$ group. Displacement of the water molecule requires the assistance of the attacking bromide ion, and reaction is bimolecular ($S_N2$).

$$\overset{\ominus}{{:}\ddot{B}r{:}} \quad CH_2{-}\overset{\oplus}{O}H_2 \longrightarrow {:}\ddot{B}r{-}CH_2(CH_2)_3CH_3 + \ddot{O}H_2$$
$$\qquad\quad | $$
$$\qquad CH_2CH_2CH_2CH_3$$

In aqueous solution formation of the hydrogen sulphate is readily reversible, and ${:}\ddot{B}r{:}^{\ominus}$ is a better nucleophile than $\ddot{O}H_2$.

(d) Butanoic acid, $CH_3CH_2CH_2COOH$, via butanamide, $CH_3CH_2CH_2CONH_2$.

Acid catalysed hydrolysis of the nitrile.

$$CH_3CH_2CH_2C{\equiv}\ddot{N} + H\ddot{C}l{:} \rightleftharpoons CH_3CH_2CH_2\overset{\oplus}{C}{=}\ddot{N}{-}H + {:}\ddot{C}l{:}^{\ominus}$$

Bimolecular reaction with a water molecule leads to the first hydrolysis product, butanamide:

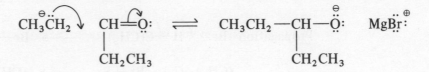

$$H_2\ddot{O} \downarrow \overset{\oplus}{C}=\ddot{N}-H \rightleftharpoons H_2\overset{\oplus}{O}-C=\ddot{N}-H \overset{-H^\oplus}{\rightleftharpoons} H\ddot{O}-C=\ddot{N}-H$$
$$\underset{CH_2CH_2CH_3}{|} \qquad \underset{CH_2CH_2CH_3}{|} \qquad \underset{CH_2CH_2CH_3}{|}$$

$$H-\overset{\frown}{\underset{..}{\ddot{O}}}-C=\overset{\curvearrowright}{\underset{..}{\ddot{N}}}-H \overset{-H^\oplus,+H^\oplus}{\rightleftharpoons} \ddot{O}=C-\ddot{N}H_2$$
$$\underset{CH_2CH_2CH_3}{|} \qquad\qquad \underset{CH_2CH_2CH_3}{|}$$

The examiner is not likely to want a second mechanism to show hydrolysis of the amide, but if you have time there is no harm in giving this! (see page 137).

(e) Pentan-3-ol, $CH_3CH_2CH(OH)CH_2CH_3$.
Nucleophilic addition.
The Grignard reagent behaves in ethereal solution as though it dissociated into $Et^\ominus$ and $MgBr^\oplus$.

$$CH_3\overset{\ominus}{\underset{..}{C}}H_2 \downarrow \overset{\frown}{CH}=\overset{\curvearrowright}{\ddot{O}}: \rightleftharpoons CH_3CH_2-CH-\overset{\ominus}{\ddot{O}}: \quad M\overset{\oplus}{g}\ddot{B}r:$$
$$\underset{CH_2CH_3}{|} \qquad\qquad\qquad \underset{CH_2CH_3}{|}$$

Then on aqueous workup,

$$CH_3CH_2-CH-\overset{\ominus}{\ddot{O}}: \quad MgBr^\oplus \overset{H_2O}{\longrightarrow} CH_3CH_2CH(\ddot{O}H)CH_2CH_3 + Mg(\ddot{O}H)\ddot{B}r:$$
$$\underset{CH_2CH_3}{|}$$

## Question 3.3

(a) The *t*-butyl group is activating and *ortho/para* directing in aromatic electrophilic substitution. This is explained by inductive stabilisation of the *ortho* and *para* sigma complex intermediates which is less effective when attack is *meta* to the substituent. You should draw representative limiting forms to illustrate this, e.g.

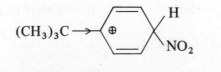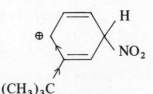

+I effect is                              +I effect operates
stabilising                                  over greater distance

Because of steric hindrance, mononitration occurs only in the *para* position. Since the nitrating mixture of $c.HNO_3/c.H_2SO_4$ is also powerfully oxidising, the final product is the 4-nitroacid.

The carboxyl group in benzoic acid is deactivating and *meta* directing. This is attributed to destabilisation of the *ortho* and *para* sigma complexes to a greater extent than the *meta* sigma complex. Illustrate this with representative limiting forms:

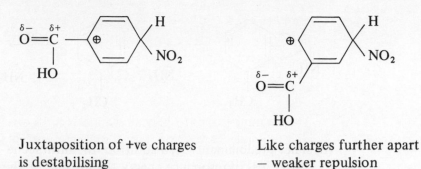

Juxtaposition of +ve charges is destabilising

Like charges further apart – weaker repulsion

Thus nitration give 3-nitrobenzoic acid.

(b) The carbon–bromine bond in bromobenzene is shorter and stronger than in bromoethane. Bimolecular attack of hydroxide ion involves the formation of a negatively charged sigma complex intermediate:

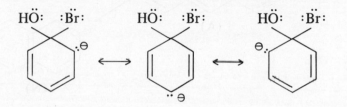

Activation energy for the formation of this intermediate is too great for this to occur under normal conditions, and at high temperature and pressure a different (benzyne) mechanism is involved. However, the presence of an electron attracting nitro group *ortho* or *para* to the bromine atom stabilises the intermediate and lowers the activation energy for its formation:

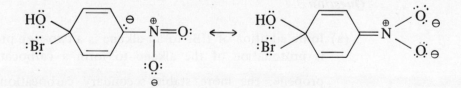

(c) This reaction involves the benzyne intermediate referred to in (b) above.

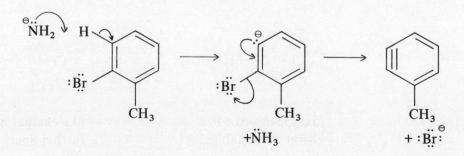

273

Reaction is completed by uptake of amide ion. Attack can occur on either of the triply bonded carbon atoms, but the +I effect of the methyl group favours the formation of 2-aminotoluene:

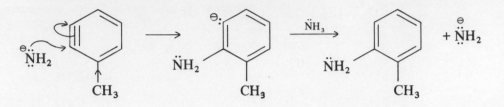

(d) The predominant 9,10-addition to anthracene is explained in terms of lower loss of resonance energy than would be the case with 1,2-addition:

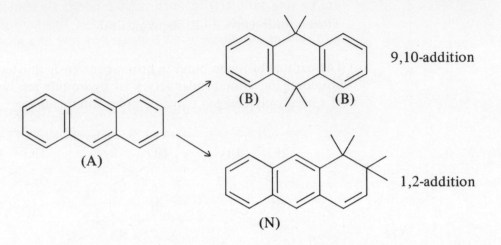

9,10-addition

1,2-addition

The 9,10-addition product has two benzenoid rings; the 1,2-addition product has one naphthalene-like ring. Since the resonance energy associated with two benzene rings is greater than that of one naphthalene ring, there is less loss of resonance energy on formation of the 9,10 product. (RE of A-2B is less than A-N.)

## Question 3.4

(a) Ionic addition of HBr to an alkene is a two-step process, the first of which is protonation of the alkene to form a carbocation intermediate. With propene, the more stable secondary carbocation, $CH_3 - \overset{\oplus}{C}H - CH_3$, is formed preferentially, leading to 2-bromopropane as the product. However, in 1,1,1-trichloropropene the carbon atom of the trichloro group is appreciably positively charged so that the secondary carbocation $CCl_3 - \overset{\oplus}{C}H - CH_3$ is less stable than the primary carbocation, $CCl_3 - CH_2 - \overset{\oplus}{C}H_2$:

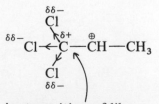

juxtaposition of like charges is destabilising

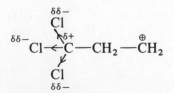

like charges now further apart

Hence the formation of 3-bromo-1,1,1-trichloropropane.

(b) Protonation of the unsaturated alcohol is followed by ionisation to give a resonance stabilised carbocation:

$$CH_2\!\!=\!\!CH-CH-CH\!\!=\!\!CH-CH_3 \; \underset{\phantom{-H_2O}}{\overset{-H_2O}{\rightleftharpoons}}$$

$$\overset{|}{\underset{\oplus OH_2}{}}$$

$$\left[ CH_2\!\!=\!\!CH-\overset{\oplus}{CH}-CH\!\!=\!\!CH-CH_3 \longleftrightarrow CH_2\!\!=\!\!CH-CH\!\!=\!\!CH-\overset{\oplus}{CH}-CH_3 \right]$$

$$\text{(I)} \qquad\qquad\qquad\qquad\qquad \text{(II)}$$

In limiting form (II) the two double bonds are conjugated and the positive charge is stabilised by the +I effect of the methyl group. Nucleophilic addition of methanol to $C^{\oplus}$ followed by loss of a proton gives the observed product.

(c) Treatment of the alcohol with thionyl chloride (sulphur oxide dichloride, $SOCl_2$) gives the chlorosulphite ester. This can ionise to an intimate ion-pair in which the carbocation has a resonance stabilised structure:

$$R-OH + SOCl_2 \longrightarrow R-O-SOCl + Cl^{\ominus} + H^{\oplus}$$

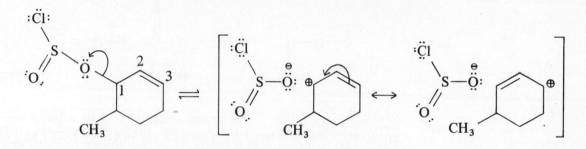

Because of the close proximity of the anion to C-1 in the intimate ion pair, it is sterically easier for $:\!\overset{..}{\underset{..}{Cl}}\!:^{\ominus}$ to attack C-3 leading to the 'rearranged' product.

(d) Protonation of the alkene occurs at the terminal carbon atom so as to give the most stable carbocation intermediate:

$$(CH_3)_3C-CH\!\!=\!\!CH_2 \xrightarrow{HCl} (CH_3)_3C-\overset{\oplus}{CH}-CH_3 + Cl^{\ominus}$$

and addition of $Cl^{\ominus}$ to this cation leads to the first product, 3-chloro-2,2-dimethylbutane, $(CH_3)_3C-CH(Cl)CH_3$.

However, the structure of the carbocation is such that migration of a methyl group can convert the secondary ion into a more stable tertiary structure:

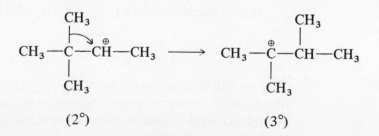

275

Reaction of the rearranged carbocation with chloride ion gives the isomeric product, 2-chloro-2,3-dimethylbutane, $(CH_3)_2C(Cl)CH(CH_3)_2$.

(e) Reaction of the tertiary halide is unimolecular, E1. Ionisation gives a carbocation with two $\beta$-H atoms:

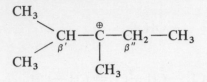

Loss of the proton occurs preferentially from the $\beta'$ position so as to give the most stable (most alkylated) alkene (Saytzeff rule).

**Question 3.5**

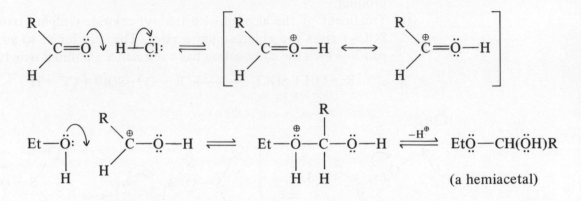

(a hemiacetal)

Protonation of the —OH group converts it into a good leaving $-\overset{\oplus}{O}H_2$ group which is displaced by a second molecule of EtOH to give the acetal (write the equations).

(b) Sodium acetate liberates the free base from semicarbazide hydrochloride and ensures a suitable pH for the nucleophilic addition reaction:

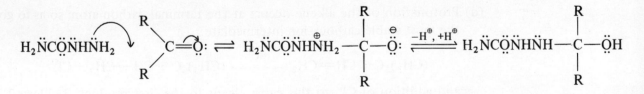

Spontaneous loss of a molecule of water gives the semicarbazone product:

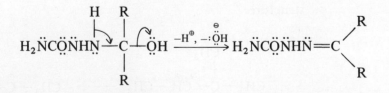

(The proton being accepted by a base, e.g. $H_2O$.)

(c) Reaction of diborane (which behaves as though it had the structure $BH_3$) involves a *cyclic* transition state, the electron-deficient boron atom receiving an electron pair from the alkene bond:

$$R-CH=CH_2 \longrightarrow \left[ \begin{array}{c} R-CH=CH_2 \\ \vdots \quad \vdots \\ H---BH_2 \end{array} \right]_{TS} \longrightarrow R-CH_2-CH_2-BH_2$$

Repetition of these steps leads to the trialkyl borane, $(RCH_2CH_2)_3B$ which is then decomposed by oxidation with $H_2O_2$:

$$(RCH_2CH_2)_3B \xrightarrow{H_2O_2/NaOH} 3 \; RCH_2CH_2OH + B(OH)_3$$

(d) This involves nucleophilic attack by the alcohol on the anhydride molecule. Pyridine is present to minimise hydrolysis of the ester product which might occur under acidic conditions.

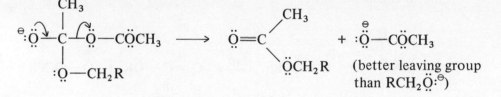

Note that the reaction involves addition with elimination, and *not* direct displacement of the acetate anion.

(e) $$RC\overset{..}{\underset{..}{O}}\overset{..}{\underset{..}{Cl}}: + AlCl_3 \rightleftharpoons [R-\overset{\oplus}{C}=\overset{..}{O} \longleftrightarrow R-C\equiv\overset{\oplus}{O}:] + Al\overset{..}{\underset{..}{Cl}}:_4^{\ominus}$$

where, for example, $B = Al\overset{..}{\underset{..}{Cl}}:_4^{\ominus}$, $BH = Al\overset{..}{\underset{..}{Cl}}:_3 + H\overset{..}{\underset{..}{Cl}}:$.

Show that the sigma complex intermediate is resonance stabilised:

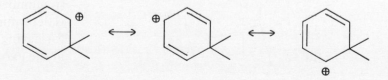

277

(f) This last reaction may not be in your syllabus, so don't worry if you were unable to do it. Reaction is acid catalysed:

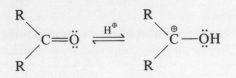

Let the 3-chloroperbenzoic acid be represented by PhCOOOH.
Nucleophilic addition of this species to the protonated ketone,

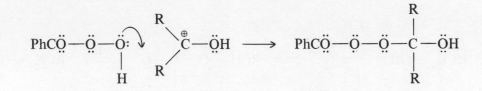

is followed by migration of an alkyl group which assists expulsion of the chlorobenzoate anion:

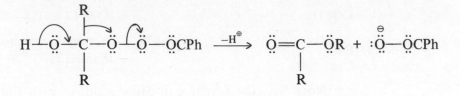

## Question 3.6

(i) The aromatic halide is able to ionise to a resonance stabilised carbocation intermediate:

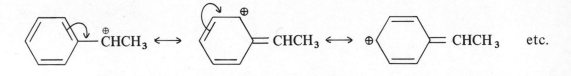

Uptake of hydroxide ion can occur from either side of the planar carbocation, giving a racemic alcohol product. The saturated halide cannot so ionise and undergoes bimolecular hydrolysis ($S_N2$) with inversion of configuration.

(ii) Reaction **A** is faster because ionisation relieves the strain resulting from close proximity of the *endo* tosylate group and hydrogen atoms on carbons 3 and 5:

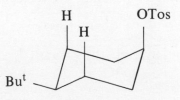

(remember that the conformation is determined by the necessity for having the bulky Bu$^t$ group in the *exo* position)

Such strain obviously is not present in the *exo* tosylate.

(iii) The *anti* conformation is the most stable for 1,2-dibromoethane as this puts the large, partially negatively charged, bromine atoms as far apart as possible.

Hydrogen bonding stabilises 1,2-ethanediol in the *gauche* conformation:

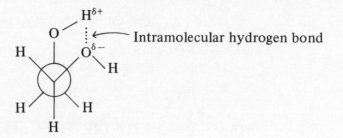

(iv) Free radical substitution.

$$Cl_2 \xrightarrow{\text{UV}} 2\ Cl^\cdot$$

The chlorine atom then preferentially abstracts the *tertiary* hydrogen atom to give a planar alkyl free radical:

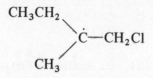

In the second propagation step (removal of $Cl^\cdot$ from $Cl_2$) the chlorine atom can add to either side of the planar radical, thus giving a racemic product.


**Question 3.7**

(a)

(i)

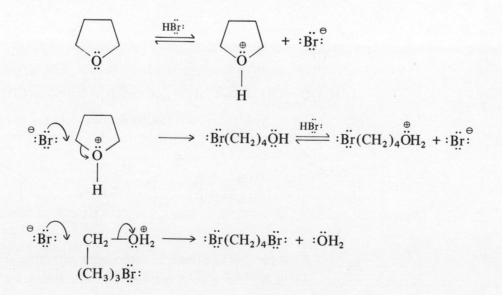

279

(ii)

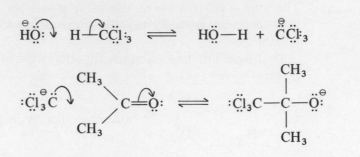

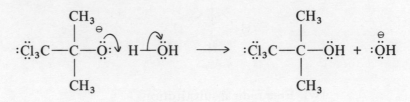

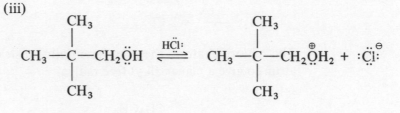

(iii)

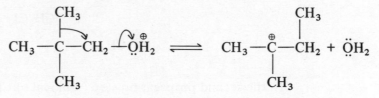

Neighbouring group participation – rearrangement to 3° structure.

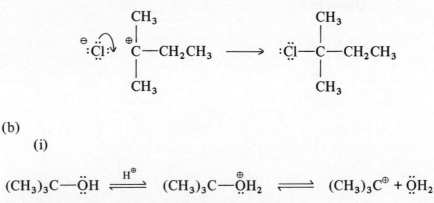

(b)

    (i)

$(CH_3)_3C\ddot{-}OH$ ⇌ $(CH_3)_3C\overset{\oplus}{-}OH_2$ ⇌ $(CH_3)_3C^{\oplus} + \ddot{O}H_2$

3°, readily ionises

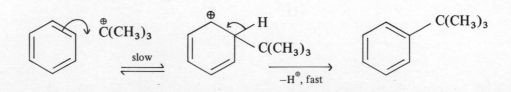

The tert-butyl group is *ortho/para* directing. For steric reasons the main product will be 1,4-ditert-butylbenzene.

(ii) This one is a bit more tricky – hence the higher allocation of marks! Treatment of a chloroalkane with sodium ethoxide will bring about elimination of HCl, and this requires an *anti*-coplanar arrangement of the chlorine and β-hydrogen atoms.

The compound shown is menthyl chloride, for which the preferred conformation (I) has the chlorine in an equatorial position:

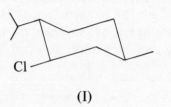

(I)

There are several possible (higher energy) conformations which have the chlorine in the required axial position. The most likely of these are the boat arrangement (II) or the chair arrangement (III):

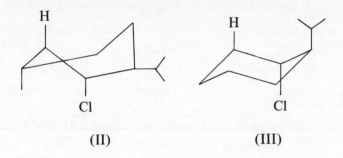

(II)                                    (III)

Elimination from either of these will give menth-2-ene, (IV):

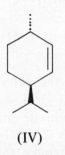

(IV)

To really clinch the 7 marks, you could finish with an outline of the bimolecular mechanism, E2:

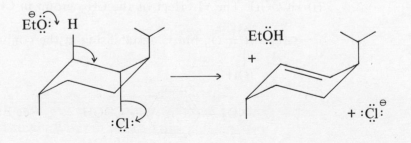

281

## Chapter 4

## Question 4.1

Least acidic is $CH_3CH_2CH_3$. The conjugate anion $CH_3CH_2\ddot{C}H_2^\ominus$ has the -ve charge localised on one carbon atom and is very unstable. Next, and still very weakly acidic, is $CH_3CH_2OH$, where the -ve charge in the conjugate alkoxide ion is localised on the oxygen atom, $CH_3CH_2\ddot{O}^\ominus$. Delocalisation of the -ve charge in the conjugate anion of cyclopentadiene puts it next in the series. Draw limiting forms to show that each of the five carbon atoms can share the -ve charge:

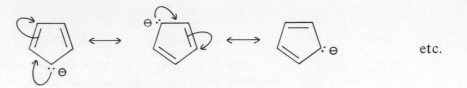

etc.

Phenol is the next most acidic. Draw limiting forms to show resonance stabilisation is more effective in the phenoxide ion than in the parent molecule,

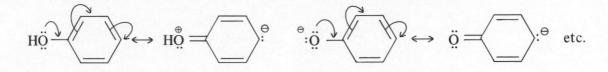

etc.

In the next most acidic compound, $CH_3COCH_2COCH_3$, resonance stabilisation of the conjugate anion is particularly effective as the -ve charge is shared by two oxygen atoms:

$$CH_3-\underset{\underset{:O:}{\|}}{C}-\overset{\overset{\ominus}{}}{\ddot{C}H}-\underset{\underset{:O:}{\|}}{C}-CH_3 \quad\longleftrightarrow\quad CH_3-\underset{\underset{:O:}{\|}}{C}-CH=\underset{\underset{:\underset{\ominus}{\ddot{O}}:}{|}}{C}-CH_3 \quad \text{etc.}$$

Most acidic are the carboxylic acids, where in the corresponding conjugate anions the -ve charge is again shared by two oxygen atoms. Draw limiting forms for the acid and its conjugate anion, and to illustrate the effect of the nitro group. The strongest acid is $CH_3CCl_2COOH$, where the two chlorine substituents destabilise the acid, but stabilise the conjugate anion (see Section 4.10 and Fig. 4.2, also Example 4.1).

## Question 4.2

(i) HCOOH. The +I effect of the $CH_3$ group in $CH_3COOH$ stabilises the acid,

$$CH_3 \overset{}{\rightarrow}\underset{\underset{:\underset{}{\ddot{O}}H}{|}}{\overset{\delta+}{C}}=\overset{\delta-}{\ddot{O}}:, \text{ but is destabilising in the conjugate anion, } CH_3\overset{}{\rightarrow}C\ddot{O}\ddot{O}^\ominus:.$$

(ii)

$$NO_2 - \langle\!\!\bigcirc\!\!\rangle - COOH \qquad \text{(see Example 4.1)}$$

(iii) $CH_3NH_2$. The methyl group (+I effect) stabilises the conjugate cation, $CH_3 \rightarrow \overset{\oplus}{N}H_3$, whereas the methoxy group (−I effect) is destabilising in the conjugate cation $CH_3\overset{\delta-}{\underset{..}{\overset{..}{O}}} \leftarrow \overset{\delta+}{C}H_2 - \overset{\oplus}{N}H_3$.

(iv) $NH_2$  $OCH_3$ The methoxy group here takes part in resonance stabilisation of the conjugate cation, $\overset{\oplus}{N}H_3$ —⟨⟩— $OCH_3$. Draw limiting forms to illustrate this:

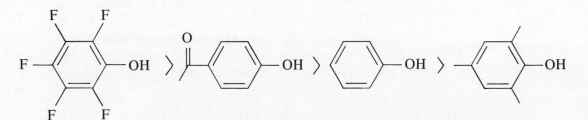

$CH_3\overset{..}{\underset{..}{O}}$ —⟨⟩— $\overset{\oplus}{N}H_3 \longleftrightarrow CH_3\overset{\oplus}{\underset{..}{O}}$ =⟨⟩= $\overset{\ominus}{}-\overset{\oplus}{N}H_3$ etc.

### Question 4.3

Phenol is the least acidic. Give reasons as outlined in Section 4.8. Next most acidic is butanoic acid (Section 4.9). The other examples illustrate how the effect of halogen substituents varies with the nature and position of the substituent (Example 4.1). Fluorine is more electronegative than chlorine and so 2-fluorobutanoic acid is stronger than 2-chlorobutanoic acid. Chlorine in the 3-position will be less effective and bromine (exerting a smaller −I effect) will have minimal effect from the more remote 4-position. Thus the order of decreasing acidity is 2-fluorobutanoic acid > 2-chloro- > 3-chloro- > 4-bromo- > butanoic acid > phenol.

### Question 4.4

(i)

F—⟨F,F,F,F⟩—OH > $\underset{O}{CH_3C}$—⟨⟩—OH > ⟨⟩—OH > CH₃—⟨CH₃,CH₃⟩—OH

The powerfully electron-withdrawing fluorine atoms will stabilise the conjugate phenoxide anion relative to the undissociated molecule. The $CH_3CO$ group has a similar but smaller effect. Methyl groups on the other hand are destabilising, and 2,4,6-trimethylphenol is the least acidic compound.

(ii) In N,N-dimethylphenylamine (3) the dimethylamino group is involved in resonance stabilisation of the molecule:

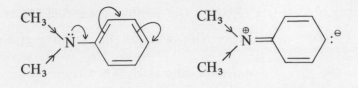

This 'contribution' is not possible when the nitrogen atom bonds with a proton,

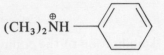

Thus, loss of resonance energy forms a barrier to the uptake of a proton and the compound is only weakly basic. For resonance to occur, the dimethylamino group has to be coplanar with the ring (see Section 4.10 and Example 4.1). In molecular orbital terms this means that the nitrogen 2p atomic orbital has to be parallel with the ring carbon 2p atomic orbital for maximum overlap to occur:

The presence of two bulky methyl substituents adjacent to the dimethyl-amino group in compound (4) prevents this necessary coplanarity and so resonance with the group is not possible. Consequently there is no loss of resonance energy on protonation of the molecule, and compound (4) is a stronger base than compound (3). (Note, however, that it is still a relatively weak base because of steric factors discussed in Section 4.12.)

**Chapter 5**

*Question 5.1*

(i) Nucleophilic strength is a measure of the ability of a species to donate an electron pair to electron-deficient carbon as in an $S_N2$ reaction. Base strength is equated with ability to donate an electron pair to a proton. However, the distinction is blurred in reaction of a Lewis base (such as ethoxyethane) with a Lewis acid (such as boron trifluoride).

(ii) Reaction is unimolecular ($S_N1$) because ionisation of the 'allylic' substrate gives a resonance-stabilised intermediate carbocation:

$$X-CH=CH-CH_2-I \rightleftharpoons [X-CH=CH-\overset{\oplus}{C}H_2 \longleftrightarrow X-\overset{\oplus}{C}H-CH=CH_2]$$

i.e. $[X-\overset{\delta+}{C}H=CH=\overset{\delta+}{C}H_2]^{\oplus}$

Thus, in addition to the expected products, $CH_3CH=CHCH_2OH$, $CH_3OCH=CHCH_2OH$ and $N\equiv CCH=CHCH_2OH$, varying amounts of rearranged products will be formed by attack of the nucleophile on the other electron-deficient carbon atom: $CH_3CH(OH)CH=CH_2$, $CH_3OCH(OH)CH=CH_2$ and $N\equiv CCH(OH)CH=CH_2$.

284

The order of decreasing reactivity is $CH_3O > CH_3 > N{\equiv}C$, related to the stability of the intermediate carbocations (or more strictly, the transition states for their formation):

$CH_3O$ is stabilising (+M effect) $CH_3\overset{..}{\underset{..}{O}}{-}\overset{\oplus}{C}H{-}CH{=}CH_2 \longleftrightarrow CH_3\overset{\oplus}{\underset{..}{O}}{=}CH{-}CH{=}CH_2$

$CH_3$ is weakly stabilising (+I effect) $CH_3{\rightarrow}\overset{\oplus}{C}H{-}CH{=}CH_2$

$N{\equiv}C$ is destabilising (−I effect) $\overset{\delta-}{\overset{..}{N}}{\equiv}\overset{\delta+}{C}{\leftarrow}\overset{\oplus}{C}H{-}CH{=}CH_2.$

Your energy profile should be labelled as follows:

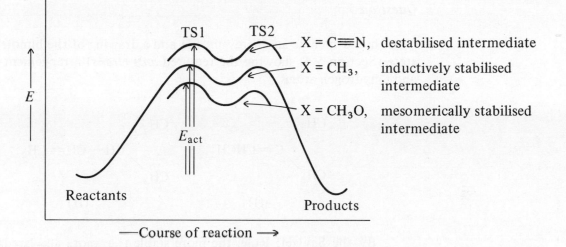

Course of reaction →

*Question 5.2*

(a) The rate determining step is ionisation of the substrate to give $(CH_3)_3C^{\oplus}$. This intermediate can then react with either $HO^{\ominus}$ or $N_3^{\ominus}$.

(b) The two chloroalkenes each readily ionise to give the same resonance-stabilised carbocation:

$$CH_3CH{=}CH{-}\overset{\oplus}{C}H_2 \longleftrightarrow CH_3\overset{\oplus}{C}H{-}CH{=}CH_2$$

i.e.     $[CH_3\overset{\delta+}{C}H{=}CH{\cdots}\overset{\delta+}{C}H_2]^{\oplus}$

Uptake of the nucleophile can occur at either positively charged carbon atom to give the two unsaturated alcohols. Aqueous silver oxide is a very mild reagent and $S_N2$ hydrolysis of the corresponding chloroalkanes would be very slow.

(c) In aqueous alkali, it will be the *carboxylate anion* which undergoes hydrolysis. Retention of configuration can be explained by the formation of a cyclic intermediate:

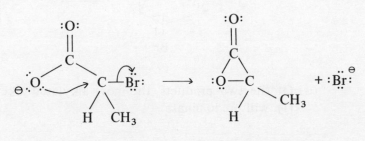

('Internal $S_N2$' – an example of 'neighbouring group participation'.) Ring-opening attack by hydroxide ion gives a hydroxyacid (carboxylate anion) with the same configuration:

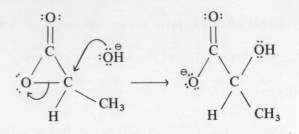

## Question 5.3

You should include an energy profile and a drawing of the bimolecular transition state (Section 5.4) showing the required *anti* (*trans*) arrangement of the halogen and β-hydrogen atoms.

(a)

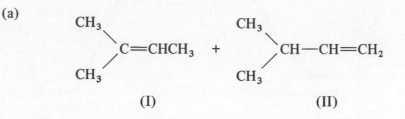

By the Saytzeff Rule, the more stable (i.e. more alkylated) alkene, I, will predominate.

(b)

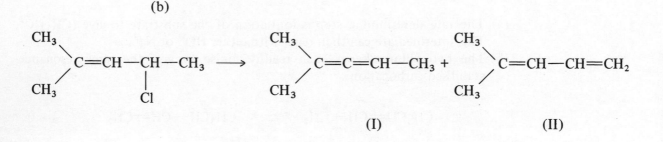

The more stable conjugated diene, II, will predominate.

(c) Because of the requirement for an *anti* (*trans*) arrangement of the displaced Br and H atoms, elimination of the proton will occur from carbon atom 6 to give 3-methylcyclohexene:

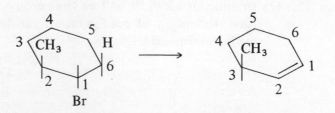

(d) Of the two products, the more alkylated alkene, 1-methylcyclohex-1-ene, (I), will predominate:

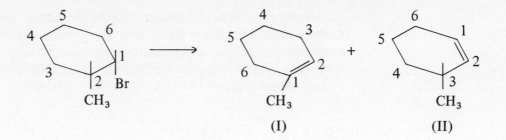

(I)          (II)

*Question 5.4*

(a) Bookwork! Mention the limitations of direct bromination of alkanes.

(b) The primary haloalkane 1-iodo-2-phenylethane reacts by $S_N2$. However, steric crowding in the bimolecular transition state for substitution in 1-iodo-1-phenylethane

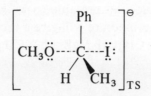

increases the activation energy for the process and it becomes energetically more favourable for the methoxide ion to attack the $\beta$-hydrogen atom

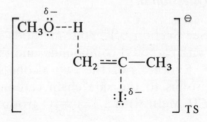

giving rise to the alkene

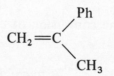

1-Iodo-1-phenylethane may also react unimolecularly, since ionisation gives a resonance-stabilised cation:

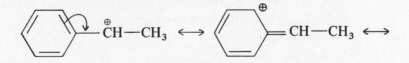

But again, the transition state for deprotonation is less sterically crowded than that for substitution, and furthermore the $\beta$-protons (on the methyl group) are more acidic in the cation. Hence elimination predominates over substitution.

(c) 2-Iodo-2-methylpentane is a *tertiary* haloalkane with a good leaving group. Thus it readily ionises, and a $\beta$-proton is preferentially lost from carbon atom 3 as this gives the more alkylated of the two possible alkenes.

The fluorine atom in 2-fluoro-2-methylpentane is a relatively poor leaving group and it is therefore likely that this compound undergoes bimolecular elimination. The controlling factor now is steric crowding in the transition state, and so attack occurs on the more accessible β-hydrogen atom of the methyl group (carbon 1) to give 2-methylpent-1-ene:

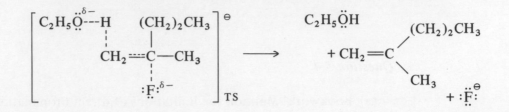

(d) At lower temperatures, bimolecular substitution will give an ether with inverted configuration. As the temperature is increased, more substrate molecules are able to ionise, and random attack on the planar carbocation intermediate will give a racemic ether.

(e) J = $CH_3CH_2MgBr$, L = $CH_3CH_2CN$ and K = $CH_3CH_2COOH$.

## Chapter 6

### Question 6.1

See Section 6.1. Your methods should include the reaction of Grignard reagents with carbonyl compounds, and the hydroboration/oxidation of alkenes.

Synthesis of the two alcohols requires the successive addition of two methyl groups to the side chain carbon atom, and this can be achieved by reaction of $CH_3MgBr$ with a $\diagdown C = O$ group.

### Synthesis of Compound (1), 2-phenylpropan-2-ol

Begin with the conversion of benzoic acid into benzaldehyde. This requires two steps:

$$PhCOOH \xrightarrow{\text{LiAlH}_4} PhCH_2OH \xrightarrow{\text{[O]}} PhCHO$$

Although $K_2Cr_2O_7$ in $CH_3COOH$ could be used, it is difficult to avoid some further oxidation to benzoic acid. Better yields are obtained with $CrO_3$ in pyridine, or with 'pyridinium chlorochromate', $C_5H_5\overset{\oplus}{N}HCrO_3\overset{\ominus}{Cl}$, prepared from $K_2Cr_2O_7$ and pyridinium chloride.

Continue:

$$PhCHO \xrightarrow{\text{CH}_3\text{MgBr}} PhCH(OH)CH_3 \xrightarrow{\text{[O]}} PhCOCH_3 \xrightarrow{\text{CH}_3\text{MgBr}} Ph-\overset{\overset{\displaystyle CH_3}{|}}{\underset{\underset{\displaystyle CH_3}{|}}{C}}$$

(Oxidation of the secondary alcohol is less difficult as ketones resist further oxidation.)

## Synthesis of Compound (2), 2-phenylpropan-1-ol

Compound (1) is the starting point, and the steps are

$$Ph-\underset{\underset{CH_3}{|}}{\overset{\overset{CH_3}{|}}{C}}-OH \xrightarrow[-H_2O]{H^{\oplus}} Ph-\underset{}{\overset{\overset{CH_3}{|}}{C}}=CH_2 \text{ (good yield, only one alkene possible)}$$

followed by

$$3\ Ph-\overset{\overset{CH_3}{|}}{C}=CH_2 \xrightarrow{\text{'(BH_3)'}} \left(Ph-\underset{\underset{H}{|}}{\overset{\overset{CH_3}{|}}{C}}-CH_2-\right)_3 B \xrightarrow{H_2O_2,\ \overset{\ominus}{OH}} 3\ Ph\ \overset{\overset{CH_3}{|}}{CH}-CH_2OH$$

The hydroboration/oxidation of alkenes complements other methods in that it gives the anti-Markownikov product.


### Question 6.2

There are many parallels in the chemistry of alcohols and amines, and most deviations which occur in behaviour can be attributed to the difference in electronegativities of oxygen and nitrogen.

Your answer could be structured according to the headings given in the question.

### Cleavage of O—H and N—H Bonds

Acid strength is linked to the stability of the conjugate anions $R\overset{..}{\underset{..}{O}}:^{\ominus}$ and $R\overset{..}{N}H^{\ominus}$ (see Chapter 4). Both compounds are only weakly acidic because in each case the negative charge is localised on the heteroatom.

Oxygen, being more electronegative than nitrogen, is better able to accommodate a negative charge, and so alcohols are slightly more acidic than amines. Neither compound will react with aqueous NaOH, but both form salts with sodium metal:

$$2\ R\overset{..}{O}H + 2\ Na^{\cdot} \longrightarrow 2\ R\overset{..}{\underset{..}{O}}:^{\ominus} Na^{\oplus} + H_2 \uparrow$$

$$2\ R\overset{..}{N}H_2 + 2\ Na^{\cdot} \longrightarrow 2\ R\overset{..}{N}H^{\ominus} Na^{\oplus} + H_2 \uparrow$$

The resulting alkoxide and alkylamide ions are powerful bases and nucleophiles. The greater acidity of alcohols is demonstrated by the reaction

$$(Na^{\oplus})\ R\overset{\ominus}{\underset{..}{N}}H \overset{\frown}{\ } H-\overset{..}{O}R \longrightarrow R\overset{..}{N}H_2 + :\overset{..}{\underset{..}{O}}R\ (Na^{\oplus})$$

### Lone Pair Donation

Two aspects should be considered here, **basicity** and **nucleophilicity**. Amines are more basic than alcohols because the (less electronegative) nitrogen atom can

289

more readily accept a positive charge. Thus, amines form stable salts with strong mineral acids, e.g.

$$R\ddot{N}H_2 \quad H—\ddot{\underset{..}{Cl}}: \longrightarrow R\overset{\oplus}{N}H_3 \quad :\ddot{\underset{..}{Cl}}:^{\ominus}$$

(an alkylammonium chloride)

Alcohols are protonated in acid solution

$$R\ddot{O}H \quad H—\ddot{\underset{..}{Cl}}: \rightleftharpoons R\overset{\oplus}{\underset{..}{O}}H_2 + :\ddot{\underset{..}{Cl}}:^{\ominus}$$

but no corresponding salts can be isolated.

Similarly, amines are stronger nucleophiles than alcohols. Amines will displace halogen from haloalkanes,

$$R\ddot{N}H_2 \quad \underset{\underset{}{}}{\overset{R}{\underset{|}{CH_2}}}—\ddot{\underset{..}{X}}: \longrightarrow \overset{R}{\underset{|}{R\overset{\oplus}{N}H_2}}—CH_2 + :\ddot{\underset{..}{X}}:^{\ominus}$$

but alcohols are unable to do this. They can displace water from a protonated alcohol, which is the basis of the Williamson etherification reaction:

$$R—\underset{\underset{H}{|}}{\ddot{O}:} \quad R—\overset{\oplus}{\ddot{O}}H_2 \longrightarrow R—\underset{\underset{H}{|}}{\overset{\oplus}{\ddot{O}}}—R + \ddot{O}H_2 \longrightarrow R—\ddot{\underset{..}{O}}—R + \overset{\oplus}{\ddot{O}}H_3$$

Both compounds react with carbonyl compounds, and, for example, show parallel behaviour (acylation) with acid chlorides:

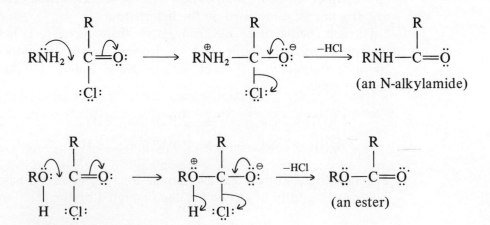

(an N-alkylamide)

(an ester)

Addition of an amine to an aldehyde or ketone is followed by elimination to give an **imine**:

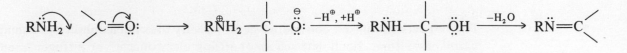

290

Aliphatic imines are unstable and tend to polymerise. Aromatic imines (Schiff bases) are more stable.

Alcohols react reversibly with aldehydes and ketones, often slowly and incompletely:

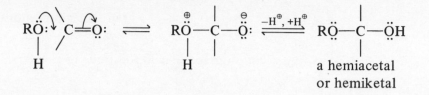

a hemiacetal
or hemiketal

The **hemiacetals** or **hemiketals** so formed are usually unstable and exist only in solution. If an alcohol is reacted with an aldehyde under anhydrous conditions, using dry hydrogen chloride as catalyst, the hemiacetal is converted into a more stable **acetal**, which can be isolated (the water formed being removed by azeotropic distillation with benzene)

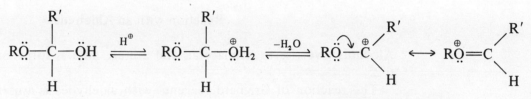

(resonance stabilised)

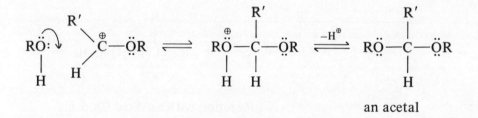

an acetal

(Ketals are usually prepared indirectly.)

### Cleavage of C—O and C—N Bonds

HO and $NH_2$ are both poor leaving groups. Protonation of OH converts this into a good-leaving water molecule — illustrated by the Williamson reaction referred to above.

The ammonia molecule is still a poor leaving group, and an analogous reaction does not occur with amines. There *are* a few examples of reactions in which $NH_3$ is displaced, e.g.

$$\ominus\ :\ddot{B}r: \quad CH_2 - \overset{\oplus}{N}H_3 \quad \overset{\Delta}{\longrightarrow} \quad :\ddot{B}r - CH_2 + \ddot{N}H_3$$

with $CH_3$ groups above.

but various side reactions occur and the reaction is of little practical value.

Alcohols, particularly tertiary alcohols, readily undergo acid-catalysed dehydration to an alkene. Such elimination is seldom encountered in amines. A notable

291

exception is the action of heat on quaternary ammonium hydroxides (Hofmann elimination):

$$\overset{\ominus}{H\ddot{O}:} \quad H\!-\!CH_2\!-\!CH_2\!-\!\overset{\oplus}{N}(CH_2CH_3)_3 \xrightarrow{\Delta} H\ddot{O}\!-\!H + CH_2\!=\!CH_2 + \ddot{N}(CH_2CH_3)_3$$

(the 3° amine is a
good leaving group)

### Question 6.3

Alcohols, amines and Grignard reagents are all nucleophilic substances, and aldehydes, acid chlorides and carboxylic acids are potential substrates for nucleophilic attack.

One approach to this question would be to compare the reaction of all three nucleophiles with each substrate in turn.

### Reaction with an Aldehyde

Alcohols give rise to hemiacetals and acetals, amines form imines. See Question 6.2.

The reaction of Grignard reagents with aldehydes is a useful method for the preparation of secondary alcohols:

$$\overset{\oplus}{(MgBr:)} \quad \overset{\ominus}{R} \overset{R'}{\underset{CH=\ddot{O}:}{\downarrow}} \longrightarrow R\!-\!\overset{R'}{\underset{|}{CH}}\!-\!\overset{\ominus}{\ddot{O}}:\!\overset{\oplus}{MgBr} \xrightarrow{H_2O} R\!-\!\overset{R'}{\underset{|}{CH}}\!-\!\ddot{O}H$$

### Reaction with an Acid Chloride

The substrate now has a good leaving group, Cl, and nucleophilic addition is followed by elimination of $:\overset{..}{\underset{..}{Cl}}:^{\ominus}$.

Alcohols are esterified, and amines form N-substituted amides (see Question 6.2).

With excess Grignard reagent there is a two-stage reaction to give first a ketone, and then a tertiary alcohol:

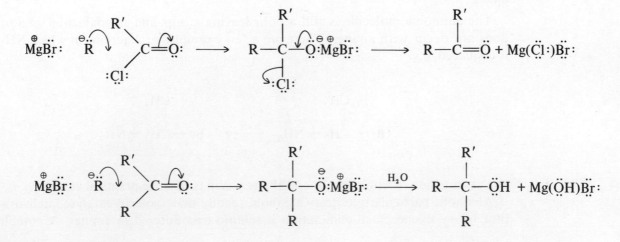

292

Carboxylic acids undergo few of the nucleophilic reactions characteristic of other carbonyl compounds, because most nucleophiles are simply protonated by the acid.

Reaction with alcohols is reversible and very slow unless acid catalysed.

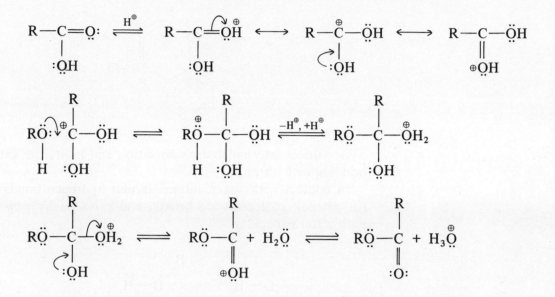

Amines are protonated by carboxylic acids

$$RNH_2 \quad H-\ddot{O}-C=\ddot{O} \longrightarrow RNH_3^\oplus + :\ddot{O}-C=\ddot{O}$$

Grignard reagents decompose with the formation of the corresponding alkane

$$MgBr^\oplus: \quad R^\ominus \quad H-\ddot{O}-C=\ddot{O} \longrightarrow RH + MgBr^\oplus :\ddot{O}-C=\ddot{O}$$

***Question 6.4***

(a) The main effects are on **volatility** and **water-solubility**. Because of *inter-molecular* hydrogen bonding,

$$\underset{O-H}{\overset{R}{\diagdown}} \cdots \underset{O-H}{\overset{R}{\diagdown}} \cdots \underset{O-H}{\overset{R}{\diagdown}}$$

alcohols and phenols have higher boiling points than isomeric ethers. For example, ethanol (b.pt. 78°C) is liquid at room temperature, whereas methoxymethane (b.pt. −25°C) is gaseous. Likewise, 4-methylphenol (m.pt. 35°C, b.pt. 202°C) is solid at room temperature, but methoxybenzene (m.pt. −38°C, b.pt. 154°C) is liquid.

Branching in alcohols gives a more compact structure, and 2-methyl-propan-2-ol (b.pt. 83°C) is more volatile than butan-1-ol (b.pt. 117°C).

Relevant to the next part of the question is the fact that 2-nitrophenol is less soluble and more volatile than 4-nitrophenol. The former compound is *intra*molecularly hydrogen bonded

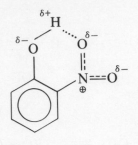

This reduces *inter*molecular association and limits the extent of hydrogen bonding with water.

In contact with water, intermolecular hydrogen bonds between alcohol (or phenol) molecules are broken and replaced by new hydrogen bonds with water molecules

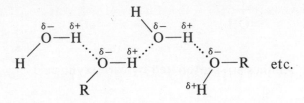

Lower alcohols are completely miscible with water. Solubility decreases as the size of the (hydrophobic) hydrocarbon skeleton increases, and phenols are only slightly soluble at room temperature.

Branching of alcohols slightly increases solubility (a more compact alkyl group is less hydrophobic than a long chain), and 2-methylpropan-2-ol is more soluble than butan-1-ol.

(b) Phenol is treated with dilute nitric acid at room temperature or below. Poor to moderate yields of 2-nitrophenol and 4-nitrophenol are obtained. The isomers may be separated by steam distillation, the 2-isomer being the more steam volatile (see previous section).

(c) Cyclohexanol is too weak an acid to react with sodium hydroxide. The conjugate anion, $R\ddot{O}\colon^{\ominus}$, is a very powerful base in which the negative charge remains localised on the oxygen atom.

Phenol readily 'dissolves' in sodium hydroxide solution because of reaction to form the resonance stabilised phenoxide ion:

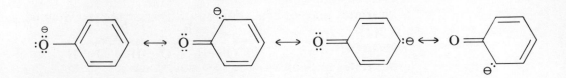

However, phenol is a weaker acid than carbonic acid and is unable to displace $CO_2$ from its salts.

Stabilisation of the phenoxide ion by three electron withdrawing nitro groups increases the acidity of trinitrophenol, which does therefore react with sodium carbonate solution. Complete your answer by drawing at least one limiting form to show resonance stabilisation, e.g.

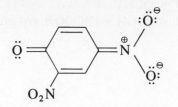

## Chapter 7

*Question 7.1*

    (1) $CD_3COPh$ ($Ph = C_6H_5$).
       Illustrates acidity of α—H

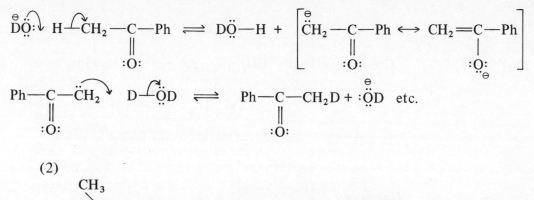

    (2)

$$\underset{Ph}{\overset{CH_3}{>}}C=N-NHPh \qquad \text{(N–phenyl hydrazone derivative)}$$

    Illustrates nucleophilic addition with elimination.

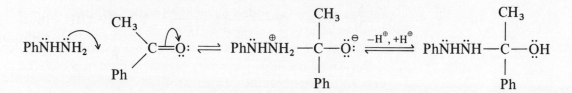

    Dehydration is acid or base catalysed (see Example 7.2(b)).
(3) $CH_3CH(OH)Ph$
    Illustrates nucleophilic addition of hydride ion (see Summary and Example 7.2(d)).
(4) $Ph_2C(OH)CH_3$
    Illustrates nucleophilic addition of carbanion

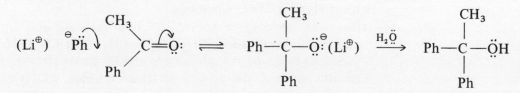

295

(5) PhCOCH=CHPh

Illustrates crossed Aldol reaction.

Note that PhCHO has no α-hydrogen and is attacked by the enolate ion of PhCOCH₃. A good answer would mention that PhCHO may give some PhCH₂OH + PhCOOH by-product from a competing Cannizzaro reaction.

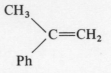

Illustrates nucleophilic addition of an ylide having carbanion character (Wittig reaction).

$$Ph_3\overset{\oplus}{P}CH_3\overset{\ominus}{I} \quad \xrightarrow[-HI]{NaOH} \quad [Ph_3\overset{\oplus}{P}=CH_2 \longleftrightarrow Ph_3\overset{\oplus}{P}-\overset{\ominus}{\underset{\cdot\cdot}{C}}H_2]$$

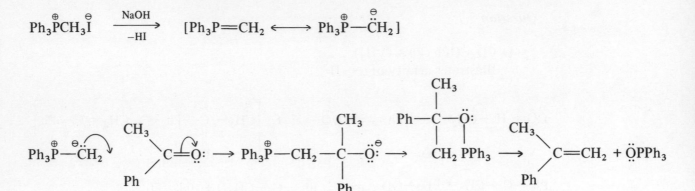

All parts carry equal marks, so the choice is yours!

**Question 7.2**

(a)

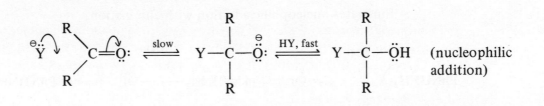

(nucleophilic addition)

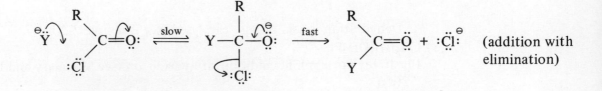

(addition with elimination)

(b) For example, oxidation of hexan-2-ol and reaction of CH₃CH₂CH₂COCl with (CH₃)₂Cd. Write equations.

(c)  (i)

3-methylheptan-3-ol, CH₃CH₂CH₂CH₂C(OH)CH₂CH₃
                                                      |
                                                      CH₃

(ii) Phenylpropanone, $CH_3CH_2COPh$, which could further react to give 1,1-diphenylpropan-1-ol,

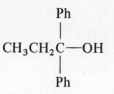

(d) (i) hexan-2-ol, $CH_3CH_2CH_2CH_2CH(OH)CH_3$
(ii) butylamine, $CH_3CH_2CH_2CH_2NH_2$

## Question 7.3

Illustrate by reference to benzaldehyde unless other examples have been given in your lecture course, e.g.

1. $PhCH_3 \xrightarrow{Cl_2} PhCHCl_2 \xrightarrow{H_2O} PhCHO$

2. $PhCH_3 \xrightarrow[Ac_2O]{CrO_3} PhCHO$

3. $PhH + CO + HCl \xrightarrow{AlCl_3} PhCHO + HCl$ (Gattermann–Koch)

4. $PhCN \xrightarrow[(ii) H_2O]{(i) SnCl_2, HCl} PhCHO$ (Stephens)

The second part says 'discuss' so you should attempt some *explanation*, as well as stating the differences. You could quote some or all of the following:
1. No $\alpha$-hydrogen, does not form an aldol, but undergoes Cannizzaro reaction.
2. Forms some cyanohydrin with HCN, but with alcoholic KCN undergoes the benzoin condensation.
3. Does not form a simple addition product with ammonia, but gives 'hydrobenzamide', $PhCH{=}NCHPhN{=}CHPh$.
4. Reduction with sodium amalgam or Zn/HCl gives phenylmethanol (benzyl alcohol) + 1,2-diphenylethan-1,2-diol, $PhCH(OH)CH(OH)Ph$, 'hydrobenzoin'.
5. Chlorination in absence of a 'halogen carrier' gives $PhCOCl$.
6. Perkin reaction with ethanoic (acetic) anhydride and sodium ethanoate (acetate):

$$PhCHO + (CH_3CO)_2O \xrightarrow{NaOOCCH_3} PhCH{=}CHCOOH$$

Outline a mechanism for (1), (2) and possibly (6).
Consult Chapter 13 for the last part of the question.
Since A,B and C all give benzene-1,2-dicarboxylic (phthalic) acid on oxidation, A contains a benzene ring which remains intact throughout the various reactions.

The sequence is:

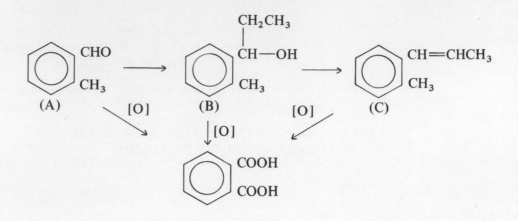

**Question 7.4**

The four main methods are:

1. **Catalytic hydrogenation**

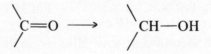

For esters a better catalyst is $CuO\text{-}CuCr_2O_4$
Reaction is often slow, and may require high T and P.

Other groups present (e.g. $\overset{\diagdown}{\underset{\diagup}{C}}=\overset{\diagup}{\underset{\diagdown}{C}}$ , $NO_2$) may also be reduced.

2. **Metal hydrides**

$$\overset{\diagdown}{\underset{\diagup}{C}}=O \longrightarrow \overset{\diagdown}{\underset{\diagup}{C}}H—OH$$

$LiAlH_4$
Will reduce all four types of compound. Does not reduce $\overset{\diagdown}{\underset{\diagup}{C}}=\overset{\diagup}{\underset{\diagdown}{C}}$ , but
will reduce $NO_2$. Expensive, highly reactive and potentially hazardous.
$NaBH_4$
Less reactive, can be used in aqueous solution (but not with RCOCl!).
Very slow to reduce esters, hence can be used to reduce keto esters to esters.
  Note also that diborane will reduce aldehydes and ketones, but not acyl
chlorides, to alcohols.

3. **Clemmensen** reduction

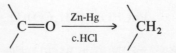

Usually only applied to aryl ketones (synthesised by Friedel–Crafts acylation).
Aldehydes often polymerise. May be useful for base sensitive compounds.

## 4. Wolff–Kishner

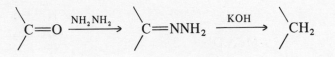

Good yields. Useful for acid sensitive aldehydes and ketones. Other methods include the **Meerwein–Ponndorf–Verley** reduction, specific for CO group, hence useful for unsaturated or nitro compounds:

$$3\ R_2CO + Al(OCH(CH_3)_2)_3 \longrightarrow (R_2CHO)_3Al + 3\ CH_3COCH_3$$

$$\Big\downarrow H^\oplus, H_2O \quad \text{distil off}$$

$$3\ R_2CHOH$$

Ketones are reduced to 'pinacols' ($\alpha$-$\beta$ diols) with Na–Hg or Mg–Hg.
Reagents which could be used are:
  (i)  Mg–Hg (pinacol reduction).
 (ii)  $K_2Cr_2O_7$ (ketones are resistant to further oxidation).
(iii)  $NaBH_4$ (will not reduce the COOH group).
 (iv)  $LiAlH_4$ (reduces both carbonyl and ester groups).
  (v)  $LiAlH(OtBu)_3$ (Unlike $LiAlH_4$, will not reduce RCHO).

## Chapter 8

### Question 8.1

Although direct esterification is possible, it is an equilibrium process, and even taking steps to drive this to the right, the yield may not be very good. The preferred methods for methyl esters would be
(1) via the acid chloride

$$RCOOH \xrightarrow{\ SOCl_2\ } RCOCl \xrightarrow{\ CH_3OH\ } RCOOCH_3$$

and

(2) by the use of an ethereal solution of diazomethane

$$RCOOH + CH_2N_2 \longrightarrow RCOCH_3 + N_2$$

The reagent is expensive, but good yields are obtained.

Ionisation of the ester ($B_{AL}1$) is possible in water (a highly polar solvent) because both the tertiary carbocation and the trifluorocarboxylate anion are reasonably stable ions:

$$Me_3C\!-\!\ddot{O}\ddot{O}C\ddot{F}_3 \rightleftharpoons Me_3C^\oplus + \underset{\underset{\ddot{\underset{\ominus}{O}}}{}}{\overset{\overset{\ominus\ddot{O}}{}}{C}}\!-\!CF_3 \longleftrightarrow \underset{\underset{\ddot{\underset{\ominus}{O}}}{}}{\overset{\overset{\ddot{O}}{\|}}{C}}\!-\!CF_3$$

alkyl-oxygen
fission, AL

In dilute alkaline solution, the more powerful hydroxide ion is able to bring about bimolecular reaction ($B_{AC}2$):

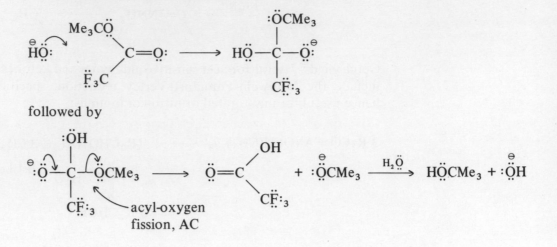

followed by

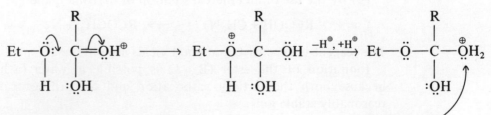

acyl-oxygen fission, AC

## Question 8.2

(a) $NaOH + Br_2 \longrightarrow Na^{\oplus} + Br^{\ominus} + H^{\oplus} + \overset{\ominus}{O}Br$

$RCONH_2 + \overset{\ominus}{O}Br \longrightarrow RCONHBr + \overset{\ominus}{O}H$

$RCONHBr + \overset{\ominus}{O}H \longrightarrow RCON\overset{\ominus}{B}r + H_2O$

followed by rearrangement

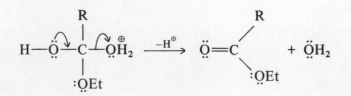

(b)

R—C=Ö $\xrightarrow{HCl}$ $\left[ R—C=\overset{\oplus}{O}H \leftrightarrow R—\overset{\oplus}{C}—\overset{..}{O}H \leftrightarrow R—C—\overset{..}{O}H \right]$ $+ :\overset{\ominus}{C}l:$
|                    |                |             ‖
:OH                 :OH             :OH          :OH
                                                  $\oplus$

Et—Ö: ⤵ C=ÖH$^{\oplus}$ $\longrightarrow$ Et—$\overset{\oplus}{O}$—C—ÖH $\xrightarrow{-H^{\oplus}, +H^{\oplus}}$ Et—Ö—C—$\overset{\oplus}{O}H_2$
|        |                    |   |                                |
H       :OH                  H  :OH                              :OH

now a good
leaving group

H—Ö ⤵ C⤴ÖH$_2$ $\xrightarrow{-H^{\oplus}}$ Ö=C + ÖH$_2$
|                                    \
:OEt                                 :OEt

300

(c)

$$(Et\ddot{O}\ddot{O}C)_2CH_2 \xrightarrow{\text{NaOEt}} (EtOOC)_2\overset{\ominus}{\ddot{C}}H \overset{\oplus}{Na}$$

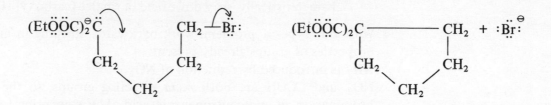

Treatment with another equivalent of NaOEt removes the second activated hydrogen atom, and this is followed by ring closure:

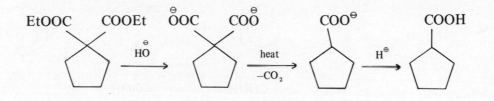

The final step involves hydrolysis of the ester groups and heating to bring about decarboxylation:

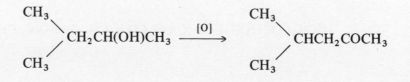

## Question 8.3

1. There are several possible solutions to this sort of problem!
One of the shorter routes is

$$C_2H_5OH \xrightarrow{[O]} CH_3CHO \xrightarrow{HO^\ominus} CH_3CH(OH)CH_2CHO \xrightarrow{[O]}$$

$$CH_3COCH_2COOH$$

followed by selective reduction of the carbonyl group by $NaBH_4$.

2. $PhCOCl \xrightarrow{\text{PhMgBr}} Ph_2CO \; (+ \; MgBrCl) \xrightarrow[\text{2. } H_2O]{\text{1. PhMgBr}} Ph_3C\text{—OH} \; (+ \; Mg(OH)Br)$

3.

301

The methylketone is then treated with iodine in aqueous alkali (haloform reaction), yielding, after acidification, the required carboxylic acid:

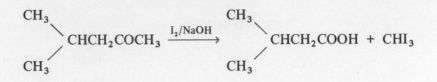

Your $^1$H NMR spectrum should show:

doublet ($6 \times 1°$ H), doublet ($2 \times 2°$ H), multiplet ($3°$ H), and characteristically, way downfield, a singlet (carboxyl H).

4. With this type of problem, work backwards, keeping in mind the directive properties of groups already present.

$NH_2$ is introduced by reduction of $NO_2$.

$NO_2$ and COOH are both *meta* directing groups, so the previous step is bromination of *meta*-nitrobenzoic acid. The step prior to that must have been nitration of benzoic acid. Thus the complete sequence is

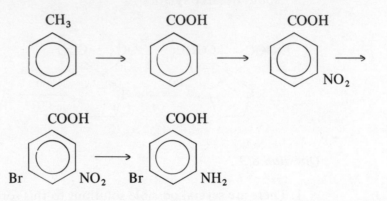

### Question 8.4

1. (a) Start with the Claisen condensation.

$$2 \, CH_3CO_2Et \xrightarrow[EtOH]{NaOEt} CH_3CO\overset{\ominus}{\underset{..}{C}}HCO_2Et \; \overset{\oplus}{Na} \xrightarrow{H^{\oplus}} CH_3COCH_2CO_2Et$$

Then make use of the 'acetoacetic ester' synthesis.

$$CH_3COCH_2CO_2Et \xrightarrow[EtOH]{NaOEt} CH_2CO\overset{\ominus}{\underset{..}{C}}HCO_2Et \; \overset{\oplus}{Na}$$

The sodio derivative is reacted with 'allyl chloride'

$$CH_3CO\overset{\ominus}{\underset{..}{C}}HCO_2Et + ClCH_2CH{=}CH_2 \longrightarrow CH_3COCHCO_2Et + Cl^{\ominus}$$
$$\qquad\qquad\qquad\qquad\qquad\qquad\qquad | $$
$$\qquad\qquad\qquad\qquad\qquad\qquad CH_2CH{=}CH_2$$

'Ketonic hydrolysis' (heat with KOH solution) gives the required product

$$CH_3COCHCO_2Et \longrightarrow CH_3COCH_2CH_2CH{=}CH_2 + CO_2 + EtOH$$
$$\underset{CH_2CH{=}CH_2}{|}$$

(b)

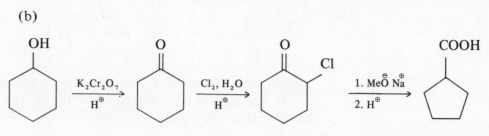

(Favorski reaction)

2. (c)

$$RCOOH \xrightarrow{SOCl_2} RCOCl \xrightarrow{RNH_2} RCONHR \xrightarrow{Na, EtOH} RCH_2NHR$$

(d)

$$\underset{R}{\overset{R}{\diagdown}}C{=}O \xrightarrow{NH_2OH} \underset{R}{\overset{R}{\diagdown}}C{=}N{-}OH \xrightarrow[\text{rearrangement}]{\text{Beckmann}} RCONHR$$

Various catalysts can be used for the rearrangement, including $H_2SO_4$, $SOCl_2$, $PCl_5$, etc.

(e)

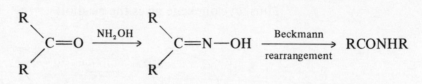

$$PhCH_2Br \xrightarrow[\text{ether}]{Mg} PhCH_2MgBr \xrightarrow{CH_2{-}CH_2} Ph(CH_2)_3OH \xrightarrow{[O]} PhCH_2CH_2COOH$$

**Chapter 9**

*Question 9.1*

(1) *Syn*-addition of two hydroxyl groups to an alkene double bond can be brought about by the use of alkaline potassium manganate(vii) (potassium permanganate), $KMnO_4$, or osmium(viii) oxide (osmium tetroxide or osmic acid), $OsO_4$.

(a) Treatment of a cyclic alkene with dilute aqueous alkaline potassium permanganate solution gives the *cis*- or *syn*-diol by a mechanism involving the formation of an intermediate cyclic manganate ester:

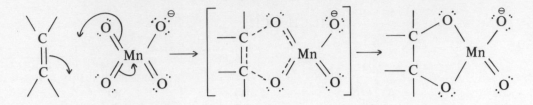

Such esters are unstable and are spontaneously hydrolysed. The use of $^{18}O$ labelled $KMnO_4$ indicates that the process involves cleavage of the Mn—O bonds

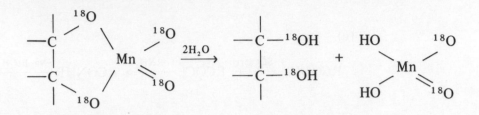

Thus, cyclohexene gives the cis-diol:

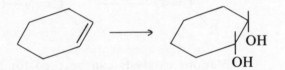

(b) Osmium tetroxide is a yellow crystalline solid that is soluble in both aqueous and non-aqueous solvents. The mechanism is analogous to that for hydroxylation with $KMnO_4$, but the intermediate ester is more stable and has to be hydrolysed by warming with aqueous $Na_2SO_3$ solution

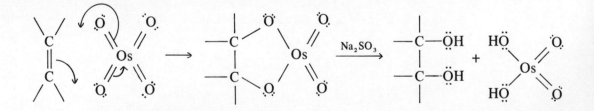

*Anti* dihydroxylation of an alkene is achieved in two stages through the use of peroxides or peroxyacids. For example, you could quote the reaction which occurs on leaving an alkene in contact with a mixture of methanoic acid and hydrogen peroxide for a few hours, when the epoxide is produced:

$$H—\underset{\underset{O}{\|}}{C}—OH + H_2O_2 \rightleftharpoons H—\underset{\underset{O}{\|}}{C}—O—OH + H_2O$$

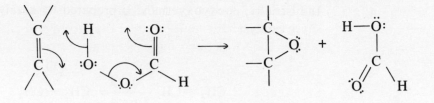

The second stage, involving ring-opening of the epoxide, is brought about by warming the compound with dilute acid or alkali:
(i) Acid catalysis

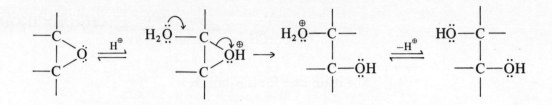

(ii) Base catalysis

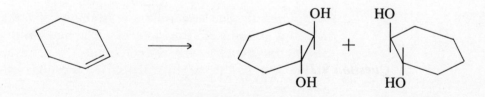

By this means, cyclohexene is converted into the *trans* 1,2-diol, which is enantiomeric:

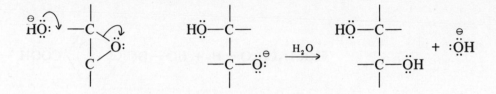

Similarly, *anti*-addition to a symmetrical Z-alkene will result in a racemic modification (mixture):

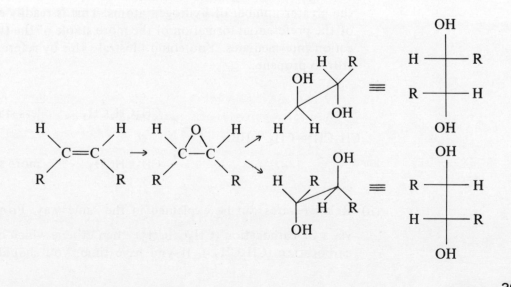

305

(2) Industrially, epoxyoxyethane is prepared by catalytic air oxidation using a silver catalyst:

$$CH_2\!=\!CH_2 \xrightarrow[250°C]{O_2/Ag} CH_2\!-\!CH_2 \ (\text{epoxide})$$

Subsequent hydrolysis to the diol requires heating with superheated steam or treatment at lower temperatures with aqueous acid:

$$CH_2\!-\!CH_2 \xrightarrow[\text{or aqueous acid, 60°C}]{\text{steam, 200°C}} CH_2(OH)CH_2OH$$

The main uses for the diol are:

(i) as a component of antifreeze mixtures for cars and aircraft wings, and
(ii) in the manufacture of the polyester, terylene, by condensation with *tere*-phthalic acid:

$$HOCH_2CH_2O\!-\!H \ + \ HO\!-\!OC \langle\bigcirc\rangle COOH \xrightarrow{-H_2O}$$

$$HOCH_2CH_2\!-\!O\!-\!OC \langle\bigcirc\rangle COOH, \text{ etc.}$$

### Question 9.2

For the mechanisms, see Example 9.2.

(i) Markownikov's Rule states that in addition of HX to a non-symmetrical alkene, the hydrogen atom adds to the carbon atom which already has the greater number of hydrogen atoms. This is readily explained in terms of the preferential formation of the more stable of the two possible carbocation intermediates. You could illustrate this by reference to addition of HBr to propene:

$$CH_3CH\!=\!CH_2 + HBr \nearrow\!\!\!\!\times \ CH_3CH_2\overset{\oplus}{C}H_2 \qquad \text{less stable } 1°$$
$$\searrow \ CH_3\overset{\oplus}{C}HCH_3 \qquad \text{more stable } 2°$$

(ii) Relative rates can be explained in the same way. Propene reacts faster, via a 2° carbocation ($CH_3\overset{\oplus}{C}HCH_3$), then ethene which can only form a 1° carbocation ($CH_3CH_2^{\oplus}$). If you have time, you should illustrate this by

306

drawing comparative energy profiles for the first (slow, rate determining) step:

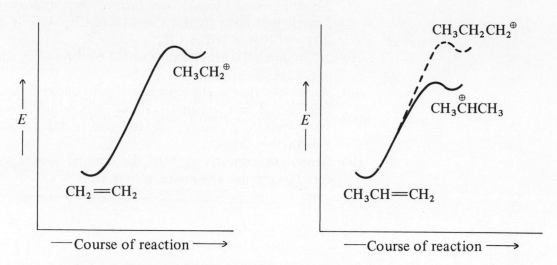

—Course of reaction ⟶                    —Course of reaction ⟶

(iii) With certain structures, the first formed carbocation may rearrange to a more stable carbocation before reacting with $Br^{\ominus}$ to form the addition product. Rearrangement may involve the shift of either a hydrogen atom with the bonding electron pair ('**hydride ion shift**'), or an alkyl group with the bonding electrons. Your answer should contain at least one illustration, such as reaction of 3-methylbut-1-ene:

$$
\begin{array}{c}
CH_3 \\
\diagdown \\
\phantom{x}CH-CH=CH_2 + HBr \\
\diagup \\
CH_3
\end{array}
\longrightarrow
\begin{array}{c}
CH_3 \\
\diagdown \\
\phantom{x}CH-\overset{\oplus}{C}H-CH_3 + Br^{\ominus} \\
\diagup \\
CH_3
\end{array}
$$

(rather than the less stable
$$
\begin{array}{c}
CH_3 \\
\diagdown \\
\phantom{x}CH-CH_2\overset{\oplus}{C}H_2 \;)\\
\diagup \\
CH_3
\end{array}
$$

followed by

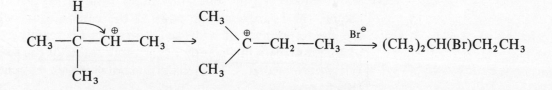

i.e. the main product is 2-bromo-2-methylbutane rather than the 'expected' 2-bromo-3-methylbutane.

## Question 9.3

(a) Obviously you will need to be able to quote the principal methods outlined at the beginning of this chapter, together with any other methods mentioned in your particular course.

(b) You should give some indication of how you arrive at the proposed structures.

(i) 2-methylhexane would be formed on hydrogenation of either 2-methylbut-2-ene $((CH_3)_3C{=}CHCH_2CH_2CH_3)$ or 2-methylbut-1-ene $(CH_2{=}C(CH_3)CH_2CH_2CH_3$.

(ii) 2,3-dibromo-5-methylhexane would be formed on addition of bromine to 5-methylhex-2-ene.

(iii) Remember that in the presence of peroxides HBr adds to alkenes by a free radical mechanism (Example 9.2) which gives rise to an anti-Markownikov product. The required starting material is therefore 3-methylhex-2-ene.

(iv) Ozonolysis converts each of the doubly bonded carbon atoms into carbonyl groups. The products here are

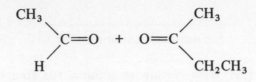

from which the structure of the original alkene is deduced to be 3-methylpent-2-ene, $CH_3CH{=}CH(CH_3)CH_2CH_3$.

## Chapter 10

### Question 10.1

Heat benzene with 'mixed acid' $(c.HNO_3 + c.H_2SO_4)$, keeping the temperature below about 50°C.

Check your answers to the next two sections against Example 10.1.

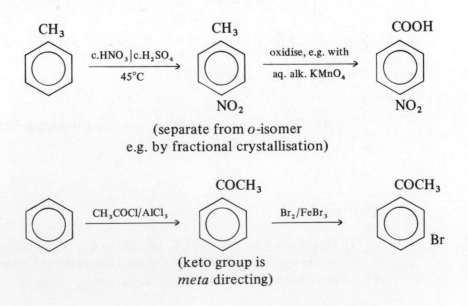

(a) Inductive stabilisation of sigma complex intermediates by the methyl substituent (Section 10.5), and powerful deactivation by the positively charged nitrogen atom of the nitro group. Difference in activation energies more significant at low temperature.

(b) The fluoro substituent is *ortho/para* directing (+M effect), but weakly deactivating (strong −I effect). Reaction is controlled by the powerfully activating methoxy group (strong +M effect), and nitration occurs in positions *ortho* and *para* to that substituent.

(c) Low temperature ensures mononitration. Steric hindrance by the bulky *tert*-butyl group accounts for preferential formation of the indicated product.

(d) In fluorobenzene the powerful −I effect of the fluorine atom reduces the amount of *ortho* product. The −I effect decreases from fluorine to iodine, allowing a higher proportion of *ortho* substitution. However, the *size* of the substituents increases in the same direction, and steric hindrance keeps the ratio of *ortho:para* less than the statistical 2:1.

*Question 10.3*

(i) Your answer should begin with a brief explanation of the effect of substituents in terms of their ability to stabilise or destabilise the sigma complex intermediates for *ortho/para* and *meta* attack (see Example 10.2).

The ammonium ($-\overset{\oplus}{N}H_3$) and nitro ($-\overset{\oplus}{N}\overset{\nearrow O^{\ominus}}{\underset{\searrow O}{}}$) groups are linked to the ring through their positively charged nitrogen atoms, and both substituents are powerfully deactivating and *meta* directing. Acetamino ($-NHCOCH_3$) is strongly activating and *ortho/para* directing (strong +M effect). However, the group is quite bulky, as is the electrophile in bromination, $\overset{\delta+}{Br}\text{----}\overset{\delta-}{Br}\text{----}\overset{\delta-}{FeBr}$. Consequently, steric hindrance to *ortho* attack (steric compression destabilising the TS for formation of the *ortho* sigma complex intermediate) would lead you to predict monobromination predominately in the *para* position. Finally, the fluoro substituent is *ortho/para* directing, but weakly deactivating. The strong negative effect will be most apparent at the adjacent *ortho* position, and so again mono-bromination will occur mainly in the *para* position.

(ii)

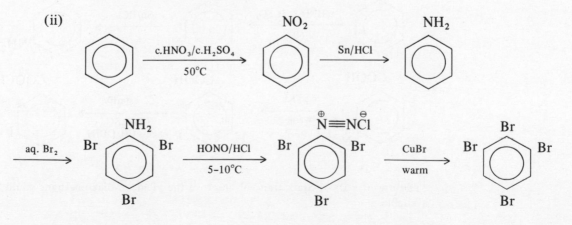

(a) Your answer should begin with a brief description of the mechanism of monobromination of benzene and the structure of the sigma complex intermediate. The carboxyl group in benzoic acid has a mesomeric polar structure in which the carbon atom carries positive charge. Juxtaposition of like charges destabilises the *ortho* and *para* sigma complexes

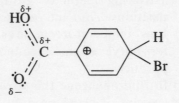

and consequently the carboxyl group is deactivating and *meta* directing. The effects of the other substituents, chlorine (*ortho/para* directing and weakly deactivating, $-I > +M$), N,N-dimethylamino (powerfully activating and *ortho/para* directing, $+M$ supported by $+I$ of the two methyl groups $\gg -I$) and methyl (moderately activating and *ortho/para* directing, $+I$) have all been discussed above.

(b)

(i)

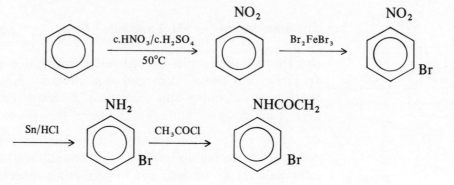

(ii)

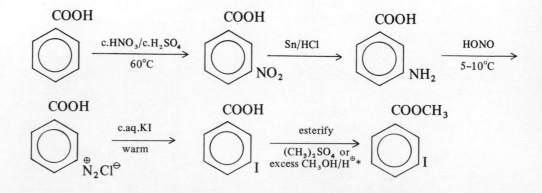

*The use of a Dean–Stark trap will improve the yield, or diazomethane could be used if this is available.

(iii)

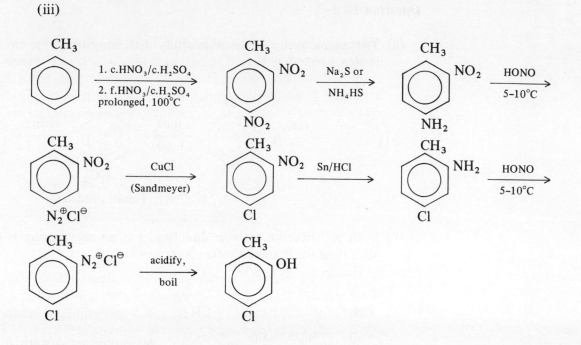

## Question 10.5

(a)

(b)

(c)

(d)

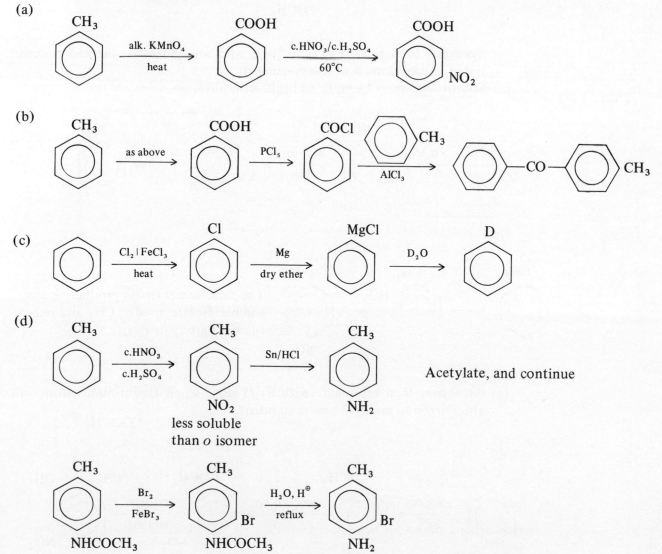

**Question 10.6**

(a) The amino group is too powerfully activating for facile monobromination. Hence, brominate first.

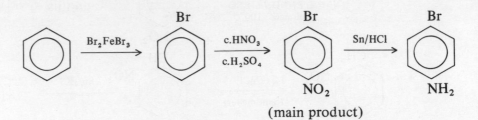

(main product)

(b) Both substituents are *o/p* directing, but an alkyl group is too powerfully activating for Friedel–Crafts alkylation.
Hence,

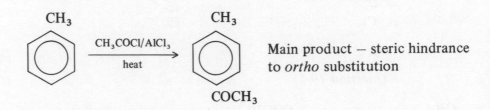

Main product – steric hindrance to *ortho* substitution

Synthesis is completed by reduction with amalgamated sinc and concentrated hydrochloric acid (Clemmensen).

(c) Substituents must be *meta*, so begin with nitration.

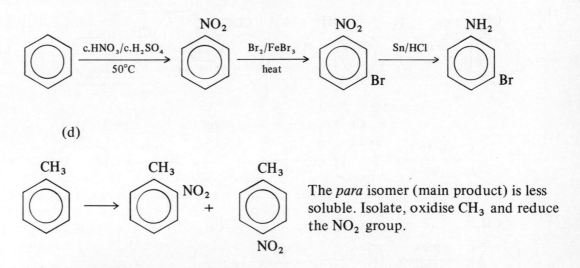

(d)

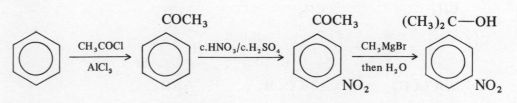

The *para* isomer (main product) is less soluble. Isolate, oxidise $CH_3$ and reduce the $NO_2$ group.

(e) As above, then brominate with $Br_2/FeBr_3$, when the bromine atom will enter *ortho* to methyl (= *meta* to nitro).

(f)

## Chapter 11

### Question 11.1

See Example 11.1. There is no simple economical way of separating components of the mixture. The next two parts are also covered in Example 11.1.

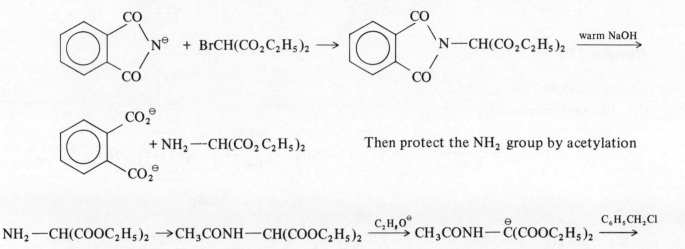

$$NH_2—CH(COOC_2H_5)_2 \rightarrow CH_3CONH—CH(COOC_2H_5)_2 \xrightarrow{C_2H_5O^{\ominus}} CH_3CONH—\overset{\ominus}{C}(COOC_2H_5)_2 \xrightarrow{C_6H_5CH_2Cl}$$

$$\underset{\overset{|}{CH_2C_6H_5}}{CH_3CONHC(COOC_2H_5)_2}$$   which is warmed in acid solution to hydrolyse the

acetamino and ester groups. Decarboxylation follows,

$$\underset{\overset{|}{CH_2C_6H_5}}{CH_3CONHC(COOC_2H_5)_2} \longrightarrow \underset{\overset{|}{CH_2C_6H_5}}{NH_2—CH—COOH} \qquad (+ CO_2 + CH_3COOH + 2C_2H_5OH)$$

### Question 11.2

(a) See Example 11.4.
(b) Coupling will only occur with activated substrates such as tertiary aromatic amines and phenols. Rate of reaction is pH dependent. At low pH unionised phenol is relatively unreactive and protonation deactivates the amine. At high pH the diazonium salt is converted into the unreactive hydroxide, $ArN=N—OH$. Coupling is most rapid at moderate pH where the free amine or the phenoxide ion is present in high concentration.

(c)
(i)

313

(ii)

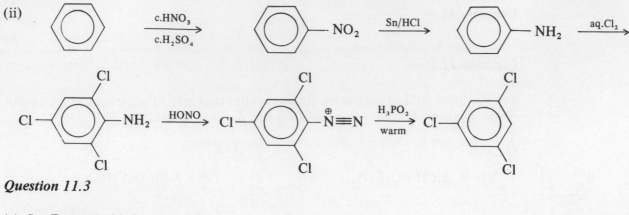

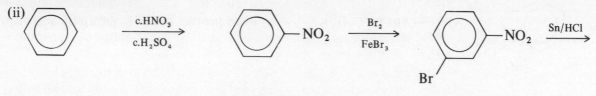

**Question 11.3**

(a) See Example 11.4.

(b) (i) See Question 11.2

(ii)

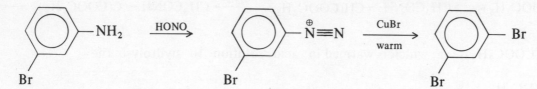

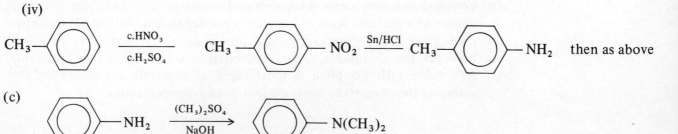

(iii) as above to give

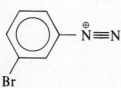

then

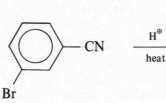

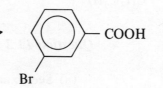

(iv)

CH₃—⬡—NO₂ ... then as above

(c)

⬡—NH₂ $\xrightarrow[\text{NaOH}]{(CH_3)_2SO_4}$ ⬡—N(CH₃)₂

then couple with diazotised sulphanilic acid.

*Question 11.4*

(a) See Chapter 4, Section 4.12.

   Basic strength of an amine relates to position of equilibrium in the reaction

$$R\ddot{N}H_2 + H_2\ddot{O} \;\rightleftharpoons\; R\overset{\oplus}{N}H_3 + :\overset{\ominus}{\ddot{O}}H$$

Aliphatic amines are more basic than ammonia because of inductive stabilisation of the conjugate cation by alkyl groups:

$$R\!\rightarrow\!\overset{\oplus}{N}H_3 \quad \text{and} \quad \overset{\displaystyle R}{\underset{\displaystyle R}{>}}\!\overset{\oplus}{N}H_2$$

However, tertiary amines are less basic than secondary because steric crowding around the positive nitrogen atom impedes solvation of the cation by water molecules (i.e. increased inductive stabilisation is outweighed by loss of solvation energy).

Protonation of aromatic amines can only occur at the expense of resonance energy associated with conjugation of the amino group with the aromatic ring

Thus phenylamine is less basic than ammonia.

(b) Amines are only weakly acidic, but with a very strong base such as sodium metal, hydrogen can be displaced. Aliphatic amines are less acidic than ammonia which in turn is less acidic than phenylamine. The latter compound gives rise to a resonance-stabilised anion:

(c) Amines are moderately strong nucleophiles and will, for example, displace halogen (X) from haloalkanes (RX)

$$R\!-\!\overset{\frown}{\ddot{N}H_2} \quad R\!-\!\overset{\frown}{\ddot{X}}: \;\longrightarrow\; R\!-\!\overset{\oplus}{N}H_3R:\overset{\ominus}{\ddot{X}}:$$

Further reaction leads to $R_3NH_2X$ and $R_4NX$, and the reaction is of limited preparative value.

Amines also react with electrophilic species such as $NO^{\oplus}$, present in strongly acidic solutions of nitrous acid. You should briefly indicate dif-

315

ferences in behaviour of primary, secondary and tertiary aliphatic and aromatic amines (see Example 11.3).

(d) The amino group, even when protonated is a poor leaving group, and amines do not exhibit the nucleophilic substitution and elimination reactions observed with alcohols. One of the few examples of nucleophilic substitution you could quote is thermal decomposition of tetramethylammonium bromide:

$$:\overset{\ominus}{\underset{..}{Br}}: \quad CH_3 - \overset{\oplus}{N}(CH_3)_3 \xrightarrow{\Delta} :\underset{..}{Br} - CH_3 + \overset{..}{N}(CH_3)_3$$

β-elimination occurs on heating quaternary ammonium hydroxides, e.g.

$$H\overset{\ominus}{\underset{..}{O}}: \quad \begin{array}{c} H \\ | \\ CH_2 - CH_2 \\ | \\ \underset{\oplus}{NR_3} \end{array} \xrightarrow{\Delta} H\underset{..}{O} - H + CH_2 = CH_2 + \overset{..}{N}R_3$$

**(Hofmann elimination)**

## Chapter 12

### Question 12.1

(a) (i) 3-methylhexane, (ii) 1-(1-methylcyclohexyl)-2-methylpropane or 1-methyl-1-(2-methylpropyl)cyclohexane, (iii) 4-ethyl-2,2-dimethy-5-propylheptane.

(b) 3-bromo-3-methylhexane.
   The bromine atom, Br·, is a relatively weak radical and is therefore very **selective**. It preferentially abstracts the tertiary hydrogen atom to give the most stable of the possible free radical intermediates.

(c) See Example 12.1.

### Question 12.2

You could begin by stating the four basic types of radical reaction: **radical pairing** (dimerisation), **radical transfer** (including **hydrogen abstraction**), **addition** to a pi bond and **disproportionation** (a special case of hydrogen abstraction) (see Section 12.8).

Examples of synthetically useful processes with which to illustrate these terms are halogenation of alkanes, thermal cracking and alkene polymerisation.

Chlorination of methane is summarised in Section 12.6, and Example 12.1. You could use monochlorination of propane as a lead in to a discussion of radical stability, quoting inductive stabilisation (or hyperconjugation, if you are familiar with this) for alkyl radicals and mesomeric (resonance) stabilisation for allylic and benzylic radicals. Make clear (give an energy profile as in Question 12.3) that the reference point for stability of radicals in each case is the parent compound.

The basic principles of thermal cracking can be outlined for decomposition of butane, where thermolysis gives three different radicals:

$$CH_3-CH_2-CH_2 \overset{\frown}{|} CH_3 \longrightarrow CH_3-CH_2-\dot{C}H_2 + \dot{C}H_3$$

$$CH_3-CH_2 \overset{\frown}{|} CH_2-CH_3 \longrightarrow CH_3-\dot{C}H_2 + \dot{C}H_2-CH_3$$

Attack by these radicals on butane will give both primary and secondary butyl radicals. For example with the methyl radical

$$\dot{C}H_3 \quad H \overset{\frown}{|} CH_2CH_2CH_2CH_3 \longrightarrow CH_4 + \dot{C}H_2CH_2CH_2CH_3$$

and

$$\dot{C}H_3 \quad H \overset{\frown}{|} \underset{\underset{CH_3}{|}}{C}HCH_2CH_3 \longrightarrow CH_4 + \underset{\underset{CH_3}{|}}{\dot{C}}HCH_2CH_3$$

These butyl radicals may themeselves undergo homolytic fission, (known as 'β-cleavage') to form an alkene and another alkyl radical:

$$CH_3CH_2 \overset{\frown}{|} CH_2 \overset{\frown}{|} \dot{C}H_2 \longrightarrow CH_3\dot{C}H_2 + CH_2{=}CH_2$$

$$\dot{C}H_3 \overset{\frown}{|} CH_2 \overset{\frown}{|} \dot{C}HCH_3 \longrightarrow \dot{C}H_3 + CH_2{=}CHCH_3$$

In this way, propagating chains are set up, leading to a mixture of methane, ethane, ethene and propene. Termination results from disproportionation,

e.g.

$$\dot{C}H_3 \quad H{-}\underset{\underset{\dot{C}H_2}{|}}{C}HCH_2CH_3 \longrightarrow CH_4 + CH_2{=}CHCH_2CH_3$$

$$\dot{C}H_3 \quad H{-}\underset{\underset{CH_3}{|}}{C}H{-}CHCH_3 \longrightarrow CH_4 + CH_3CH{=}CHCH_3$$

or radical combination

e.g.

$$\dot{C}H_3 \quad \dot{C}H_2CH_2CH_2CH_3 \longrightarrow CH_3CH_2CH_2CH_2CH_3$$

$$\dot{C}H_3 \quad \underset{\underset{CH_3}{|}}{\dot{C}}HCH_2CH_3 \longrightarrow CH_3\underset{\underset{CH_3}{|}}{C}HCH_2CH_3$$

Radical polymerisation is considered in Section 12.7.


Question 12.3

(a) H = 3-bromo-3-methylpentane

$$Br_2 \underset{UV}{\rightleftharpoons} 2\,\dot{B}r \qquad \text{(initiation)}$$

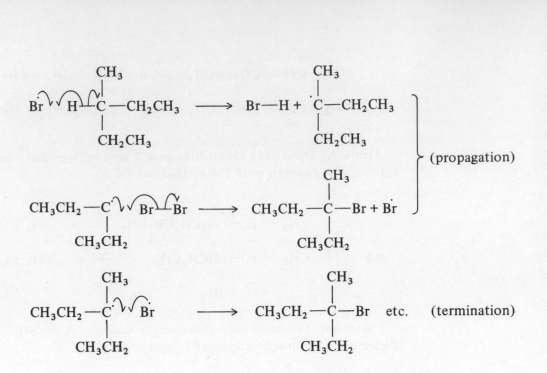

$$\} \text{ (propagation)}$$

$$\} \text{ etc. (termination)}$$

$Br\dot{}$ is a relatively unreactive radical and preferentially removes the tertiary hydrogen atom so as to give the most stable of the possible radical intermediates.

(b) Uptake of a chlorine atom by the ring carbon can only occur with perturbation of the aromatic ring (i.e. loss of resonance energy).

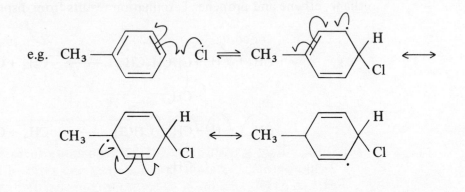

Thus although the intermediate radical is stabilised by delocalisation of the unpaired electron (Section 12.5), activation energy for its formation is quite large. At elevated temperatures the process is readily reversible because loss of $Cl\dot{}$ regenerates the aromatic system (i.e. restores the 'lost' resonance energy).

On the other hand, hydrogen abstraction from the methyl group by a chlorine atom results in a resonance stabilised benzylic radical intermediate:

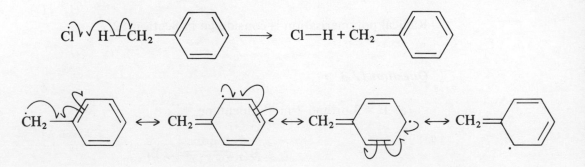

Thus, activation energy is much lower for this process, and the benzyl radical takes part in a chain reaction which eventually leads to trichloromethylbenzene:

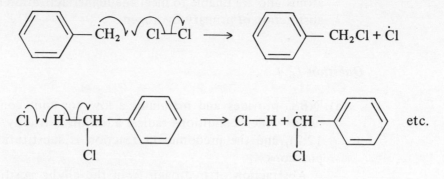

Illustrate your answer with an energy profile,

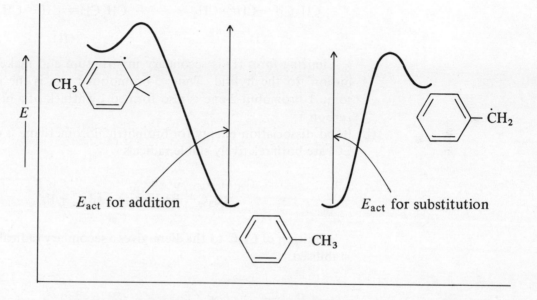

$E_{act}$ for addition          $E_{act}$ for substitution

(c) On a purely statistical basis (the **probability factor**) 1-chloropropane and 2-chloropropane should be formed in the ratio of 3:1. The higher than expected proportion of 2-chloropropane is due to the greater ease of removal of secondary hydrogen atoms compared with primary hydrogen atoms (the **reactivity factor**).

Activation energy is less for the removal of secondary hydrogen as this leads to the more stable secondary propyl radical. Illustrate this with an energy profile,

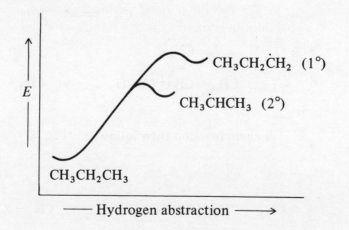

and outline the free radical mechanism.

Bromine molecules dissociate more readily than chlorine molecules, but the resulting bromine atoms are correspondingly less reactive than chlorine atoms and are unable to meet the higher activation energy requirements for abstraction of primary hydrogen.

### Question 12.4

(a) 'NBS' provides and maintains a low constant concentration of bromine, under which conditions radical addition is readily reversible (see Example 12.2), and the predominant reaction is substitution in allylic or benzylic positions.

Abstraction of hydrogen from the allylic position in but-1-ene gives a resonance stabilised radical intermediate:

$$CH_3\dot{C}H-CH=CH_2 \longleftrightarrow CH_3CH=CH-\dot{C}H_2$$

(I)                                    (II)

Limiting form (I) is secondary in structure and makes the larger 'contribution' to the hybrid. Thus, 3-bromobut-1-ene is the major product, but some 1-bromobut-2-ene is also formed by attack of a bromine molecule on carbon 1.

(b) Bond dissociation energy for bromotrichloromethane is quite low as $\dot{B}r$ and $\dot{C}Cl$ are both relatively stable radicals

$$Cl_3C-Br \xrightarrow{UV} Cl_3\dot{C} + \dot{B}r$$

Addition of $Cl_3\dot{C}$ to the diene gives a secondary radical which is resonance stabilised:

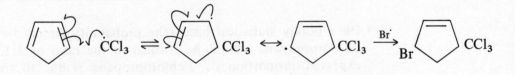

The bromine atom is quite large and probably adds only to the observed position because of steric hindrance adjacent to the large $CCl_3$ group.

(c)

$$Cl_2 \rightleftharpoons 2\dot{C}l$$

$$CH_3-CH=CHCH_3 \longrightarrow CH_3-\underset{\underset{Cl}{|}}{CH}-\dot{C}HCH_3$$

A chain reaction then follows

$$CH_3-\underset{\underset{Cl}{|}}{CH}-\dot{C}HCH_3 + Cl_2 \longrightarrow CH_3-\underset{\underset{Cl}{|}}{CH}-\underset{\underset{Cl}{|}}{CH}-CH_3 + \dot{C}l \text{ etc.}$$

The minor product is formed by a disproportionation step:

$$CH_3-CH-CH-CH_2-H \quad Cl \longrightarrow CH_3-CH-CH=CH_2 + HCl$$

with Cl substituents on the second carbon of each structure.

## Chapter 13

### *Question 13.1*

This is a nice straightforward question to begin with!

(A) = $C_4H_{10}O$     oxidised to a carbonyl compound, so probably an alcohol, $C_4H_9-OH$

↓ $CrO_3$

(B) = $C_4H_8O$     a carbonyl compound, $C_3H_8-CO$

↓ $I_2$/NaOH     the iodoform test for $CH_3CO$ group

$CHI_3 + CH_3CH_2CO\overset{\ominus}{O} \overset{\oplus}{Na}$

(B) = $CH_3CH_2COCH_3$, and

(A) = $CH_3CH_2CH(OH)CH_3$

Mechanism of the iodoform reaction

$$\overset{\ominus}{H\ddot{O}:} \quad H-CH_2C\ddot{O}CH_2CH_3 \rightleftharpoons H\ddot{O}-H + \overset{\ominus}{\ddot{C}H_2}C\ddot{O}CH_2CH_3$$

$$CH_3CH_2C\ddot{O}\overset{\ominus}{CH_2} \quad :\ddot{I}-\ddot{I}: \longrightarrow CH_3CH_2C\ddot{O}CH_2\ddot{I}: + :\ddot{I}:^{\ominus}$$

Repetition leads to $CH_3CH_2C\ddot{O}C\ddot{I}_3$.
Then

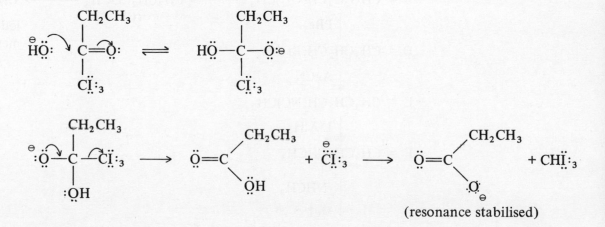

(resonance stabilised)

**Question 13.2**

Compound (A) is synthesised by aldol condensation of propanone with benzaldehyde.

$$CH_3COCH_3 + C_6H_5CHO \xrightarrow{HO^{\ominus}} CH_3COCH=CHC_6H_5 + H_2O$$

4-phenylbut-3-en-2-one
(benzalacetone)

in agreement with the molecular formula $C_{10}H_{10}O$.
    Then

| | | |
|---|---|---|
| (A)= $CH_3COCH=CHC_6H_5$ | | (+ve Brady's, iodoform, $Br_2/H_2O$) |

$\downarrow$ $LiAlH_4$  (reduces $\diagdown C=O$ to $-CH(OH)-$)

(B) = $CH_3CH(OH)CH=CHC_6H_5$    (+ve iodoform, $Br_2/H_2O$)

$\downarrow$ $H_2/Pd$  (double bond hydrogenated)

(C) = $CH_3CH(OH)CH_2CH_2C_6H_5$    (+ve iodoform)

$\downarrow$ c.$H_2SO_4$  (alcohol dehydrated)

(D) = $CH_3CH=CHCH_2C_6H_5$    (Saytzeff, most stable alkene; +ve $Br_2/H_2O$)

$\downarrow$ $O_3$/workup

$CH_3CHO + C_6H_5CH_2CHO$

**Question 13.3**

Compound A is a carbonyl compound, and with only 3 carbon atoms it is either propanal or propanone. Reaction with a Grignard reagent gives an oxidisable alcohol, and therefore A must be propanal. (Propanone would give a tertiary alcohol, resistant to oxidation.)

A = $CH_3CH_2CHO \xrightarrow{2,4\text{-DNP}} CH_3CH_2CH=NNH$ ⬡ $NO_2$ with $NO_2$

$\downarrow$ MeMgI, hydrolytic work-up

B = $CH_3CH_2CH(OH)CH_3 \xrightarrow{\text{chromic acid}} CH_3CH_2COCH_3 \xrightarrow{I_2/KOH} CH_3CH_2COOH + CHI_3$

(C)    iodoform from a methyl ketone

$\downarrow$ $PBr_3$

D = $CH_3CH_2CH(Br)CH_3$

$\downarrow$ AgCN

E = $CH_3CH_2CH(NC)CH_3$

$\downarrow$ $LiAlH_4$

F = $CH_3CH_2CHCH_3$
        |
      $NHCH_3$
$\downarrow$ MeI

322

$G = CH_3CH_2CHCH_3$
      |
      $\overset{\oplus}{N}(CH_3)_3 \ \overset{\ominus}{I}$
      |
      ↓ $Ag_2O$

$H = CH_3CH_2CHCH_3$
      |
      $\overset{\oplus}{N}(CH_3)_3 \ \overset{\ominus}{O}H$
      |
      ↓ dry distillation

$I = CH_3CH_2CH{=}CH_2 + (CH_3)_3N$

Cyanide ion is an ambident ion, and when associated with the $Ag^{\oplus}$ ion, reacts by donation of an electron pair from the *nitrogen* atom:

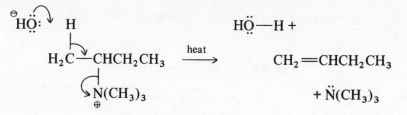

Dry distillation of the quaternary ammonium hydroxide, H, brings about Hofmann elimination:

$$\overset{\ominus}{H\ddot{O}}: \quad H$$
$$H_2C{-}CHCH_2CH_3 \xrightarrow{heat} H\ddot{O}{-}H + \quad CH_2{=}CHCH_2CH_3$$
$$\overset{\oplus}{N}(CH_3)_3 \qquad\qquad\qquad + \ddot{N}(CH_3)_3$$

Note that in order to minimise steric crowding in the transition state, Hofmann elimination proceeds so as to give the least alkylated alkene (but-1-ene rather than but-2-ene).

**Question 13.4**

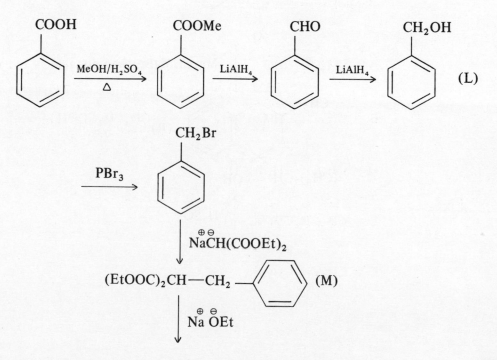

323

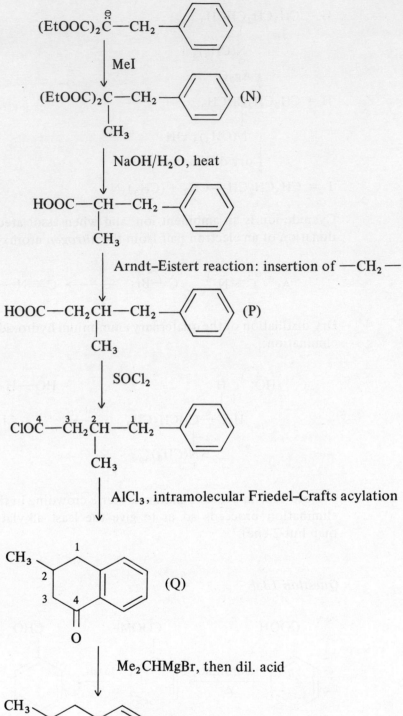

$$(EtOOC)_2\overset{\ominus}{C}-CH_2- \text{\textcopyright}$$

↓ MeI

$$(EtOOC)_2C-CH_2- \text{\textcopyright} \quad (N)$$
$$\quad\quad\quad | $$
$$\quad\quad\quad CH_3$$

↓ NaOH/H₂O, heat

$$HOOC-CH-CH_2- \text{\textcopyright}$$
$$\quad\quad\quad | $$
$$\quad\quad\quad CH_3$$

↓ Arndt–Eistert reaction: insertion of —CH₂—

$$HOOC-CH_2CH-CH_2- \text{\textcopyright} \quad (P)$$
$$\quad\quad\quad\quad\quad | $$
$$\quad\quad\quad\quad\quad CH_3$$

↓ SOCl₂

$$ClO\overset{4}{C}-\overset{3}{C}H_2\overset{2}{C}H-\overset{1}{C}H_2- \text{\textcopyright}$$
$$\quad\quad\quad\quad\quad | $$
$$\quad\quad\quad\quad\quad CH_3$$

↓ AlCl₃, intramolecular Friedel–Crafts acylation

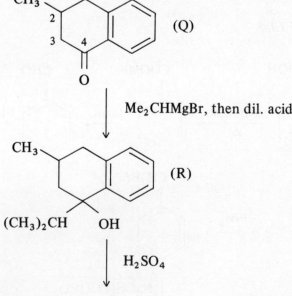

(Q)

↓ Me₂CHMgBr, then dil. acid

(R)

↓ H₂SO₄

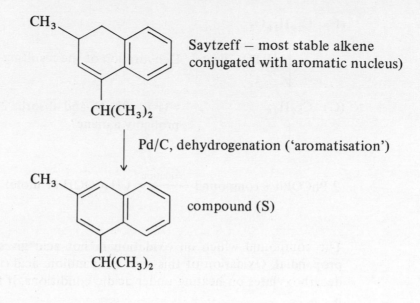

Saytzeff − most stable alkene
conjugated with aromatic nucleus)

Pd/C, dehydrogenation ('aromatisation')

compound (S)

**Question 13.5**

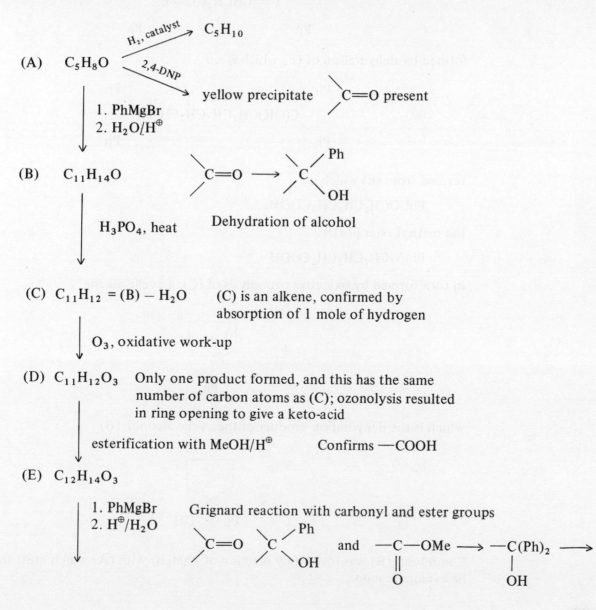

(A)  C$_5$H$_8$O  $\xrightarrow{\text{H}_2, \text{ catalyst}}$  C$_5$H$_{10}$

$\xrightarrow{\text{2,4-DNP}}$  yellow precipitate  $\diagup$C=O present

1. PhMgBr
2. H$_2$O/H$^\oplus$

(B)  C$_{11}$H$_{14}$O  $\diagup$C=O  $\longrightarrow$  $\diagup$C$\diagup^{Ph}_{OH}$

H$_3$PO$_4$, heat    Dehydration of alcohol

(C)  C$_{11}$H$_{12}$ = (B) − H$_2$O    (C) is an alkene, confirmed by
absorption of 1 mole of hydrogen

O$_3$, oxidative work-up

(D)  C$_{11}$H$_{12}$O$_3$    Only one product formed, and this has the same
number of carbon atoms as (C); ozonolysis resulted
in ring opening to give a keto-acid

esterification with MeOH/H$^\oplus$    Confirms —COOH

(E)  C$_{12}$H$_{14}$O$_3$

1. PhMgBr    Grignard reaction with carbonyl and ester groups
2. H$^\oplus$/H$_2$O

$\diagup$C=O  $\diagup$C$\diagup^{Ph}_{OH}$  and  —C—OMe $\longrightarrow$ —C(Ph)$_2$ $\longrightarrow$
$\parallel$    $\mid$
O    OH

(F)  $C_{29}H_{28}O_2$

  │ $H_3PO_4$, heat    Dehydration of the resulting diol
  ↓

(G)  $C_{29}H_{24}$           = (F) − $2H_2O$, and absorbs 2 mole hydrogen
                                probably a diene?

  │ ozonolysis
  ↓

2 PhCOPh + compound $\xrightarrow{\text{oxidation}}$ $CH_3COOH$ (1 mole) + $CO_2$ (1 mole)

The compound which on oxidation in hot acid gives ethanoic acid and $CO_2$ is propandial. Oxidation of this gives propandioic acid (malonic acid) which readily decarboxylates on heating under acidic conditions. It follows that compound (G) is

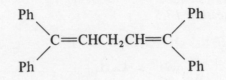

formed by dehydration of (F) which is

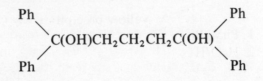

formed from (E) which is

   $PhCOCH_2CH_2CH_2COOMe$,

the methyl ester of (D),

   $PhCOCH_2CH_2CH_2COOH$

in turn formed by oxidative ozonolysis of (C), a cyclic alkene,

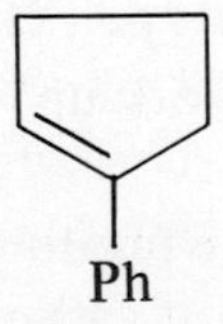

which is the dehydration product of the cyclic alcohol, (B)

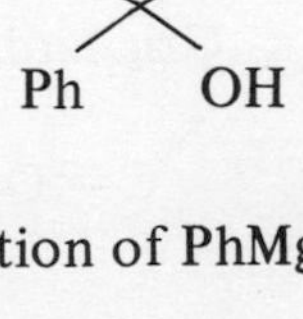

Compound (B) was formed by reaction of PhMgBr with (A) which must therefore be cyclopentanone,

Cyclopentanone is reduced by $H_2$/catalyst, and forms a derivative with 2,4-dinitrophenylhydrazine,

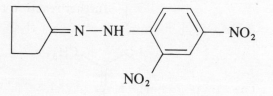

cyclopentanone 2,4-dinitrophenylhydrazone

## Question 13.6

$$\text{(A), } C_8H_{14}O, \text{ achiral} \xrightarrow{O_3, \ Zn/H_2O} \text{(B), } C_5H_8O_2 + O{=}C(CH_3)_2$$

(A) ↓ $I_2$/NaOH → CHI$_3$

(B) ↓ NaOH, warm

(C), $C_5H_6O$, = (B) − $H_2O$

CHI$_3$
+

↓ KMnO$_4$    an acid

Conditions for the aldol reaction — but only one molecule is involved, ∴ internal? In which case, (B) is a dicarbonyl compound.

(D), $C_5H_8O_3$

↓ $C_2H_5OH$/HCl

(E), $C_7H_{12}O_3$

Compound (A) gives a +ve iodoform test and is therefore a methyl ketone of the form $CH_3COR$, where R is unsaturated. (Since (A) is achiral, it cannot be $CH_3CH(OH)R$, which would also give a +ve iodoform test.) Oxidation of (A) cleaves the molecule at the double bond to give a keto-acid, (D), which is esterified with $C_2H_5OH$ and HCl to give (E), a keto-ester.

For compound (B) to readily cyclise, the resulting aldol must have a five-membered ring structure (a four-membered ring would form much less readily). Therefore (B), and consequently also (D) have linear structures. The reaction sequence is:

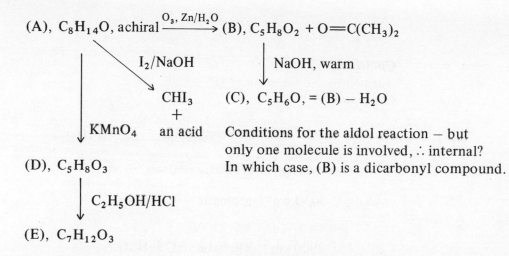

(A) = $CH_3COCH_2CH_2CH{=}C(CH_3)_2$ $\xrightarrow{\text{ozonolysis}}$ (B) = $CH_3COCH_2CH_2CHO + O{=}C(CH_3)_2$

↓ oxidation

(D) = $CH_3COCH_2CH_2COOH$

(B) ↓ aldol reaction

(C) =

↓ esterification

(E) = $CH_3COCH_2CH_2COOC_2H_5$

cyclopent-2-enone

(note additional stability form enone conjugation)

Reaction of (B) with hydrazine

(B)  = CH$_3$COCH$_2$CH$_2$CHO

$\downarrow$  NH$_2$NH$_2$

CH$_3$COCH$_2$CH$_2$CH=NNH$_2$ + H$_2$O

$\downarrow$  further reaction results in ring closure

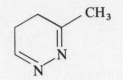

(which is readily oxidised to 3-methylpyridazine, 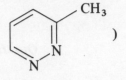 )

### Question 13.7

*Compound H*

IR data  1720 cm$^{-1}$ non conjugated $\diagdown$C=O str.

1600 cm$^{-1}$ aromatic —C==C— str.

3030 cm$^{-1}$ aromatic ==C—H str.

2900 cm$^{-1}$ aliphatic —C—H str.

$^1$H nmr  Downfield multiplet 7.3$\delta$, 5 aromatic H
downfield triplet 9.71$\delta$, aldehydic H adjacent to —CH$_2$—
upfield doublet 3.66$\delta$, —CH$_2$— adjacent to —CHO

To undergo the aldol condensation, the compound must have an activated H. The data is consistent with the structure of phenylethanal,

Compound (H) = ⬡—CH$_2$CHO (C$_8$H$_8$O)

*Compound I*

IR data  Absorptions at 1600 cm$^{-1}$ and 1695 cm$^{-1}$ indicate an aromatic compound with a conjugated carbonyl group (longer $\lambda$). (I) is isomeric with (H), but undergoes the Cannizzaro reaction in strong alkali, and therefore has no $\alpha$-H. Thus, (I) is probably a methylbenzaldehyde. Absorptions at 2820 cm$^{-1}$ and 2730 cm$^{-1}$ could be due to —C—H str. of
the aldehyde group (which often appears as a doublet).

328

<sup>1</sup>H nmr   Downfield singlet 9.51δ due to aldehydic H
upfield singlet 2.41δ due to —CH₃ group
downfield doublets 7.66δ and 7.19δ due to two pairs of aromatic H in
different environments.
Thus the data is consistent with the structure of 4-methylbenzaldehyde,

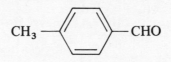

### Reactions

2 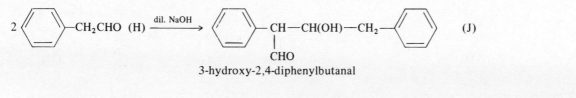 (J)

3-hydroxy-2,4-diphenylbutanal

2 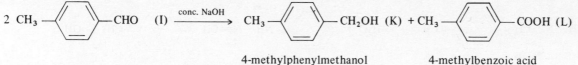 (K) + (L)

4-methylphenylmethanol          4-methylbenzoic acid

The phenylhydrazones are

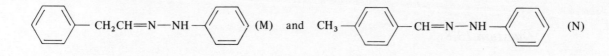 (M)   and   (N)

### Mechanisms

Aldol

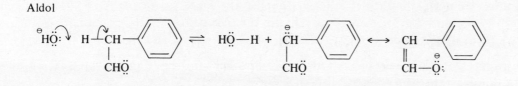

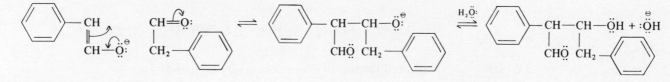

Cannizzaro

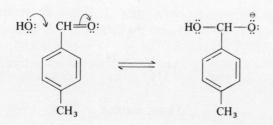

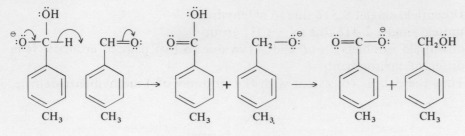

(hydride transfer)

Formation of phenylhydrazones

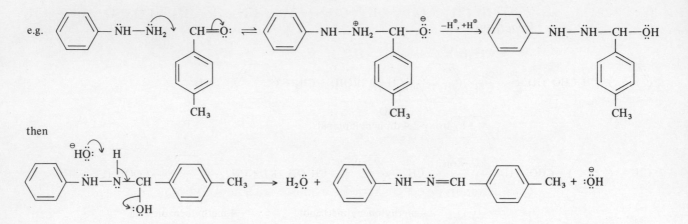

then

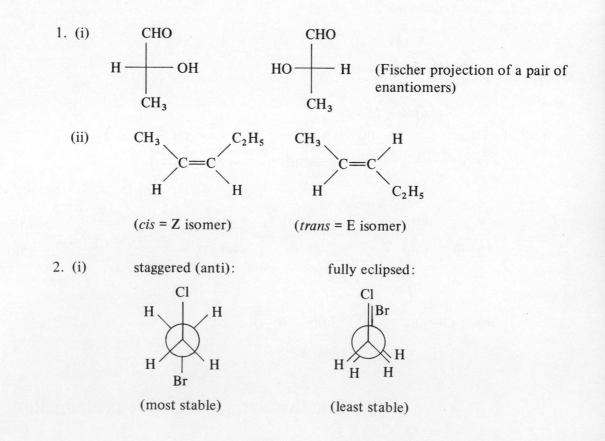

*Question 13.8*

1. (i)

| CHO | CHO |
| H —— OH | HO —— H |
| CH₃ | CH₃ |

(Fischer projection of a pair of enantiomers)

(ii)

(*cis* = Z isomer)     (*trans* = E isomer)

2. (i)    staggered (anti):     fully eclipsed:

(most stable)     (least stable)

330

(ii)

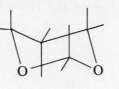 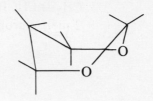

most stable           least stable

3. (i) nucleophile, (ii) free radical, (iii) none of these, (iv) nucleophile.

4. (i) $CH_3-CH=CH-CH_2-CH_2-Cl$    (ii) $CH_3-CO-O-CH=CH_2$

5. $CH_3COO^{\ominus}Na^{\oplus} + C_2H_5OH$

    $CH_3CH(OH)OC_2H_5$

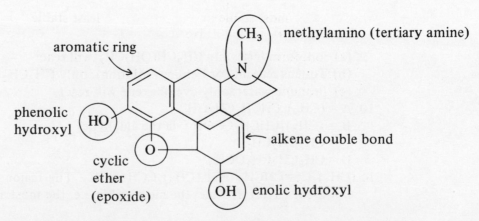

6. Phenols give a green or blue colour with neutral $FeCl_3$.

7. ⟨ ⟩$-NH_2 < NH_3 < CH_3NH_2$

## Question 13.9

1. (B), 2. (D), 3. (B), 4. (A), 5. (B), 6. (C).

## Question 13.10

1.

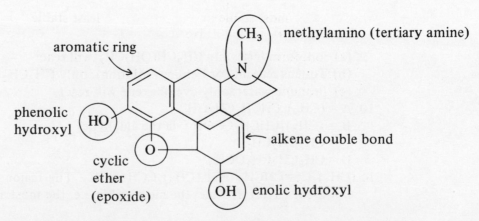

2. (a) $CH_3COCH(CH_3)COCH_3$   (b) $CH_3CH_2C≡CCH(CH_3)CH_2OH$

3. $CH_2{=}CHCH_2CH_3$, but-1-ene, $CH_3CH{=}CHCH_3$, but-2-ene; $(CH_3)_2C{=}CH_2$, 2-methylpropene; $CH_2{-}CH_2$, methylcyclopropane;

$$\underset{CH_3}{\overset{}{\underset{|}{CH}}}$$

$$\begin{array}{c} CH_2{-}CH_2 \\ |\qquad | \\ CH_2{-}CH_2 \end{array}, \text{ cyclobutane}$$

4. But-2-ene;

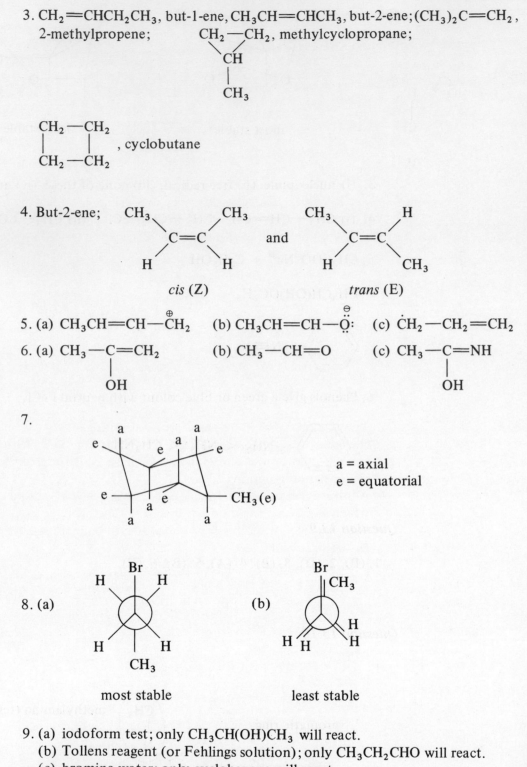

   $cis$ (Z)        and        $trans$ (E)

5. (a) $CH_3CH{=}CH{-}\overset{\oplus}{C}H_2$   (b) $CH_3CH{=}CH{-}\overset{..}{\underset{..}{O}}\text{:}$   (c) $\dot{C}H_2{-}CH_2{=}CH_2$

6. (a) $CH_3{-}\underset{\underset{OH}{|}}{C}{=}CH_2$   (b) $CH_3{-}CH{=}O$   (c) $CH_3{-}\underset{\underset{OH}{|}}{C}{=}NH$

7.

$a$ = axial
$e$ = equatorial

8. (a)   (b)

   most stable        least stable

9. (a) iodoform test; only $CH_3CH(OH)CH_3$ will react.
   (b) Tollens reagent (or Fehlings solution); only $CH_3CH_2CHO$ will react.
   (c) bromine water; only cyclohexene will react.
10. A = $(CH_3)_2CHCH_2CH_2OH$
    B = $(CH_3)_2CHCH_2COOH$ (via the aldehyde)
    C = $CH_3CH_2CH_2Br$
    D = $CH_3CH_2CH_2OCH_3$
11. $(CH_3)_2C{=}CHCH_3$ and $(CH_3)_2CCH{=}CH_2$. The major product is 2-methyl-but-2-ene (Saytzeff Rule, the most stable, i.e. the most alkylated alkene)

*Question 13.11*

1. (a) $CH_3CH_2CH=CH_2$

(b)

$$\overset{\ominus}{H\ddot{O}:}\, H$$

$$CH_2-CHCH_2CH_3 \rightarrow \left[ \begin{array}{c} \overset{\delta-}{H\ddot{O}}\text{---}H \\ \\ CH_2 = CHCH_2CH_3 \\ \\ (CH_3)_3N^{\delta+} \end{array} \right]_{TS} \rightarrow \begin{array}{c} H\ddot{O}-H\; + \\ \\ CH_2=CHCH_2CH_3 \\ \\ +\, (CH_3)_3\ddot{N} \end{array}$$

$$(CH_3)_3\underset{\oplus}{N}$$

Reacting conformation has the $\beta H$ and $N^{\oplus}$ atoms *anti* (*trans*) to one another. Hofmann elimination gives the less stable, least alkylated alkene — minimised steric hindrance in the TS.

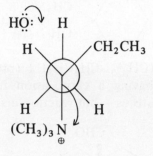

2. The secondary bromoalkane does not ionise to any extent, and with ethanol the main reaction is bimolecular substitution, $S_N2$. With sodium ethoxide, the alkoxide ion is a very powerful base, and the main reaction then is bimolecular elimination, E2.

3. Addition of $D^{\oplus}$ to *cis*-but-2-ene gives a planar carbocation intermediate which can be attacked on either side by the bromide ion, $Br^{\ominus}$. Since there are two possible structures for the carbocation, there will be four structures for the deuterobromide:

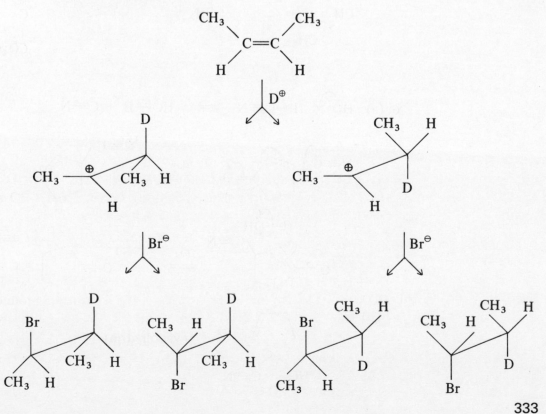

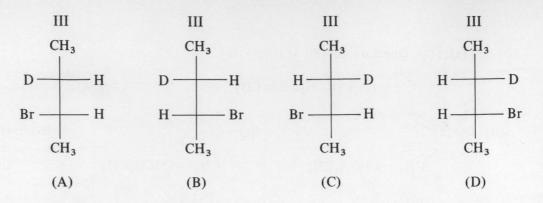

| III | III | III | III |
|---|---|---|---|
| (A) | (B) | (C) | (D) |

Compounds (A) and (D) form an enantiomeric pair, as do (B) and (C), and the two pairs are diastereoisomers.

4. (a) $(CH_3)_2CClCH_3$ (via the more stable of the two possible carbocation intermediates)

(b)  (via the more stable 3° carbocation intermediate)

(c) $(CH_3)_3\overset{\oplus}{N}CH_2CH_2I \ \overset{\ominus}{I}$ (protonation of the left-hand carbon atom avoids having a carbocation intermediate with positive charges on adjacent atoms — but reaction may be slow because of steric hindrance)

5. (a)

(i)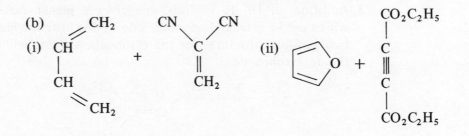

(ii)

(b)
(i) 

(ii)

6. (a) 

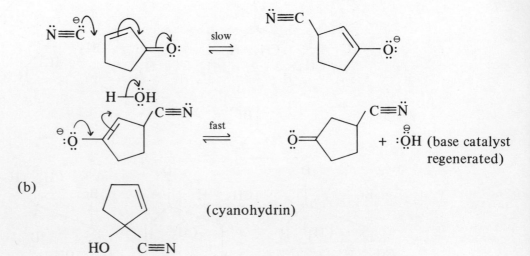

(b) (cyanohydrin)

334

7. (b) and (c); ((a) does not obey the Huckel rule)

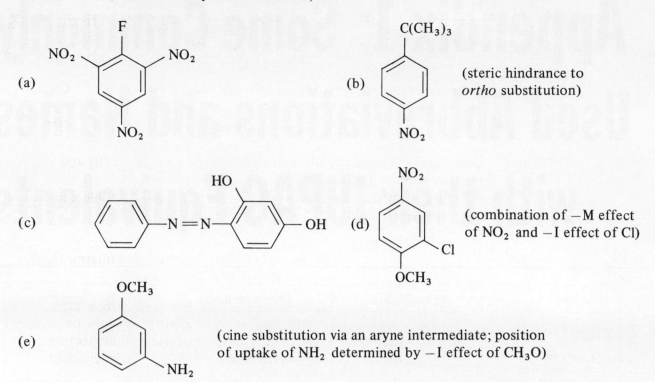

(a)

(b)    (steric hindrance to *ortho* substitution)

(c)

(d)    (combination of $-M$ effect of $NO_2$ and $-I$ effect of Cl)

(e)    (cine substitution via an aryne intermediate; position of uptake of $NH_2$ determined by $-I$ effect of $CH_3O$)

Structures for solution 13.5, page 245.

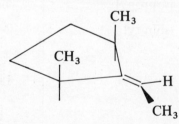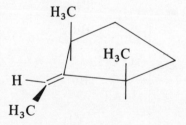

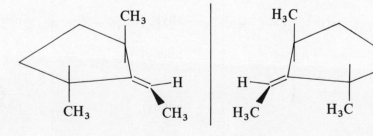

# Appendix 1: Some Commonly Used Abbreviations and Names with their IUPAC Equivalents

| Structure | Abbreviation | Common name | IUPAC name |
|---|---|---|---|
| $CH_3CO-$ | Ac | acetyl | ethanoyl |
| $CH_3COOH$ | AcOH | acetic acid | ethanoic acid |
| $(CH_3CO)_2O$ | $Ac_2O$ | acetic anhydride | ethanoic anhydride |
| $CH_3CH_2O-$ | EtO | ethoxy | nc |
| $CH_3O-$ | MeO | methoxy | nc |
| $(CH_3)_2CH-$ | iPr | isopropyl | methylethyl |
| $(CH_3)_2CHCH_2$ | iBu | isobutyl | 2-methylpropyl |
| $CH_3CH_2$ <br> $\quad\diagdown$ <br> $\qquad CH-$ <br> $\quad\diagup$ <br> $CH_3$ | sBu | sec-butyl | 1-methylpropyl |
| $(CH_3)_3C-$ | tBu | tert-butyl | 1,1-dimethylethyl |
| $CH_2\!=\!CH-$ | – | vinyl | ethenoyl |
| $CH_2\!=\!CHCH_2$ | – | allyl | propenoyl |
| ⬡— | Ph | phenyl | nc |
| ⬡—$CH_2$ | Bz | benzyl | phenylmethyl |
| ⬡—$CO$ | – | benzoyl | benzene carbonyl |
| ⬡—$NH_2$ | – | aniline | phenylamine |

For a more comprehensive list, see *Chemical Nomenclature, Symbols and Terminology for Use in School Science,* Third Edition, 1985, published by The Association for Science Education.

# Appendix 2: A Short List of Books on How to Study

Francis Casey, *How to Study*, Macmillan Education, 1985.
Maddox, Harry, *How to Study*, Pan, 2nd edn, 1988.
Rowntree, Derek, *Learn how to Study*, MacDonald, 1984.

# Index